Carbon-Based Materials

Carbon-Based Materials

Editor

Julia A. Baimova

MDPI • Basel • Beijing • Wuhan • Barcelona • Belgrade • Manchester • Tokyo • Cluj • Tianjin

Editor
Julia A. Baimova
Institute for Metals
Superplasticity Problems of RAS
Russia

Editorial Office
MDPI
St. Alban-Anlage 66
4052 Basel, Switzerland

This is a reprint of articles from the Special Issue published online in the open access journal *Materials* (ISSN 1996-1944) (available at: https://www.mdpi.com/journal/materials/special_issues/carbon_Mater).

For citation purposes, cite each article independently as indicated on the article page online and as indicated below:

LastName, A.A.; LastName, B.B.; LastName, C.C. Article Title. *Journal Name* **Year**, *Volume Number*, Page Range.

ISBN 978-3-0365-3439-8 (Hbk)
ISBN 978-3-0365-3440-4 (PDF)

© 2022 by the authors. Articles in this book are Open Access and distributed under the Creative Commons Attribution (CC BY) license, which allows users to download, copy and build upon published articles, as long as the author and publisher are properly credited, which ensures maximum dissemination and a wider impact of our publications.

The book as a whole is distributed by MDPI under the terms and conditions of the Creative Commons license CC BY-NC-ND.

Contents

About the Editor vii

Preface to "Carbon-Based Materials" ix

Alenka Vesel, Rok Zaplotnik, Gregor Primc and Miran Mozetič
Synthesis of Vertically Oriented Graphene Sheets or Carbon Nanowalls—Review and Challenges
Reprinted from: *Materials* 2019, 12, 2968, doi:10.3390/ma12182968 1

Nonjabulo P. D. Ngidi, Moses A. Ollengo and Vincent O. Nyamori
Effect of Doping Temperatures and Nitrogen Precursors on the Physicochemical, Optical, and Electrical Conductivity Properties of Nitrogen-Doped Reduced Graphene Oxide
Reprinted from: *Materials* 2019, 12, 3376, doi:10.3390/ma12203376 19

Mariano Palomba, Gianfranco Carotenuto, Angela Longo, Andrea Sorrentino, Antonio Di Bartolomeo, Laura Iemmo, Francesca Urban, Filippo Giubileo, Gianni Barucca, Massimo Rovere, Alberto Tagliaferro, Giuseppina Ambrosone and Ubaldo Coscia
Thermoresistive Properties of Graphite Platelet Films Supported by Different Substrates
Reprinted from: *Materials* 2019, 12, 3638, doi:10.3390/ma12213638 45

Massimo Calovi, Emanuela Callone, Riccardo Ceccato, Flavio Deflorian, Stefano Rossi and Sandra Dirè
Effect of the Organic Functional Group on the Grafting Ability of Trialkoxysilanes onto Graphene Oxide: A Combined NMR, XRD, and ESR Study
Reprinted from: *Materials* 2019, 12, 3828, doi:10.3390/ma12233828 57

Elena A. Korznikova, Leysan Kh. Rysaeva, Alexander V. Savin, Elvira G. Soboleva, Evgenii G. Ekomasov, Marat A. Ilgamov and Sergey V. Dmitriev
Chain Model for Carbon Nanotube Bundle under Plane Strain Conditions
Reprinted from: *Materials* 2019, 12, 3951, doi:10.3390/ma12233951 71

Xiwei Wang, Peng Duan, Zhenzhong Cao, Changjiang Liu, Dufu Wang, Yan Peng and Xiaobo Hu
Homoepitaxy Growth of Single Crystal Diamond under 300 torr Pressure in the MPCVD System
Reprinted from: *Materials* 2019, 12, 3953, doi:10.3390/ma12233953 85

Alexander V. Savin and Yuriy A. Kosevich
Modeling of One-Side Surface Modifications of Graphene
Reprinted from: *Materials* 2019, 12, 4179, doi:10.3390/ma12244179 97

Xiwei Wang, Peng Duan, Zhenzhong Cao, Changjiang Liu, Dufu Wang, Yan Peng, Xiangang Xu and Xiaobo Hu
Surface Morphology of the Interface Junction of CVD Mosaic Single-Crystal Diamond
Reprinted from: *Materials* 2019, 13, 91, doi:10.3390/ma13010091 111

Konstantin P. Katin, Mikhail M. Maslov, Konstantin S. Krylov and Vadim D. Mur
On the Impact of Substrate Uniform Mechanical Tension on the Graphene Electronic Structure
Reprinted from: *Materials* 2020, 13, 4683, doi:10.3390/ma13204683 123

Yuzhe Wu, Yuntong Li, Conghui Yuan and Lizong Dai
A Facile Method for the Generation of Fe_3C Nanoparticle and $Fe-N_x$ Active Site in Carbon Matrix to Achieve Good Oxygen Reduction Reaction Electrochemical Performances
Reprinted from: *Materials* **2020**, *13*, 4779, doi:10.3390/ma13214779 . . . . **143**

Qiang Wu, Can Sun, Zi-Zong Zhu, Ying-Dong Wang and Chong-Yuan Zhang
Effects of Boron Carbide on Coking Behavior and Chemical Structure of High Volatile Coking Coal during Carbonization
Reprinted from: *Materials* **2021**, *14*, 302, doi:10.3390/ma14020302 . . . . **155**

Liliya R. Safina, Karina A. Krylova, Ramil T. Murzaev, Julia A. Baimova, Radik R. Mulyukov
Crumpled Graphene-Storage Media for Hydrogen and Metal Nanoclusters
Reprinted from: *Materials* **2021**, *14*, 2098, doi:10.3390/ma14092098 . . . . **167**

About the Editor

Julia A. Baimova Prof., Dr. Sci (Dr. Hab.) is the Head of Laboratory for "Mechanics and Physics of Carbon Nanostructures" at the Institute for Metals Superplasticity Problems at the Russian Academy of Sciences. Her research focuses on studying the stability and mechanical and physical properties of various carbon nanomaterials, such as graphene, carbon nanotubes, and crumpled graphene. She obtained a Ph.D. in Condensed Matter (2014) and Habilitation in Condensed Matter (2016). Her scientific interests are centered around the search for links between structure peculiarities and physical and mechanical properties and ways of improving the properties of carbon nanostructures through external treatment. Now, she is working on the study of graphene–metal composites, understanding fabrication techniques and the effect of heat and temperature treatment on their physical and mechanical properties.

Preface to "Carbon-Based Materials"

This book contains publications included in the Special Issue of *Materials* entitled "Carbon-Based Materials", which appeared in print in 2021. The aim of this Special Issue was to present the state-of-the-art research progress in the field of carbon-based nanomaterials and to demonstrate their growing importance in many fields. The goal of this book is to present different studies on the fabrication and physical and mechanical properties of various carbon nanostructures, which will further demonstrate the importance of this subject. Both experimental and simulation works are included to the book. We believe that this book will provide a useful update of the newest results, trends, and lines of research in this field and will assist in planning further studies.

This book is targeted toward professionals, researchers, and graduate students working in a field of nanomaterials.

As a Guest Editor, I would like to express thanks to the publisher, MDPI, and particularly to Ms. Fannie Xu, Editor, for providing extensive help, guidance, and encouragement in preparing this publication.

Julia A. Baimova
Editor

Review

Synthesis of Vertically Oriented Graphene Sheets or Carbon Nanowalls—Review and Challenges

Alenka Vesel, Rok Zaplotnik, Gregor Primc and Miran Mozetič *

Department of Surface Engineering, Jozef Stefan Institute, Jamova cesta 39, 1000 Ljubljana, Slovenia; alenka.vesel@guest.arnes.si (A.V.); rok.zaplotnik@ijs.si (R.Z.); gregor.primc@ijs.si (G.P.)
* Correspondence: miran.mozetic@ijs.si

Received: 23 August 2019; Accepted: 11 September 2019; Published: 12 September 2019

Abstract: The paper presents a review on the current methods for deposition of vertically oriented multilayer graphene sheets (often called carbon nanowalls—CNWs) on solid substrates. Thin films of CNWs are among the most promising materials for future applications in capacitors, batteries, electrochemical devices, and photovoltaics, but their application is currently limited by slow deposition rates and difficulties in providing materials of a desired structure and morphology. The review paper analyzes results obtained by various groups and draws correlations between the reported experimental conditions and obtained results. Challenges in this scientific field are presented and technological problems stressed. The key scientific challenge is providing the growth rate as well as morphological and structural properties of CNWs thin films versus plasma parameters, in particular versus the fluxes of reactive plasma species onto the substrate surface. The technological challenge is upgrading of deposition techniques to large surfaces and fast deposition rates, and development of a system for deposition of CNWs in the continuous mode.

Keywords: carbon nanowalls; plasma synthesis; growth mechanism; deposition speed; deposition parameters; deposition temperature

---

1. Introduction

Nanocarbon has attracted enormous attention in the past two decades. It can exist in various configurations such as graphene, carbon nanotubes (CNTs), or carbon nanowalls (CNWs). Robert F. Curl Jr., Sir Harold W. Kroto and Richard E. Smalley received the Noble prize in chemistry in 1996 for discovery of fullerenes, while Andre Geim and Konstantin Novoselov in physics in 2010 for ground-breaking experiments regarding the two-dimensional material graphene. S. IIjima has been a candidate for the prize, too, for discovery of carbon nanotubes, and he received the first Kavli prize in Nanotechnology in 2008. Many research groups are nowadays involved in research on nanocarbon worldwide. In Figure 1, a comparison of publications per year on a synthesis of CNTs and CNWs is shown. Nanocarbon in the form of CNWs has attracted less attention but represents a promising material for application in fuel cells, lithium ion batteries, photovoltaic devices, thin-film transistors, sensors of specific gaseous molecules, field-emission devices, batteries, light absorbers, enhanced detectors for electrochemical and gas sensors, supercapacitors and scaffolds for tissue engineering [1–8]. The unique property of carbon nanowalls versus any other known material is a combination of stability, chemical inertness, electrical conductivity, and huge surface-to-mass ratio. Carbon nanowalls are often referred to as "multilayer graphene sheets stretching perpendicularly to the substrate surface". Such vertically oriented graphene sheets have a high density of atomic-scale graphitic edges that are potential sites for electron field emission [9]. Due to their high surface-to-mass ratio they are also good candidates for biosensors and energy storage applications [9]. Since awarding the Nobel prize, tens of thousands research groups have been involved in basic research as well as in application of graphene worldwide and promising results were reported; however, mass application of this type of carbon is yet to be realized. A way to implement this material in a mass production is depositing graphene perpendicular to a substrate surface and thus taking full advantage of its unique properties. This review paper intends to present current state-of-the-art on methods for deposition of carbon nanowalls as well as their properties where they are reported. Exact growth mechanisms are far from being well-understood; therefore, both theoretical and experimental study is yet to be performed.

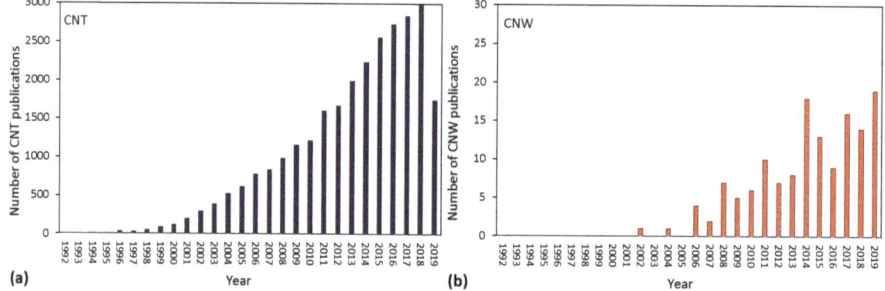

Figure 1. Number of publications per year regarding carbon nanotube (CNT) synthesis (**a**) and carbon nanowall (CNW) synthesis (**b**). Source: Web of Science.

A typical SEM image of CNWs is shown in Figure 2. CNWs were first synthesized more than 10 years ago and can be deposited onto a substrate using a classical plasma-enhanced chemical vapor deposition (PECVD) method [10–13]. A carbon-containing gas (usually methane CH_4 or acetylene C_2H_2) is partially atomized and ionized upon plasma conditions and the resultant radicals condensate on a substrate surface. Upon limited range of experimental conditions, carbon in the form of nanowalls (multilayer graphene sheets) grows on the substrate surface. The commonly accepted growth mechanism for CNWs is illustrated in Figure 3 and can be summarized as follows [9,14]:

1. Adsorption of CH_3 radicals and formation of amorphous carbon layer on the substrate.
2. Formation of defects and dangling bonds because of ion irradiation leading to the formation of nucleation sites.
3. Migration of carbon species and formation of nanoislands with dangling bonds.
4. Nucleation of small graphene nanosheets on dangling bonds followed by a two-dimensional growth.
5. Formation of nanographene sheets with a random orientation.
6. Bonding of reactive carbon species to the edge of graphene sheets. Nanosheets that are standing almost vertically preferably grow faster and shadow low-lying graphene sheets, therefore their growth is suppressed.

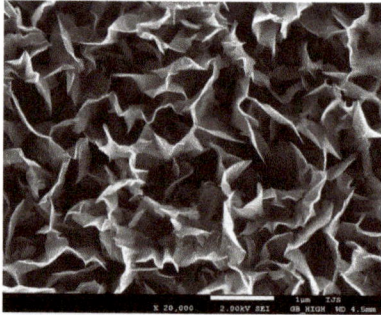

Figure 2. An example of carbon nanowalls grown on the surface of a titanium foil.

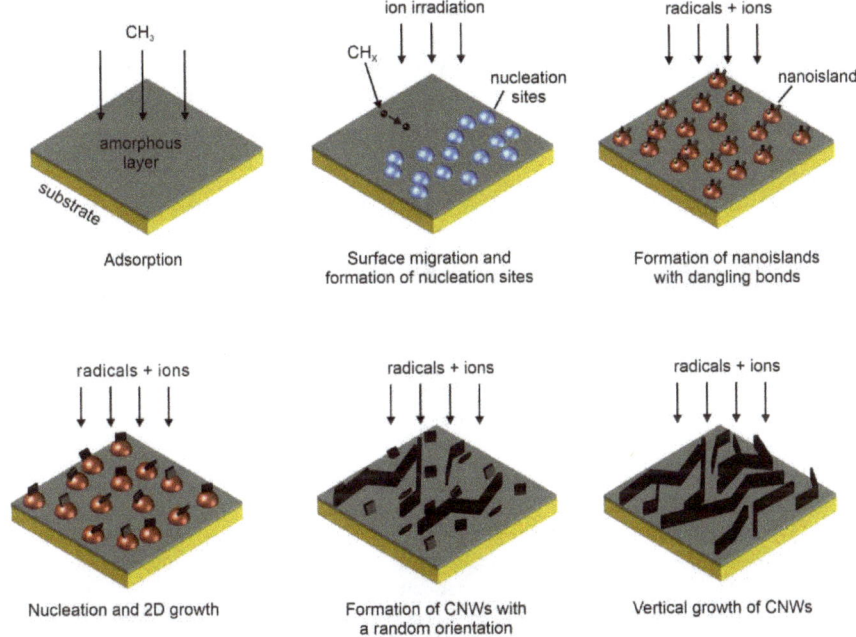

Figure 3. Schematic presentation of CNWs growth mechanism as suggested by Hiramatsu [9].

Images taken by scanning electron microscopy (SEM) at various deposition times confirm the above mechanism, at least for experimental conditions adopted by [13,15,16].

Hydrogen was found to play a significant role in the growth process of CNWs by numerous authors [17–19]. Hydrogen is needed to etch and remove any amorphous carbon that is formed on the substrates, furthermore, it prevents formation of additional graphene layers by etching weakly bounded carbon atoms and it was found to enhance migration of carbon precursors [14]. Therefore, addition of hydrogen can greatly improve the quality of CNWs. The quality of CNWs can be also enhanced by addition of argon and even oxygen. By addition of argon and/or oxygen it was found to be possible to control the density of nanoislands in the initial stage of growth and thus consequently also the density of CNWs. This is the reason why CNWs are most frequently synthesized using a mixture of CH_4 with H_2 and Ar, where researchers usually use different pressures and gas flows to influence the deposition rate as well as quality of CNWs.

CNWs can be also formed by employing C_2F_6 gas in a mixture with hydrogen or oxygen [16]. In this case, C and CF_3 radicals serve as building blocks, whereas hydrogen atoms are needed for abstraction of fluorine from the growing film. A mechanism for CNWs growth in a non-equilibrium C_2F_6 gas environment was proposed by Kondo et al. [16]. Another example of CNWs synthesis is an application of a CO/H_2 gas environment upon heavily non-equilibrium conditions [20]. CNWs were observed only if H_2 was added, whereas formation of nanofibers was observed if CO was mixed with Ar/O_2 [21]. In recent years, CNWs were also successfully synthesized by using other precursors like organic precursor (p-xylene [22], ethanol or hexane [23,24]) or metal-organic precursors (aluminum acetylacetonate [25]).

A drawback of the currently known techniques for synthesizing CNWs is a low growing rate and inability to obtain uniform coatings on large substrates. Therefore, they are inappropriate for industrial application at this stage of the scientific knowledge. So far, researchers have managed to obtain uniform coatings on surfaces measured in square centimeters and the growth is accomplished in a time scale of minutes if not hours. The key problem arises from the fact that deposition rates using

CH$_x$ radicals cannot be enhanced because carbon agglomerates at elevated pressure (forming dusty plasma); therefore, the structure of the deposit is not appropriate—instead of nanowalls, carbon of various morphological shapes including soot or hydrogenated amorphous carbon grows at elevated pressure. So far, few alternatives to CH$_x$ radicals have been reported except for CO and C$_2$F$_6$ as mentioned above.

Another problem limiting the application of carbon nanowalls on industrial scale is associated with deposition of CNWs at rather high substrate temperatures. Currently available methods allow deposition of CNWs only at temperatures in the range of 600–800 °C that are not appropriate for deposition of CNWs on polymer substrates [10]. Temperatures even higher than 800 °C were also reported. The researchers observed that the quality of CNWs is increasing with increasing temperature and also their growth rate [17,19,25–29]. Substrate temperature has therefore an important effect on their size and density. If the temperature was too low, CNWs did not grow and only an amorphous carbon layer was observed, or carbon in other morphological nanostructures [26,29]. Tii et al. found that addition of Ar to N$_2$/CH$_4$ gas system could lower the substrate temperature for CNWs deposition to approximately 650 °C [30]. Contrary, Park et al. reported deposition of CNWs at temperatures a bit lower than 400 °C (depending on the substrate material) [31]. Additionally, Singh [32] managed to synthesize CNWs on a glass surface at a temperature of approximately 400 °C, however he used hot-wire chemical vapor deposition. Although it was reported that graphene films have been deposited at low temperatures such as 240 °C [33], and carbon nanotubes (CNT) at just 120 °C [34], it is still a big challenge for researchers working on deposition of vertically oriented graphene sheets to minimize the substrate temperature. In one experiment, CNWs were deposited onto SiO$_2$ substrates at various substrate temperatures ranging from 600 to 800 °C [26]. The deposition was performed using ECR-PECVD (electron-cyclotron resonance plasma-enhanced chemical vapor deposition) in CH$_4$/Ar environment. The authors found an important effect of the substrate temperature on the vertical growth of CNWs through nanoscale graphitic islands. In this paper, 600 °C was found to be a minimum temperature where formation of nano-graphitic islands was observed. These two-dimensional nanoislands changed to three-dimensional structures when the substrate temperature was increased to 625 °C. However, further increase of the sample temperature led to formation of a higher density of CNWs. Their height and growth rate increased with increasing temperature and a formation of nest-like structure was observed [26]. In another experiment, Gentoiu et al. [29] found strong dependence of a structure, morphology, and graphitization on deposition temperature. At low temperatures ~200 °C carbon nanotubes (CNTs) were formed, at temperatures ~300–400 °C rather amorphous carbon nanoparticles appeared, whereas at temperatures ~500–700 °C formation of CNWs was observed. These experiments clearly show that the temperature has an important effect on the morphology and growth of CNWs. Deposition of CNWs is thus currently still limited to materials that can withstand high temperatures.

Besides temperature, also the choice of a substrate material may influence the initial stage of the carbon cluster nucleation because of a different lattice matching with graphite and consequently different quality of CNWs may be formed [10]. In addition, carbon solubility in the substrate material may have a strong influence on the nucleation and growth of CNWs as reported by Giese at al. [25]. The authors investigated deposition of CNWs on various substrates including stainless steel, aluminum, nickel and silicon which strongly differ in carbon solubility. The effect of a bias voltage and substrate temperature was investigated as well. With increasing bias voltage and temperature, the morphology changed from nanorods which were formed at low bias voltage and temperature, to thorny structures, followed by straight CNWs, whereas curled CNWs were formed at high bias voltages and temperatures. These growth regimes were shifted for different materials. On stainless steel and aluminum all mentioned structures were formed, however they appeared at different bias voltage and temperature. Whereas for Ni and Si, no nanorods were formed and only straight and curled CNWs were found. This was explained by difference in carbon bulk and surface diffusion for these materials and different affinity to form carbides at the surface. Additionally, Vizireanu et al. [35] investigated the effect of substrate material on CNWs synthesis and their morphology. CNWs were deposited on various substrates including SiO_2/Si, titanium, stainless steel, Quartz, MgO, and carbon paper, that were previously covered with clustered nickel catalyst. The authors found that the type, morphology, and electrical characteristics (conductive, insulator, or semiconductor) were not important for CNWs growth.

2. Early Scientific Documents of Plasma Synthesis of Carbon Nanowalls

An excellent and comprehensive review summarizing earlier achievements in CNWs deposition was prepared by Himatsu and Hori in 2010 [14]. Appropriate references of the earlier papers are summarized in this classical monograph. The first report on the synthesis of CNWs structures appeared in scientific literature in 2002 by Wu et al. [36]. The gases used were methane and hydrogen of flow rates 40 and 10 sccm, respectively. Such a gas mixture is a natural choice for depositing any carbon nanomaterials, because methane partially dissociates and ionizes upon plasma conditions and the radicals such as C, CH, CH_2, and CH_3 stick to the substrate surface. A rather high substrate temperature of about 700 °C was used to favor decomposition of hydrogenated carbon radicals to almost pure carbon suitable for growing carbon structures almost free from hydrogen. Additional DC biasing was applied to deliver more energy to the substrates upon growing of CNWs. A catalyst (typically NiFe) was applied to stimulate the nucleation. Addition of hydrogen was essential because atomic hydrogen and positively charged ions caused removal of weaker-bonded carbon what was found beneficial for appropriate structure of the CNWs. Wu et al. found their CNWs suitable for application in batteries, light-emitting and conversion devices, catalysts, and other areas requiring high surface area materials.

In 2005 the group of Shiji [37] reported fabrication of CNWs by capacitively coupled radio-frequency plasma enhanced chemical vapor deposition (CCP-PECVD) employing fluorocarbon/hydrogen mixtures. Correlation between CNWs growth and fabrication conditions, such as the carbon source gases, was investigated. In addition, the influence of H-atom density in the plasma was measured using vacuum ultraviolet absorption spectroscopy to discuss the growth mechanism of CNWs.

Additionally, also in 2005, Tanaka et al. [38] reported growth of CNWs on a SiO_2 substrate by microwave plasma enhanced chemical vapor deposition. They investigated the growth process and revealed that the CNWs grew at the fine-textured structure on SiO_2 and the growth process did not require the catalyst (as opposite to Wu et al.). It was found that the height of CNWs as a function of growing time obeyed the square-root law. Rather high growth rates of approximately 10 micrometers per hour were achieved. They also used hydrocarbons with hydrogen as a useful gas mixture.

Dikonimos et al. [39] reported CNWs with a maximum longitudinal dimension ranging from 10 to 200 nm and a wall thickness lower than 5 nm. Such structures were grown in a high-frequency chemical vapor deposition reactor on Si substrates. The growth precursor was methane diluted with a noble gas (He). The growth rate and film morphology were explored. The experimental setup consisted of a two-grid system which allowed to vary the voltage and current density on the substrate surface independently. An increase of growth rate was observed as the film thickness increased from a few nanometers to about 200 nm when the substrate current density was increased.

The importance of hydrogen in the gas mixtures was elaborated by Cui et al. [17]. Without addition of H_2, graphite sheets were difficult to produce, and the film contained other forms of carbon. At small H_2 fluxes (40 sccm), the carbon nano-sheets were not clearly distinguished. When H_2 flux was increased the vertical graphene sheets became more obvious (80–120 sccm). However, if the H_2 flow was too high (150 sccm) the density of the vertical sheets decreased. At low H_2 flow rates, the supply of hydrogen was insufficient to etch away the amorphous part, however at high H_2 flow rates also CNWs were etched by excessive hydrogen [17]. Therefore, at optimal conditions a mixture should contain just the right density ratio of CH_x radicals acting as a source of carbon species and hydrogen atoms needed to etch away the amorphous part. Similar findings were also found by Jiang et al. [18].

Teii et al. [30] revealed the importance of Ar in the production of C_2 dimers, which were found to be the most important radicals responsible for CNWs growth. He performed synthesis of CNWs in ASTex microwave plasma using $Ar/N_2/CH_4$ or $Ar/N_2/C_2H_2$ gas mixtures with various Ar concentrations. The amount of C_2 dimers was increasing by adding Ar. Furthermore, addition of Ar reduced the substrate temperature needed for CNWs deposition to 650 °C. Rather high deposition rates of approximately 1 µm/min were obtained.

Vizireanu et al. [35] synthesized CNWs structures in Ar/H$_2$/C$_2$H$_2$ mixture on various substrates including SiO$_2$/Si, Ti, stainless steel, Quartz glass, MgO, carbon paper, that were previously covered with clustered Ni catalyst. SEM images of CNWs on various substrates revealed that the type, morphology and electrical characteristics of the substrates (conductive, insulator, or semiconductor) were not important for CNWs growth. A deposition rate was approximately 1 µm per 30 min. The authors also investigated the influence of the pressure and gas flows. It was found that the quality of CNWs could be altered by changing the pressure or Ar flow. CNWs with large length-to-thickness ratio and well-isolated between themselves were deposited at low pressure and high carrier flow rates, whereas poor quality of CNWs was obtained at high pressure or low Ar flow.

Jiang et al. [18] also investigated the morphology of CNWs grown in CH$_4$/H$_2$ mixture at various CH$_4$ flow rates and CH$_4$ to H$_2$ ratios. CH$_4$ flow rate was changed from 5 to 100 sccm whereas H$_2$ flow rate was kept constant. It was found that the size of graphene sheets first increased with increasing CH$_4$ flow rate, reached a maximum in the range of flows 10–30 sccm, and then decreased with further increase of the CH$_4$ flow rate. This result was explained by higher density of nucleation sites, faster nucleation, and sufficient density of carbon radicals with increasing flow rate. However, if the flow rate was too high, too high density of the nucleation sites was reported, thus hindering the nucleus from growing into large sizes of graphene sheets because of insufficient interspace between the neighbouring nuclei. Moreover, when CH$_4$ flow rate was manipulated, also CH$_4$ to H$_2$ ration was changed which influenced the etching effect of hydrogen radicals. Too high H$_2$ content led to a small size of the graphene sheets because of the excessive etching according to Jiang. Therefore, it was concluded that controlling the dynamic competition between growth and etching was the key factor for obtaining good quality of CNWs.

Davami et al. [40] investigated the morphology of CNWs grown in CH$_4$/H$_2$ systems on various substrates including Cu, Si, or Si coated with a thin layer of Ni or Au. The authors found that CNWs on pure Si substrates were denser and thinner in comparison to CNWs deposited on Si/Ni or Si/Au substrates, whereas CNWs on Cu were much finer than on all other substrates.

The growth rate of PECVD techniques is usually limited to tens of nanometers per minute that is insufficient for practical applications. Zhang et al. [41] used "high density meso-plasma CVD" and obtained fast growth rate of the order of ~10 µm/min, depending on a power of a radio-frequency (RF) generator and CH$_4$ flow rate. The meso-plasma system was actually a modified ICP-jet plasma in combination with a planar-coiled antenna. In such a configuration, they obtained fast deposition because of a high dissociation rate of CH$_4$. The CNWs deposition was performed in CH$_4$/H$_2$/Ar mixture. A deposition rate was increasing with increasing RF power (12–18 kW) and increasing CH$_4$ flow rate (10–80 sccm), when keeping H$_2$ flow constant. The highest growth rate (18 µm/min) was observed, when the flow of H$_2$ was zero, what was explained by a lower etching effect of hydrogen. An increase in the plasma power and CH$_4$ flow did not only change the growth rate but it also had an effect on CNWs morphology and structure. Different morphological forms including petal-like, cauliflower-like, maze-like, or floc-like structures were observed.

3. A Brief Review of Patents

As already mentioned, there is a great commercial interest in application of carbon nanowalls in different devices. In order to make this review rather complete, the most relevant patents on deposition of carbon nanowalls are listed below and briefly described.

Probably the first patent application on growth of CNWs was filed in 2007 by Hiramatsu and Hori [42]. They disclosed a method and a device for producing thin films of CNWs on solid substrates. A source gas containing carbon was introduced into a reaction chamber where plasma was sustained with a capacitively coupled generator. The authors disclose also a second radical-generating chamber which was disposed outside the reaction chamber. Hydrogen radicals were generated by decomposing radical source gas containing hydrogen using RF or another method. The hydrogen radicals were introduced into the plasma, whereby CNWs were formed on a substrate disposed on the second electrode of the CCP. The growth of the CNWs with this method was found to be rather slow (about 1 µm high-quality CNWs in approximately 5 h). The key innovative step in this patent application was application of a remote source of atomic hydrogen, which was essential for the growth of high-quality CNWs. The drawback of the method is a very long treatment time. This drawback was suppressed in the patent [43] which discloses a method and a device for deposition of carbon nanostructures where the base materials forming carbon nanostructures can be continuously fed, thus mass-production could be facilitated. The method described in [43] is actually based on the method revealed in previous patent [42] by the same group. The patent [44] further improves the method described in [42], in particular to improve the crystallinity of CNWs. However, the improvement of crystallinity had a negative effect on the growth rate, since it was reported to decrease from about 60 to 20 nm/min.

A method for growing CNWs on a solid substrate is disclosed also in the patent by Ghoanneviss et al. [45]. In this patent, a method is described which comprises mixing a predetermined amount of a hydrocarbon gas with a predetermined amount of at least one non-hydrocarbon gas, placing the solid substrate into a reaction chamber; creating gaseous radicals in the reaction chamber which comprises hydrocarbon and non-hydrocarbon radicals; applying the radicals to the solid substrate; and growing CNWs on said solid substrate exposed to said radical. This invention comprises a method where CNWs are created under atmospheric pressure. The CNWs growth with this method usually takes tens of minutes. No fluxes nor fluences of said radicals are disclosed in this patent application.

CNWs were also formed as a product in a CO_2 reduction device with the CO_2 reduction method disclosed in a recent patent application by Ohmae et al. [46]. This CO_2 reduction method produces CNWs by transforming CO_2 gas into carbon using microwave (MW) plasma chemical vapor deposition and, essentially, using water vapor as a carrier gas. In the preferred embodiment of this patent, the method based on MW plasma chemical vapor deposition is used to reduce CO_2 gas in carbon oxide-containing gas flowing through the inside of an U-shaped reaction tube made from glass. The water vapor is used just as a carrier gas of the carbon oxide-containing gas according to Ohmae. Unlike all previously cited documents the methods disclosed in this patent application do not rely on injection of hydrocarbons into gaseous plasma. The CO_2 gas is dissociated upon plasma conditions and CNWs are produced on a solid substrate positioned inside the glass tube. The inventors claim a CO_2 reduction system which has the U-shaped CO_2 reduction device whose gas exhaust tube is connected with the gas introduction tube. The inventors also claim a CO_2 reduction method which produces CNWs by conversion of CO_2 gas into carbon source using MW plasma CVD method and water vapor as a carrier gas. In fact, the tube is mounted into a MW waveguide of such a shape that an extremely large electromagnetic field is obtained right at a bend of the tube, therefore the power density is extremely large. Unfortunately, the authors of this patent do not report the exact value of the power density, nor the substrate temperature, but both should be large in such a configuration. The scalability of the method is questionable, though. The decomposition rate of CO_2 increases with increasing discharge power. The electric power generated by photovoltaic power generation is used for powering the MW plasma generator in one embodiment, thus making the device highly economical. The sediment (i.e., the CNWs film) as deposited by the methods of Ohmae also contains other morphological forms of carbon.

CNWs can be used for fuel cells, lithium ion batteries, diodes, and photovoltaic devices, etc. In another patent by Hori's group [47] a method for manufacturing a catalyst layer for a fuel cell is disclosed. Here, CNWs are refined in order to enhance the power generation efficiency of a fuel cell by improving the contact of hydrogen molecules and oxygen molecules which take part in a reaction with a metal catalyst and an electrolyte in the fuel cell to sufficiently form a three-phase interface.

The method that simplifies the process for manufacturing an electrode layer for fuel cells and improves the dispensability of the catalyst component and the electrolyte, whereby the generation efficiency of a fuel cell can be improved, is also revealed in yet another patent by Hori et al. [48].

The CNWs could be also used as a material for the negative electrode in a lithium battery. Tachibana and Tanaike [49] disclose the negative electrode material for a lithium ion battery. This material is prepared using as minute graphite material, flaky CNWs constituted of aggregates in which crystallites having a 10 to 30 nanometer range are highly oriented. A thin lithium battery which uses the innovative material is also provided. There are four other patents on CNWs for negative electrodes for lithium batteries [49–51] and a patent disclosing application of CNWs for a positive electrode [52].

CNWs can be also used as a part of a sample substrate for laser desorption ionization mass spectrometry (LDI-MS) as described in a patent [53]. CNWs are known as excellent absorbents because of their morphology, structure, and composition. Therefore, their possible application can be for a saturable absorbing element with a wide absorption band, a high light absorbance, and a high modulation depth as disclosed in [54]. CNWs can be also used in medical applications, for example when they are deposited on a substrate of an implantable medical device [55]. They are also used as a raw material for producing other materials, such as graphene nanoribbons [56,57] or metal-supported nano-graphite [58].

The patents do not provide details about the particular setups or just disclose the preferred embodiments, so it is difficult to extract the deposition parameters.

4. Summary of Literature Review on PECVD Deposition of CNWs

In Table 1 comparison of conditions used for deposition of CNWs by PECVD is shown. According to data in Table 1, CNWs are usually synthesized by various PECVD methods. These can be microwave plasma enhanced chemical vapor deposition (MW PECVD), capacitively coupled radio-frequency plasma enhanced chemical vapor deposition (CCP PECVD), inductively coupled radio-frequency plasma enhanced chemical vapor deposition (ICP PECVD), direct current plasma enhanced chemical vapor deposition (DC PECVD), and electron cyclotron resonance plasma enhanced chemical vapor deposition (ECR PECVD). A combination of these methods is sometimes used as well as additional biasing of the substrates. Especially in the case of RF plasmas, CCP configuration is often combined with ICP or an external H radical injection [13,16]. Deposition was usually performed at low pressures, however, there were also reports on the deposition at atmospheric pressure giving much higher deposition rates [23,59]. Another way to synthesize graphene sheets was also the application of a discharge in a liquid where the carbon precursor can be either the electrode material or the liquid medium [60,61]. A solution containing graphene sheets was then filtered to collect graphene sheets. Li et al. synthesized graphene sheets by pulsed arc discharge in water with petroleum asphalt as a carbon source [61]. Typical synthesis time was 20 min. On the contrary, Lee et al. synthesized graphene flakes by plasma generated between two carbon electrodes which were immersed in distilled water [60]. Recently, Amano et al. synthesized graphene flakes in ethanol with added iron phthalocyanine [62]. The synthesis time was only 5 min.

As already mentioned in the introduction and also shown in Table 1, the growth rate and quality of CNWs can be controlled by increasing gas pressure or/and flow, discharge power, and substrate temperature. Especially, addition of H_2 and Ar has an important influence on the quality of CNWs; therefore, the right proportion of gasses is needed for optimal CNWs deposition. Higher gas flow rates usually give higher growth rates, but also higher etching rates and loss of a desired morphology; therefore, flows and ratios should be optimized for particular applications of CNWs thin films.

As shown in Table 1, CNWs were successfully deposited to various substrates, electrically conductive and nonconductive. When first invented, deposition was performed with the help of the catalysts [36]. Nowadays, PECVD deposition is usually performed without any catalyst. As reported in the literature, deposition was successfully performed on materials such as Cu, GaAs, Si, SiO_2, sapphire, Al_2O_3, Mo, Zr, Ti, Hf, Nb, W, Ta, stainless steel, MgO, TiN, Quartz glass, carbon paper, and even on non-flat surfaces such as carbon fibres and Ni foam. Yu et al. [63] managed to synthesize patterned CNWs. CNWs were grown on a gold pattern made of a network of squares and other geometrical structures that were coated on the SiO_2 substrate before the deposition by plasma methods.

A rather high temperature is required for deposition of CNWs. Temperatures reported in the literature are usually in the range of 600–800 °C for PECVD methods. Sometimes also temperatures higher than 800 °C were reported (up to about 1000 °C). In some cases, authors managed to deposit CNWs also at temperatures lower than 600 °C (see Table 1), depending on the substrate material. Temperatures required for CNWs deposition on glass (~400–500 °C) were usually lower than for metals (~600–800 °C) [28]. Nevertheless, despite high temperatures that are still needed for PECVD methods, they still enable deposition at temperatures lower than conventional thermal CVD methods. It is interesting, however, that temperatures of approximately 500 °C were reported for hot-wire CVD deposition of CNWs on a stainless steel substrate or quartz glass with Ni catalyst [64,65].

Table 1. Methods and conditions for CNWs deposition.

Ref.	Gas	Temperature (°C)	Growth Rate or Time	Method	Substrate Material	Important Findings
[36]	CH_4/H_2	650–700	-	MW PECVD with catalyst and DC bias	Cu, GaAs, Si, SiO_2, sapphire	-
[19]	CH_4/H_2	600–900	~several m/h	ICP PECVD	Si, SiO_2, Al_2O_3, Mo, Zr, Ti, Hf, Nb, W, Ta, Cu, stainless steel 304	The growth rate was increasing with increasing temperature and CH_4 concentration. CNWs on all substrates showed the same general morphology.
[37]	C_2F_6, CH_4, CF_4, CHF_3, or C_4F_8 with H_2	500	~180 nm/h	CCP PECVD + ICP for H radical injection	Si	The growth rate depended on the type of gas and it was the highest for C_2F_6/H_2 and the lowest for CF_4/H_2: $C_2F_6/H_2 > CHF_3 > CH_4 > CF_4/H_2$. CNWs did not grow in C_4F_8/H_2 gas.
[38]	CH_4/H_2	-	~8 m/h	MW PECVD with DC bias	SiO_2	The height of CNWs as a function of time obeyed the square root law.
[39]	CH_4/He	1000	~7 nm/min	DC PECVD	Si	The average size and film thickness were increasing with increasing total plasma current.
[30]	$Ar/N_2/CH_4$ $Ar/N_2/C_2H_2$	min. 650	1 µm/min	ASTex MW PECVD	Si or silica	Addition of Ar gas reduced the deposition temperature and increased the production of C_2 dimers.
[20]	CO/H_2	700	1 µm/min	ASTex MW PECVD	Si	High growth rate was obtained at a relatively low MW power of 60 W.
[31]	CH_4/H_2	~400	up to 180 s	ECR-MW PECVD	SiO_2, glass, Cu	Deposition temperature depended on the substrate material.
[26]	CH_4/Ar	625–800	~10 nm/min	ECR PECVD	SiO_2/Si	The growth rate and quality of CNWs could be enhanced by increasing the substrate temperature, decreasing the distance between the MW source and the substrate, and increasing the MW power. Below 625 °C CNWs did not grow.
[16]	C_2F_6/H_2 w/o O_2	580	~25 nm/min	Radical injection CCP PECVD	Si	O_2 gas addition reduced the amorphicity and disorder of CNWs and assisted in nucleation of CNWs.
[40]	CH_4/H_2	680	1 µm/20 min	RF PECVD	Cu, Si, and Si with a film of Ni or Au	Morphology of CNWs depended on the type of a substrate
[18]	CH_4/H_2	-	1.5 m/2 min	MW PECVD	Cu	The size of graphene sheets depended on a flow rate. A maximum was observed at 10–30 sccm.
[35]	$Ar/H_2/C_2H_2$	700	1 µm/30 min	RF plasma beam PECVD	SiO_2/Si, Ti, stainless steel, Quartz, MgO, carbon paper (all substrates covered with clustered Ni catalyst)	Type of the substrate material was not critical for CNWs growth. Quality of CNWs depended on pressure and Ar flow rate. Low pressure and high carrier flow rate was found to be optimal.

Table 1. Cont.

Ref.	Gas	Temperature (°C)	Growth Rate or Time	Method	Substrate Material	Important Findings
[41]	$Ar/H_2/CH_4$	-	~10 µm/min	Mesoplasma (CCP+ICP) PECVD	Si	Growth rate was increasing with increasing RF power (12–18 kW) and increasing CH_4 flow rate (10–80 sccm). Various CNWs morphologies were observed.
[27]	Ar/CH_4	750–900	up to 10 min	CCP PECVD	Cu	The density of CNWs increased with substrate temperature, plasma power, and deposition time.
[28]	$Ar/H_2/CH_4$	475–550	~10 nm/min	ICP PECVD	glass	The size and density of CNWs increased with increasing temperature.
[17]	$Ar/H_2/C_2H_2$	550, 650, 750	-	RF PECVD	Si, Ni/Si, Al_2O_3, carbon fiber	CNWs did not grow at 550 °C. Morphology of CNWs depended on temperature, pressure, and gas flow.
[59]	Ar/CH_4	700	~300 nm/min in lateral size	Atmospheric DC PECVD	Polished stainless steel	Growth rate is much higher compared to low-pressure synthesis.
[23]	$Ar/H_2/$ethanol or hexane vapor	800	100 nm/min	Atmospheric DC PECVD	Ni	Growth rate is much higher compared to low-pressure synthesis.
[24]	$Ar/H_2/$ethanol vapor	700	>15 min	Atmospheric DC PECVD	Si, Cu, stainless steel	-
[63]	Ar/CH_4 or Ar/C_2H_2	-	Several min	Low-pressure PECVD	SiO_2/Si with Au pattern	CNWs were grown on a substrate with a designed pattern.
[66, 67]	H_2/CH_4	~1000	~50–55 nm/min	DC PECVD	Glassy carbon, Si	Substrate temperature depended on the film thickness. An increase in temperature of the substrate surface resulted in an increase in the nanowall average linear size.
[22]	p-xylene	450	20 min	ICP PECVD	Si coated with TiN	Three types of carbon nanostructured were formed depending on the flow rate: fibers, free standing nanowalls, or interconnected nanowalls.
[29]	$Ar/H_2/C_2H_2$	200–700	60 min	RF plasma beam PECVD	Si	Strong dependence of morphology on temperature: CNTs were observed at 200 °C, amorphous carbon nanoparticles in the range of 300–400 °C and CNWs at 500–700 °C.
[25]	aluminum acetyl-acetonate + Ar	350, 425, 500	50 min	ICP PECVD	Stainless steel, Ni, Al, Si	Strong influence of the bias voltage, substrate temperature, and substrate material on the morphology of CNWs. Nanorods or thorny, straight, or curled CNWs were found.
[68]	H_2/CH_4	600	40 min	RF PECVD	Ni foam, copper, glass	-
[2]	$Ar/H_2/CH_4$	520–550	12 nm/min	ICP PECVD	SiO_2	Quality of CNWs increased with plasma power and temperature.

5. Comparison of Available Literature

The prior state-of-the-art can be summarized as follows:

- Either gaseous plasma or hot wires are used for production of reactive carbon-containing molecules that stick to the surface substrate and cause growing of CNWs on said substrate;
- Reactive carbon containing molecules are usually produced from hydrogenated carbon precursors, sometimes fluorinated, or from carbon oxide
- Precursors are essentially gaseous and are continuously leaked into a reaction chamber to facilitate growing of CNWs. The gases are continuously removed from the reaction chamber;
- Hydrogen is leaked into the reaction chamber simultaneously with hydrogenated carbon precursors in order to obtain good quality nanowalls. Noble gases are often added into the gas mixture leaked into the reaction chamber to alter the quality of CNWs

- Metallic catalysts were applied in early documents but have been omitted later;
- Elevated temperatures of the substrates (usually in the range of 600–800 °C) are needed for CNWs growth.

Different authors used different experimental setups so any comparison of results might not be scientifically perfect. Still, it is interesting to draw at least some correlations. Of particular importance is the growth rate versus the parameters reported in literature cited in Table 1. Figure 4 reveals the growth rates (where reported) versus the substrate temperature (where reported), Figure 5 the growth rates for different gas mixtures, and Figure 6 the growth rates for different discharges. As mentioned above the results summarized in Figures 4–6 are based on statistical evaluation of available literature from different authors as presented in Table 1.

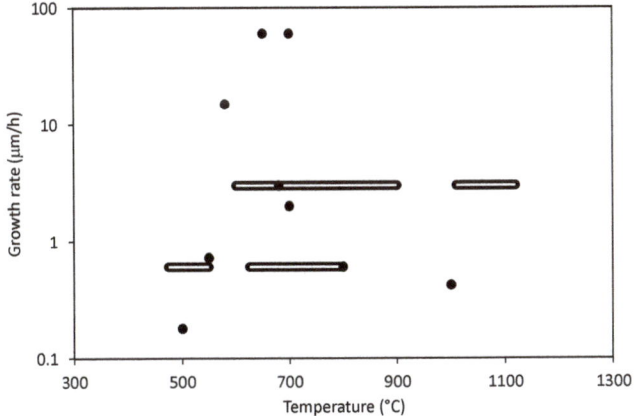

Figure 4. The growth rate versus the temperature as reported in literature shown in Table 1. The dots represent results in the cases when the authors performed experiments at a constant temperature. Some authors reported a range of temperatures during deposition—these results are represented with longitudinal bars.

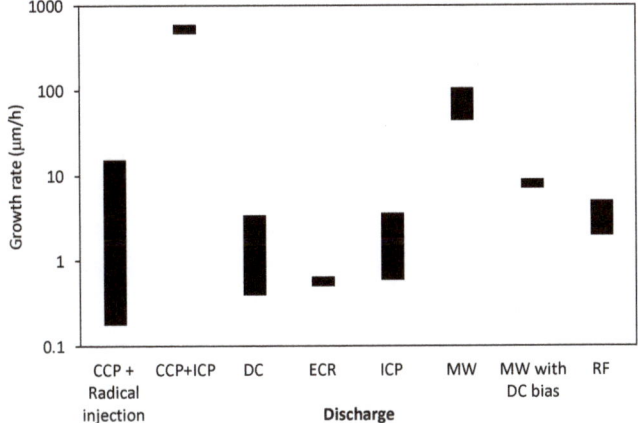

Figure 5. The growth rate for different types of discharges as reported in literature shown in Table 1. The height of the bars indicates the range of growth rates found in the literature.

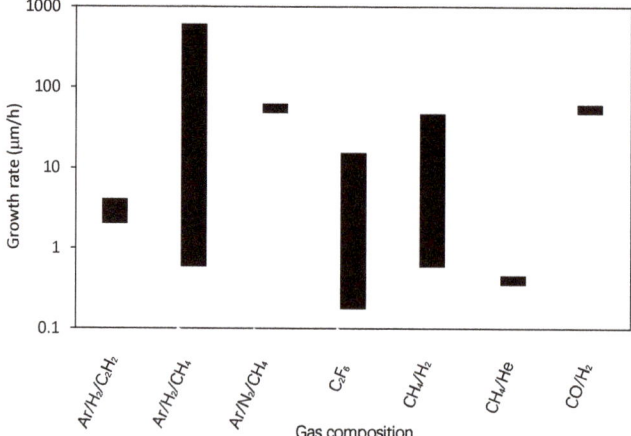

Figure 6. The growth rate for different gases as reported in literature shown in Table 1. The height of the bars indicates the range of growth rates found in the literature.

Let us first examine Figure 4 which represents the growth rate of CNWs versus reported substrate temperature. The results are scattered widely, which is explained by different experimental conditions adopted by different authors. It seems that the surface temperature alone is not a decisive parameter regarding the growth rate of CNWs. Obviously, other parameters play a more significant role as long as the growth rate is the merit. There are a couple of dots in Figure 4 that stretch from others: i.e., the measured growth rate at about 60 µm/h. Both results were obtained using microwave discharge for plasma sustenance. The MW plasma adopted by Teii [30] and Mori [20] is known for the high power density, so this parameter may be more important than the substrate temperature. The power density, of course, influences the heat dissipated on the sample upon plasma treatment, and the heat in turn influences the substrate temperature. Any sample exposed to plasma is heated by bombardment with positive ions, neutralization of charged particles, recombination of radicals (in particular atoms), accommodation of any metastables, and absorption of light quanta. The prevailing mechanism depends on fluxes of reactive species and biasing. Unfortunately, only a few authors mention these parameters, so it is difficult to deduce the heating power. In any case, the fluxes usually increase with increasing power density of the discharge until the saturation is reached. For example, the atom density next to the sample surface (and thus the atom flux onto the sample) increases with the power density, but it also depends on the properties of any material facing plasma. Carbon nanowalls should represent an almost perfect sink for atoms because they are trapped in gaps between neighboring walls, therefore they experience numerous collisions with walls before being able to escape. At each collision there is a certain probability for recombination to parent molecules and because the collisions are numerous, only few atoms are able to avoid surface recombination on a material of such a rich morphology as CNWs. In fact, one of the highest recombination coefficients was recently reported by Zaplotnik et al. [69]. Unfortunately, none of the authors cited in this review reported the atom (usually hydrogen) flux on the sample surface.

The probability for surface neutralization of positive ions is close to 100% thus the heating by this mechanism could be deduced if plasma density is measured. The ions are accelerated when crossing the sheath next to the sample surface and the kinetic energy gained is often between 10 and 20 eV (depending on the plasma potential and the ion mass); therefore, this heating mechanism is easily evaluated if the plasma density and electron temperature are known. Again, only few authors reported these parameters. The heating by ions is of course enhanced if the sample is biased, but in such a case, the thermal contact between the sample and the electrode is usually good so biasing itself does not assure for a higher sample temperature.

To understand the influence of a discharge type on the CNWs growth rate we summarize the results reported by various authors in Figure 5. According to literature shown in Table 1 and description in this review paper, various discharges were adopted. Unfortunately, the discharge power density is almost impossible to deduce from a good number of papers. Still, the results summarized in Figure 5 are useful for giving a hint on the role of a discharge in growth of CNWs. The highest growth rate was observed when combining the CCP with ICP. Unfortunately, only one author reported such a large growth rate [41]. The ICP in the H-mode is known for its ability to absorb large RF powers in small volumes so it can be concluded that the large power density is beneficial for fast growing of the CNWs. Furthermore, the MW discharges also provide high growth rates. As mentioned above, these discharges are also capable of sustaining dense plasma in a rather small volume. The results summarized in Figure 5 therefore indicate that the large power density is highly beneficial for a rapid growth of CNWs.

Finally, it is worth discussing the results of Figure 6, in particular because several authors stressed the influence of a gas composition on the growth of CNWs. As in Figures 4 and 5, the results are scattered over a couple of orders of magnitude. The high growth rate using a mixture of carbon monoxide and hydrogen can be explained by a very high power density in plasma sustained by MW discharge in a small volume, despite using a relatively low power of 60 W [20]. The results obtained using other gases or gas mixtures are scattered so much that it may be concluded the gas mixture is not the key parameter governing the growth rate of CNWs.

6. Challenges and Roadmap

The binding energy of carbon atoms in a hexagonal structure is much larger than between the graphene layers so it is natural that the synthesis of wall-like structures is dominated as long as the weakly bonded atoms are removed continuously upon growth of CNWs. The removal of such "wrongly deposited" atoms is assured by using plasma species that react chemically with weakly bonded carbon atoms and most authors agree that H atoms are particularly useful. Unfortunately, the supply of H atoms onto the surface upon the growth of CNWs seems to be too small to enable immediate removal of "wrongly deposited" carbon atoms and thus high-quality CNWs so elevated temperatures are needed, because chemical etching of carbon materials by atomic hydrogen increases with increasing sample temperature. The particular morphology of CNWs depends on deposition parameters and it has been suggested by numerous authors that bombardment of the sample with positively charged ions upon CNWs deposition is beneficial for the growth of vertically oriented (as opposite to randomly oriented) graphene sheets.

The challenges in deposition of CNWs are apparent from the text in this review paper. Although numerous authors discussed the influence of various plasma species on the growth kinetics, very few reported about the fluxes or fluences of plasma species onto the substrates. The greatest immediate challenge is therefore measuring plasma parameters. The key parameters are densities of radicals next to the substrate surface and corresponding fluxes onto the surface. While current techniques for plasma diagnostics allow for measuring densities of a variety of species they have rarely been applied. An important challenge is also determination of gradients of reactive species which appear next to or within the samples due to the loss of radicals on the surface.

A great challenge for any future application of CNWs is upscaling. Best plasma parameters are usually found in small experimental reactors of sample size measured in cm^2. Upscaling plasma of the right parameters to large systems is always a scientific and technological challenge. To make CNWs useful on an industrial scale, upscaling to systems that enable deposition of CNWs on large surfaces, at least as large as wafers, is essential. Most currently reported deposition rates are prohibitively slow therefore other solutions should be considered. Preferred deposition of any thin films for industrial application is in a continuous mode: the substrate (preferably an infinite sheet) moves through a dense plasma sustained by a suitable discharge and the deposition rate is high enough to assure a rapid deposition at a reasonable speed of the substrate. Such a mode, however, has not been adopted even for "traditional" plasma industries such as microelectronics so it might take a long time to invent techniques for fast deposition of CNWs on continuous materials.

Author Contributions: Conceptualization, A.V. and M.M.; Methodology, G.P. and R.Z.; Formal Analysis, R.Z. and G.P.; Investigation, R.Z. and G.P.; Data Curation, A.V., R.Z. and G.P.; Writing—Original Draft Preparation, A.V.; Writing—Review and Editing, M.M., G.P. and R.Z.; Visualization, R.Z.; Supervision, A.V. and M.M.; Project Administration, A.V.; Funding Acquisition, A.V.

Funding: The authors acknowledge the financial support from the Slovenian Research Agency–project No. L2-1834 (Carbon nanowalls for future supercapacitors).

Conflicts of Interest: The authors declare no conflict of interest.

References

1. Kwon, S.H.; Kim, H.J.; Choi, W.S.; Kang, H. Development and performance analysis of carbon nanowall-based mass sensor. *J. Nanosci. Nanotechnol.* **2018**, *18*, 6552–6554. [CrossRef] [PubMed]
2. Li, J.H.; Zhu, M.J.; An, Z.L.; Wang, Z.Q.; Toda, M.; Ono, T. Constructing in-chip micro-supercapacitors of 3D graphene nanowall/ruthenium oxides electrode through silicon-based microfabrication technique. *J. Power Sources* **2018**, *401*, 204–212. [CrossRef]
3. Liu, L.L.; Guan, T.; Fang, L.; Wu, F.; Lu, Y.; Luo, H.J.; Song, X.F.; Zhou, M.; Hu, B.S.; Wei, D.P.; et al. Self-supported 3D NiCo-LDH/Gr composite nanosheets array electrode for high-performance supercapacitor. *J. Alloy Compd.* **2018**, *763*, 926–934. [CrossRef]
4. Sarani, A.; Nikiforov, A.Y.; De Geyter, N.; Morent, R.; Leys, C. Surface modification of polypropylene with an atmospheric pressure plasma jet sustained in argon and an argon/water vapour mixture. *Appl. Surf. Sci.* **2011**, *257*, 8737–8741. [CrossRef]
5. Shin, S.C.; Yoshimura, A.; Matsuo, T.; Mori, M.; Tanimura, M.; Ishihara, A.; Ota, K.; Tachibana, M. Carbon nanowalls as platinum support for fuel cells. *J. Appl. Phys.* **2011**, *110*, 104308. [CrossRef]
6. Krivchenko, V.A.; Itkis, D.M.; Evlashin, S.A.; Semenenko, D.A.; Goodilin, E.A.; Rakhimov, A.T.; Stepanov, A.S.; Suetin, N.V.; Pilevsky, A.A.; Voronin, P.V. Carbon nanowalls decorated with silicon for lithium-ion batteries. *Carbon* **2012**, *50*, 1438–1442. [CrossRef]
7. Takeuchi, W.; Kondo, H.; Obayashi, T.; Hiramatsu, M.; Hori, M. Electron field emission enhancement of carbon nanowalls by plasma surface nitridation. *Appl. Phys. Lett.* **2011**, *98*, 123107. [CrossRef]
8. Wei, W.; Hu, Y.H. Highly conductive Na-embedded carbon nanowalls for hole-transport-material-free perovskite solar cells without metal electrodes. *J. Mater. Chem. A* **2017**, *5*, 24126–24130. [CrossRef]
9. Hiramatsu, M.; Kondo, H.; Hori, M. Graphene Nanowalls. In *New Progress on Graphene Research*; Gong, J.R., Ed.; IntechOpen: Rijeka, Croatia, 2013.
10. Liu, R.L.; Chi, Y.Q.; Fang, L.; Tang, Z.S.; Yi, X. Synthesis of carbon nanowall by plasma-enhanced chemical vapor deposition method. *J. Nanosci. Nanotechnol.* **2014**, *14*, 1647–1657. [CrossRef]
11. Vizireanu, S.; Stoica, S.D.; Luculescu, C.; Nistor, L.C.; Mitu, B.; Dinescu, G. Plasma techniques for nanostructured carbon materials synthesis. A case study: Carbon nanowall growth by low pressure expanding RF plasma. *Plasma Sources Sci. Technol.* **2010**, *19*, 034016. [CrossRef]
12. Achour, A.; Solaymani, S.; Vizireanu, S.; Baraket, A.; Vesel, A.; Zine, N.; Errachid, A.; Dinescu, G.; Pireaux, J.J. Effect of nitrogen configuration on carbon nanowall surface: Towards the improvement of electrochemical transduction properties and the stabilization of gold nanoparticles. *Mater. Chem. Phys.* **2019**, *228*, 110–117. [CrossRef]
13. Hiramatsu, M.; Shiji, K.; Amano, H.; Hori, M. Fabrication of vertically aligned carbon nanowalls using capacitively coupled plasma-enhanced chemical vapor deposition assisted by hydrogen radical injection. *Appl. Phys. Lett.* **2004**, *84*, 4708–4710. [CrossRef]
14. Hiramatsu, M.; Hori, M. *Carbon Nanowalls: Synthesis and Emerging Applications*; Springer: Wien, Austria, 2010.
15. Li, J.; Liu, Z.; Guo, Q.; Yang, S.; Xu, A.; Wang, Z.; Wang, G.; Wang, Y.; Chen, D.; Ding, G. Controllable growth of vertically oriented graphene for high sensitivity gas detection. *J. Mater. Chem. C* **2019**, *7*, 5995–6003. [CrossRef]
16. Kondo, S.; Kawai, S.; Takeuchi, W.; Yamakawa, K.; Den, S.; Kano, H.; Hiramatsu, M.; Hori, M. Initial growth process of carbon nanowalls synthesized by radical injection plasma-enhanced chemical vapor deposition. *J. Appl. Phys.* **2009**, *106*, 094302. [CrossRef]

17. Cui, L.; Chen, J.; Yang, B.; Sun, D.; Jiao, T. RF-PECVD synthesis of carbon nanowalls and their field emission properties. *Appl. Surf. Sci.* **2015**, *357*, 1–7. [CrossRef]
18. Jiang, L.; Yang, T.; Liu, F.; Dong, J.; Yao, Z.; Shen, C.; Deng, S.; Xu, N.; Liu, Y.; Gao, H.-J. Controlled Synthesis of Large-Scale, Uniform, Vertically Standing Graphene for High-Performance Field Emitters. *Adv. Mater.* **2013**, *25*, 250–255. [CrossRef] [PubMed]
19. Wang, J.; Zhu, M.; Outlaw, R.A.; Zhao, X.; Manos, D.M.; Holloway, B.C. Synthesis of carbon nanosheets by inductively coupled radio-frequency plasma enhanced chemical vapor deposition. *Carbon* **2004**, *42*, 2867–2872. [CrossRef]
20. Mori, S.; Ueno, T.; Suzuki, M. Synthesis of carbon nanowalls by plasma-enhanced chemical vapor deposition in a CO/H_2 microwave discharge system. *Diam. Relat. Mater.* **2011**, *20*, 1129–1132. [CrossRef]
21. Mori, S.; Suzuki, M. Non-catalytic low-temperature synthesis of carbon nanofibers by plasma-enhanced chemical vapor deposition in a $CO/Ar/O_2$ DC discharge system. *Appl. Phys. Express* **2009**, *2*, 015003. [CrossRef]
22. Lehmann, K.; Yurchenko, O.; Urban, G. Effect of the aromatic precursor flow rate on the morphology and properties of carbon nanostructures in plasma enhanced chemical vapor deposition. *RSC Adv.* **2016**, *6*, 32779–32788. [CrossRef]
23. Meško, M.; Vretenár, V.; Kotrusz, P.; Hulman, M.; Šoltýs, J.; Skákalová, V. Carbon nanowalls synthesis by means of atmospheric dcPECVD method. *Phys. Status Solidi B* **2012**, *249*, 2625–2628. [CrossRef]
24. Yu, K.H.; Bo, Z.; Lu, G.H.; Mao, S.; Cui, S.M.; Zhu, Y.W.; Chen, X.Q.; Ruoff, R.S.; Chen, J.H. Growth of carbon nanowalls at atmospheric pressure for one-step gas sensor fabrication. *Nanoscale Res. Lett.* **2011**, *6*, 202. [CrossRef] [PubMed]
25. Giese, A.; Schipporeit, S.; Buck, V.; Wohrl, N. Synthesis of carbon nanowalls from a single-source metal-organic precursor. *Beilstein J. Nanotechnol.* **2018**, *9*, 1895–1905. [CrossRef] [PubMed]
26. Ghosh, S.; Polaki, S.R.; Kumar, N.; Amirthapandian, S.; Kamruddin, M.; Ostrikov, K.K. Process-specific mechanisms of vertically oriented graphene growth in plasmas. *Beilstein J. Nanotechnol.* **2017**, *8*, 1658–1670. [CrossRef] [PubMed]
27. Yang, Q.; Wu, J.; Li, S.; Zhang, L.; Fu, J.; Huang, F.; Cheng, Q. Vertically-oriented graphene nanowalls: Growth and application in Li-ion batteries. *Diam. Relat. Mater.* **2019**, *91*, 54–63. [CrossRef]
28. Zhang, N.; Li, J.; Liu, Z.; Yang, S.; Xu, A.; Chen, D.; Guo, Q.; Wang, G. Direct synthesis of vertical graphene nanowalls on glass substrate for thermal management. *Mater. Res. Express* **2018**, *5*, 065606. [CrossRef]
29. Gentoiu, M.A.; Betancourt-Riera, R.; Vizireanu, S.; Burducea, I.; Marascu, V.; Stoica, S.D.; Bita, B.I.; Dinescu, G.; Riera, R. Morphology, microstructure, and hydrogen content of carbon nanostructures obtained by PECVD at various temperatures. *J. Nanomater.* **2017**, *2017*, 1374973.
30. Teii, K.; Shimada, S.; Nakashima, M.; Chuang, A.T.H. Synthesis and electrical characterization of n-type carbon nanowalls. *J. Appl. Phys.* **2009**, *106*, 084303. [CrossRef]
31. Park, H.J.; Ahn, B.W.; Kim, T.Y.; Lee, J.W.; Jung, Y.H.; Choi, Y.S.; Song, Y.I.; Suh, S.J. Direct synthesis of multi-layer graphene film on various substrates by microwave plasma at low temperature. *Thin Solid Films* **2015**, *587*, 8–13. [CrossRef]
32. Singh, M.; Jha, H.S.; Agarwal, P. Synthesis of vertically aligned carbon nanoflakes by hot-wire chemical vapor deposition: Influence of process pressure and different substrates. *Thin Solid Films* **2019**, *678*, 26–31. [CrossRef]
33. Kalita, G.; Wakita, K.; Umeno, M. Low temperature growth of graphene film by microwave assisted surface wave plasma CVD for transparent electrode application. *RSC Adv.* **2012**, *2*, 2815–2820. [CrossRef]
34. Hofmann, S.; Kleinsorge, B.; Ducati, C.; Ferrari, A.C.; Robertson, J. Low-temperature plasma enhanced chemical vapour deposition of carbon nanotubes. *Diam. Relat. Mater.* **2004**, *13*, 1171–1176. [CrossRef]
35. Vizireanu, S.; Mitu, B.; Luculescu, C.R.; Nistor, L.C.; Dinescu, G. PECVD synthesis of 2D nanostructured carbon material. *Surf. Coat. Technol.* **2012**, *211*, 2–8. [CrossRef]
36. Wu, Y.H.; Qiao, P.W.; Chong, T.C.; Shen, Z.X. Carbon nanowalls grown by microwave plasma enhanced chemical vapor deposition. *Adv. Mater.* **2002**, *14*, 64–67. [CrossRef]
37. Shiji, K.; Hiramatsu, M.; Enomoto, A.; Nakamura, N.; Amano, H.; Hori, M. Vertical growth of carbon nanowalls using rf plasma-enhanced chemical vapor deposition. *Diam. Relat. Mater.* **2005**, *14*, 831–834. [CrossRef]

38. Tanaka, K.; Yoshimura, M.; Okamoto, A.; Ueda, K. Growth of carbon nanowalls on a SiO_2 substrate by microwave plasma-enhanced chemical vapor deposition. *Jpn. J. Appl. Phys.* **2005**, *44*, 2074–2076. [CrossRef]
39. Dikonimos, T.; Giorgi, L.; Giorgi, R.; Lisi, N.; Salernitano, E.; Rossi, R. DC plasma enhanced growth of oriented carbon nanowall films by HFCVD. *Diam. Relat. Mater.* **2007**, *16*, 1240–1243. [CrossRef]
40. Davami, K.; Shaygan, M.; Kheirabi, N.; Zhao, J.; Kovalenko, D.A.; Rümmeli, M.H.; Opitz, J.; Cuniberti, G.; Lee, J.-S.; Meyyappan, M. Synthesis and characterization of carbon nanowalls on different substrates by radio frequency plasma enhanced chemical vapor deposition. *Carbon* **2014**, *72*, 372–380. [CrossRef]
41. Zhang, H.; Wu, S.; Lu, Z.; Chen, X.; Chen, Q.; Gao, P.; Yu, T.; Peng, Z.; Ye, J. Efficient and controllable growth of vertically oriented graphene nanosheets by mesoplasma chemical vapor deposition. *Carbon* **2019**, *147*, 341–347. [CrossRef]
42. Hiramatsu, M.; Hori, M. Method for Producing Carbon Nanowalls, Carbon Nanowall, and Apparatus For Producing Carbon Nanowalls. Patent No. US20070184190A1, 8 August 2007.
43. Hori, M.; Koaizawa, H.; Shibayama, S.; Toda, S. Production Method and Production Device for Carbon Nanostructure. Patent No. JP2008063196A, 21 March 2008.
44. Hori, M.; Hiramatsu, M.; Kano, H. Method for Producing Carbon Nanowalls. Patent No. US2011045207A1, 24 February 2011.
45. Ghoanneviss, M.; Eslami, A.P.; Laheghi, S.N. Method for Growing Carbon Nanowalls. Patent No. US2009274610A1, 5 November 2009.
46. Ohmae, N.; Toyoshima, A. CO_2 Reduction Device and CO_2 Reduction Method. Patent No. WO2016024301A1, 16 February 2016.
47. Hori, M.; Kano, H.; Hama, Y. Method for Refining Carbon Nanowall (CNW), Refined Carbon Nanowall, Method for Manufacturing Catalyst Layer for Fuel Cell, Catalyst Layer for Fuel Cell, and Polymer Electrolyte Fuel Cell. Patent No. JP2008239369A, 9 October 2008.
48. Hori, M.; Hiramatsu, M.; Kano, H.; Yoshida, S.; Katayama, Y.; Sugiyama, T. Method and Apparatus for Producing Catalyst Layer for Fuel Cell. Patent No. US2008274392A1, 6 November 2008.
49. Tachibana, M.; Tanaike, O. Negative Electrode Material for Lithium Ion Battery and Rapid Charging/Discharging Lithium Ion Battery Using the Same. Patent No. JP2010009980A, 14 January 2010.
50. Yoshida, S.; Hama, Y.; Hori, M.; Hiramatsu, M.; Kano, H. Negative Electrode for Lithium Secondary Battery, Method for Preparing the Negative Electrode, Lithium Secondary Battery having the Negative Electrode, and Vehicle having the Lithium Secondary Battery. Patent No. CN102668180A, 14 November 2012.
51. Izuhara, K.; Daifuku, M.; Miyata, Y. Lithium Secondary Battery Negative Electrode and Method For Manufacturing the Same. Patent No. US2014170490A1, 19 June 2014.
52. Yoshida, S.; Hama, Y.; Hori, M.; Hiramatsu, M.; Kano, H. Positive Electrode for Lithium Secondary Battery, Method for Preparing the Positive Electrode, Lithium Secondary Battery having the Positive Electrode, and Vehicle having the Lithium Secondary Battery. Patent No. CN102668181A, 12 September 2012.
53. Hori, M.; Sato, H.; Toyoshima, Y.; Hiramatsu, M. Sample Substrate for Laser Desorption Ionization-Mass Spectrometry, and Method and Device Both Using the Same for Laser Desorption Ionization-Mass Spectrometry. Patent No. US2012175515A1, 12 July 2012.
54. Omi, S.; Kawaguchi, N.; Kondo, M.; Harasaki, O. Saturable Absorbing Element, Saturable Absorbing Element Producing Method, and Laser Apparatus. Patent No. JP2015118348A, 25 June 2015.
55. Junkar, I.; Modic, M.; Vesel, A.; Mozetic, M.; Dinescu, G.; Vizireanu, S.I.; Kleinschek, K.S. Method of Growing Carbon Nanowalls on a Substrate. Patent No. WO2016059024A1, 21 April 2016.
56. Zhou, M.; Yuan, X.; Wang, Y.; Wu, F. Carbon Nanowall and Graphene Nanoribbon Preparation Method. Patent No. CN103935975A, 23 July 2014.
57. Zhou, M.; Yuan, X.; Wang, Y.; Wu, F. Graphene Nanoribbon Preparation Method. Patent No. CN103935982A, 15 December 2014.
58. Yoshimura, A.; Matsuo, T.; Tachibana, M.; Shin, S.C. Fabrication Method for Metal-Supported Nano-Graphite. Patent No. US2014127411A1, 8 May 2014.
59. Bo, Z.; Yu, K.; Lu, G.; Wang, P.; Mao, S.; Chen, J. Understanding growth of carbon nanowalls at atmospheric pressure using normal glow discharge plasma-enhanced chemical vapor deposition. *Carbon* **2011**, *49*, 1849–1858. [CrossRef]

60. Lee, H.; Bratescu, M.A.; Ueno, T.; Saito, N. Solution plasma exfoliation of graphene flakes from graphite electrodes. *RSC Adv.* **2014**, *4*, 51758–51765. [CrossRef]
61. Li, Y.; Chen, Q.; Xu, K.; Kaneko, T.; Hatakeyama, R. Synthesis of graphene nanosheets from petroleum asphalt by pulsed arc discharge in water. *Chem. Eng. J.* **2013**, *215–216*, 45–49. [CrossRef]
62. Amano, T.; Kondo, H.; Ishikawa, K.; Tsutsumi, T.; Takeda, K.; Hiramatsu, M.; Sekine, M.; Hori, M. Rapid growth of micron-sized graphene flakes using in-liquid plasma employing iron phthalocyanine-added ethanol. *Appl. Phys. Express* **2017**, *11*, 015102. [CrossRef]
63. Yu, K.H.; Wang, P.X.; Lu, G.H.; Chen, K.H.; Bo, Z.; Chen, J.H. Patterning vertically oriented graphene sheets for nanodevice applications. *J. Phys. Chem. Lett.* **2011**, *2*, 537–542. [CrossRef]
64. Itoh, T.; Shimabukuro, S.; Kawamura, S.; Nonomura, S. Preparation and electron field emission of carbon nanowall by Cat-CVD. *Thin Solid Films* **2006**, *501*, 314–317. [CrossRef]
65. Shimabukuro, S.; Hatakeyama, Y.; Takeuchi, M.; Itoh, T.; Nonomura, S. Effect of hydrogen dilution in preparation of carbon nanowall by hot-wire CVD. *Thin Solid Films* **2008**, *516*, 710–713. [CrossRef]
66. Krivchenko, V.; Shevnin, P.; Pilevsky, A.; Egorov, A.; Suetin, N.; Sen, V.; Evlashin, S.; Rakhimov, A. Influence of the growth temperature on structural and electron field emission properties of carbon nanowall/nanotube films synthesized by catalyst-free PECVD. *J. Mater. Chem.* **2012**, *22*, 16458–16464. [CrossRef]
67. Krivchenko, V.A.; Dvorkin, V.V.; Dzbanovsky, N.N.; Timofeyev, M.A.; Stepanov, A.S.; Rakhimov, A.T.; Suetin, N.V.; Vilkov, O.Y.; Yashina, L.V. Evolution of carbon film structure during its catalyst-free growth in the plasma of direct current glow discharge. *Carbon* **2012**, *50*, 1477–1487. [CrossRef]
68. Zhou, H.T.; Liu, D.B.; Luo, F.; Luo, B.W.; Tian, Y.; Chen, D.S.; Shen, C.M. Preparation of graphene nanowalls on nickel foam as supercapacitor electrodes. *Micro Nano Lett.* **2018**, *13*, 842–844. [CrossRef]
69. Mozetic, M.; Vesel, A.; Stoica, S.D.; Vizireanu, S.; Dinescu, G.; Zaplotnik, R. Oxygen atom loss coefficient of carbon nanowalls. *Appl. Surf. Sci.* **2015**, *333*, 207–213. [CrossRef]

© 2019 by the authors. Licensee MDPI, Basel, Switzerland. This article is an open access article distributed under the terms and conditions of the Creative Commons Attribution (CC BY) license (http://creativecommons.org/licenses/by/4.0/).

Article

Effect of Doping Temperatures and Nitrogen Precursors on the Physicochemical, Optical, and Electrical Conductivity Properties of Nitrogen-Doped Reduced Graphene Oxide

Nonjabulo P. D. Ngidi, Moses A. Ollengo and Vincent O. Nyamori *

School of Chemistry and Physics, University of KwaZulu-Natal, Westville Campus, Private Bag X54001, Durban 4000, South Africa; nonjabulongidi@gmail.com (N.P.D.N.); mosesollengo@gmail.com (M.A.O.)
* Correspondence: nyamori@ukzn.ac.za; Tel.: +27-31-2608256; Fax: +27-31-2603091

Received: 10 September 2019; Accepted: 5 October 2019; Published: 16 October 2019

Abstract: The greatest challenge in graphene-based material synthesis is achieving large surface area of high conductivity. Thus, tuning physico-electrochemical properties of these materials is of paramount importance. An even greater problem is to obtain a desired dopant configuration which allows control over device sensitivity and enhanced reproducibility. In this work, substitutional doping of graphene oxide (GO) with nitrogen atoms to induce lattice–structural modification of GO resulted in nitrogen-doped reduced graphene oxide (N-rGO). The effect of doping temperatures and various nitrogen precursors on the physicochemical, optical, and conductivity properties of N-rGO is hereby reported. This was achieved by thermal treating GO with different nitrogen precursors at various doping temperatures. The lowest doping temperature (600 °C) resulted in less thermally stable N-rGO, yet with higher porosity, while the highest doping temperature (800 °C) produced the opposite results. The choice of nitrogen precursors had a significant impact on the atomic percentage of nitrogen in N-rGO. Nitrogen-rich precursor, 4-nitro-o-phenylenediamine, provided N-rGO with favorable physicochemical properties (larger surface area of 154.02 m^2 g^{-1}) with an enhanced electrical conductivity (0.133 S cm^{-1}) property, making it more useful in energy storage devices. Thus, by adjusting the doping temperatures and nitrogen precursors, one can tailor various properties of N-rGO.

Keywords: reduced graphene oxide; nitrogen-doping; chemical vapor deposition; physicochemical properties; optical properties; electrical conductivity

1. Introduction

Functionalization of carbon-based materials, such as graphene and carbon nanotubes for different purposes, is gaining a lot of attention in the field of material science. This interest arises because of their low cost, unique stable physicochemical properties, and broad applications. The former includes energy-harvesting devices [1], supercapacitors [2], sensors [3], field-effect transistors [4] and medical uses [5]. Certain extrinsic properties, such as electrical conductivity, high chemical stability, and a zero band-gap, enable some carbon-based materials to perform as semi-metals and semi-conductors [6]. Graphene, for instance, has a zero band-gap which needs to be manipulated for use in various applications such as solar cells. When graphene is chemically doped, it can change one absorbed photon and cause an increase in power conversion efficiency of solar cells [7]. However, graphene is a transparent material with a low coefficient of light absorption. Therefore, when graphene is applied in solar cells, it tends to produce a lower power conversion efficiency than solar cells based on heteroatom-doped graphene [7]. Thus, creating a well-tuned and sizeable band-gap to improve the coefficient of light absorption of graphene is a great challenge but with enormous interest.

The band-gap of graphene can be tuned by altering the surface chemistry through substitutional doping [8]. This can be achieved using selected heteroatoms to tune and enhance the band structure and conductivity [9,10]. Various heteroatoms that have commonly been employed in substitutional doping include boron [11–13] and nitrogen [14–16]. This is because they possess similar atomic radii and sizes to carbon and impact interesting electron chemistry within the graphene framework [17]. These heteroatoms have a significant effect on the electrical properties of graphene which is shown by a p-type conductivity for boron whereas nitrogen results in n-type conductivity [18]. In the case of strong p-type doping, it can be conferred by the interaction with the environment, hence, nitrogen-doping does not always confer n-type conductivity unless graphene is encapsulated [19]. The nitrogen atom is mostly used in chemical doping of graphene or graphene oxide (GO). This is because nitrogen atom acts as defect site in the crystal structure of graphene and these defective centers can enhance the electrochemical activity of graphene or GO [20].

Nitrogen-doping suppresses the density of state of graphene near the Fermi level and results to band-gap opening. Furthermore, nitrogen-doping tends to introduce strong electron donor states and leads to n-type or p-type semiconductor behavior depending on the bonding configuration. The conductivity and carrier mobilities of nitrogen-doped graphene are lower than pristine graphene due to the presence of nitrogen atom and defects introduced during the nitrogen-doping process which are capable of functioning as scattering centers that hinder the electron or hole transport [21]. Boron-doping in graphene, results in a p-type doping and is also highly favorable. This is because B-C bond is about 0.5% longer than the C-C bond while N-C bond is about the same as C-C bond in length, enabling formation of relaxed structure of boron-doped graphene. Boron-doping tends to introduce more holes into the valence band of graphene resulting in a high carrier concentration. Boron-doped graphene is reported to have high conductivity compared to pristine graphene [22] and nitrogen-doped graphene [23], due to the large density of state near the Fermi level.

Doping GO with nitrogen, results in nitrogen-doped reduced graphene oxide (N-rGO). The ideal physicochemical properties of N-rGO for optoelectrical applications include a large surface area and high chemical stability. These physicochemical properties of N-rGO can be significantly enhanced by improving the atomic percentage of nitrogen [24] and the bonding configuration [25]. Various bonding configurations of nitrogen in N-rGO have been reported, e.g., pyrrolic-N [25], pyridinic-N [26], quaternary-N [27,28] and oxide-N [29]. These bonding configurations impart various effects on the carrier concentration which tend to produce well-defined band structures in doped GO [30].

The mechanism of formation of N-rGO is still a fascinating phenomenon because it is not well understood and there is more that can be done to manipulate it. Therefore, synthetic procedures for N-rGO need a certain level of control regarding the required extent of doping and the bonding configuration of nitrogen. Different synthetic approaches have been employed in the in-situ synthesis of N-rGO, such as arc discharge [31], plasma method [32], thermal annealing [33] and chemical vapor deposition (CVD) [34,35]. The CVD approach is mostly preferred because it is easier to scale-up and produces relatively high-quality N-rGO. Scientific reports on the synthesis of N-rGO via the CVD approach indicate that the mostly used materials are metal catalyst (Cu, Ni, Co or Fe) [36,37] and organic molecules [38].

In the CVD synthesis of N-rGO, several factors including the type of carrier gas, doping temperature and nitrogen precursor (used either as a solid, liquid or in the gaseous phase), influence the nitrogen content and properties of the final product [39,40]. Nang et al. [41] and Panchakarla et al. [42] reported the use of dimethylformamide and pyridine, respectively, as liquid nitrogen precursors for the synthesis of N-rGO, with the former achieving a very low nitrogen content of 0.64%. The drawback of liquid nitrogen precursors is that they are expensive, dangerous, and highly flammable when used in the CVD method.

The alternative to liquid and gaseous nitrogen precursors is solid nitrogen precursors. The use of solid nitrogen precursors, such as monoethanolamine [43], urea [33,44], 1,3,5-triazine [24], pentachloropyridine [36] and the combination of imidazole and melamine [45] have been reported and observed to result in high doping levels. Lu et al. [24] reported the CVD synthesis of a few-layered nitrogen-doped graphene oxide containing atomic percentages of between 2.1 and 5.6% nitrogen by making use of the carbon and nitrogen precursor 1,3,5-triazine and Cu foil as catalyst. Doped graphene films with a higher nitrogen content of approximately 5.6% were obtained at a doping temperature of 990 °C, with melamine [46] as a solid nitrogen precursor. The use of the solid nitrogen precursor, pentachloropyridine, in the synthesis of nitrogen-doped graphene was reported by Wan et al. [36] to yield a nitrogen content between 4.4 and 7.5%. Solid nitrogen precursors are cost-effective and are easy to handle compared with liquid and gaseous nitrogen precursors.

In this work, we report for the first time, the effect of different doping temperatures and solid nitrogen precursors on the physicochemical (nitrogen content, crystallinity, thermal stability and bonding configuration), optical (band-gap energy and charge recombination) and electrical conductivity properties of N-rGO. The synthesis of N-rGO (Figure 1) was achieved by liquid exfoliation of GO, high temperature vapor reduction of GO and doping it with nitrogen atoms from various solid nitrogen precursors (4-nitroaniline, 4-aminophenol and 4-nitro-o-phenylenediamine). These nitrogen precursors were chosen because they possess different number of nitrogen atoms on their structures or frameworks, and therefore the effect of the number of nitrogen atoms contained in the nitrogen precursor was also investigated.

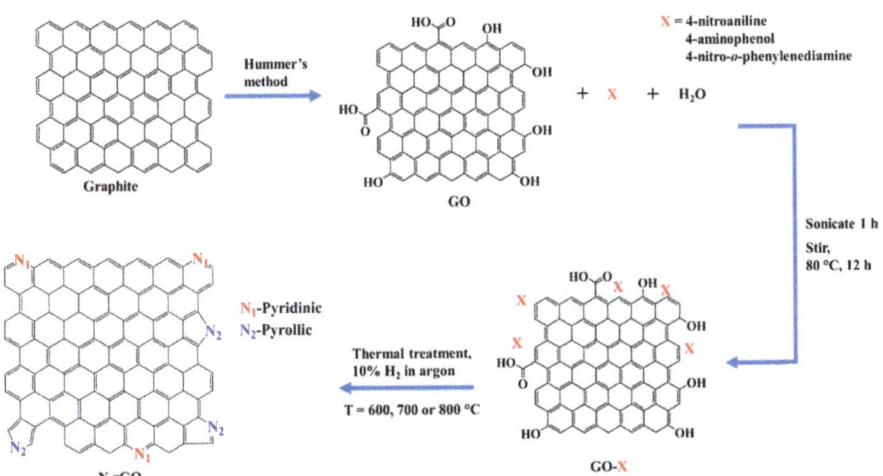

Figure 1. Schematic diagram of the conversion of graphite to N-rGO.

2. Materials and Methods

2.1. Materials and Instrumentation

Graphite powder (99.99% -trace metal basis), sodium nitrate (99%), potassium permanganate (99%), 4-nitroaniline (≥99%), 4-aminophenol (≥99%) and 4-nitro-o-phenylenediamine (≥99%) were purchased from Sigma-Aldrich, Johannesburg, South Africa. These chemicals were of analytical grade and were used without further purification. Hydrogen peroxide (100%) was purchased from Merck Laboratory Supplies, Johannesburg, South Africa. Sulfuric acid (98%) was purchased from Promark Chemicals, Johannesburg, South Africa. Double distilled water was obtained from a double distiller, Glass Chem water distiller model Ws4lcd was supplied by Shalom Laboratory Supplies, Durban, South Africa. A gas mixture of 10% hydrogen in argon (v/v) was purchased from Afrox Limited Gas Co., Durban, South Africa. Weighing of N-rGO was done on an electronic weighing balance, Mettler AE 200, Mundelein, IL USA. Ultrasonication was performed in a digital ultrasonic water bath (400 W) obtained from Shalom Laboratory Supplies, Durban, South Africa.

2.2. Synthesis of N-rGO

A modified Hummer's method was used to synthesize GO [47]. In brief, approximately 1 g of graphite powder and 1 g of sodium nitrate were mixed with 50 mL of concentrated sulfuric acid in a 500 mL round-bottom flask placed in an ice-bath and stirred for 30 min. After that, 6 g of potassium permanganate was added slowly to the mixture with the temperature kept at 5 °C to prevent explosion and excessive heating. Thereafter, the mixture was stirred for 3 h at a temperature of 35 °C and then further treated with 200 mL of 3% hydrogen peroxide while stirring for 30 min. The resulting GO was washed with double distilled water until a pH of 6 was achieved. The product was then filtered and dried in the oven for 24 h at a temperature of 80 °C.

The synthesized GO was simultaneously reduced (using 10% hydrogen in argon as a reducing agent) and doped with different nitrogen precursors, namely 4-nitroaniline, 4-aminophenol and 4-nitro-o-phenylenediamine. This was done by mixing 70 mg of GO and 30 mg of nitrogen precursor in 50 mL double distilled water, followed by sonication (25 °C) for 1 h. The mixture was further stirred and heated for 6 h at a temperature of 100 °C to remove the excess water. After drying, the resulting black solid was heat-treated in a ceramic quartz boat placed in a tube furnace (Elite Thermal Systems Ltd., Model TSH12/50/160) in a mixture of 10% hydrogen in argon (v/v) at a flow rate of 100 mL min^{-1}. The doping temperature of the furnace was set at each of 600, 700 and 800 °C, for each nitrogen precursor. The carrier gas flow rate and the doping temperature were maintained constant throughout the synthesis period of 2 h. After 2 h, the furnace was allowed to cool naturally to room temperature and N-rGO was collected and subsequently characterized.

2.3. Physicochemical Characterization

The surface morphology of N-rGO was investigated by field emission scanning electron microscopy (FE-SEM, Carl Zeiss Ultra Plus, Cambridge, UK). Briefly, the aluminum stub sample holders were coated with piece of a sticky carbon tape; after that the N-rGO was sprinkled on the carbon tape and gold coated thrice before SEM analysis. The microstructural features of the N-rGO were evaluated by means of high-resolution-transmission electron microscopy (HR-TEM, JOEL JEM model 1010, Peabody, MA, USA), and set at an accelerating voltage of 100 kV at different magnifications.

The crystallinity or graphitic nature of the N-rGO was investigated with a Delta Nu Advantage 532^{TM} Raman spectrometer (Laramie, WY, USA) equipped with NuSpec^{TM} software (1.0., Microsoft Publisher, Redmond, WA, USA) and operated at a wavelength (λ) of 514.5 nm. The functional groups present in the N-rGO were investigated with a PerkinElmer Spectrum 100 Fourier transform infrared (FTIR) spectrometer (Akron, OH, USA) equipped with an attenuated total reflectance (ATR) accessory. Approximately 0.22 g of the N-rGO was pressed into a pellet for about 2 min, under a pressure of 10 Tons. The pellets were then placed on the diamond crystal for analysis.

The thermal stability of N-rGO was measured with a TA Instruments Q series^{TM} thermal analysis instrument (DSC/thermogravimetric analysis (TGA) (SDT-Q600), New Castle, PA, USA) in air flowing at a rate of 50 mL min^{-1} and heated from room temperature up to 1000 °C at a ramping rate of 10 °C min^{-1}. N-rGO were further characterized by X-ray photoelectron spectroscopy (XPS, Quantum 2000 with an X-ray source of monochromatic Al K$_\alpha$ (1486.7 eV), Chanhassen, MN, USA) to investigate the surface chemical composition of carbon and nitrogen.

A Micromeritics Tristar II 3020 surface area and porosity analyzer (Norcross, GA, USA) was used to determine the textural properties of N-rGO. Typically, a mass of approximately 0.1 g of N-rGO was degassed at 90 °C for 1 h, the temperature was then raised to 160 °C and the sample further degassed for 12 h using Micromeritics Vacprep 061 (sample degas system), before fitting N-rGO in the Micromeritics Tristar II instrument for analysis. The textural properties of the N-rGO were investigated at a temperature of −196 °C with N$_2$ as the adsorbate. The specific surface areas were calculated with the Brunauer, Emmett and Teller (BET) model and the pore volumes were obtained by applying the Barrett-Joyner-Halenda (BJH) model.

The phase characteristics of the synthesized N-rGO were determined by X-ray powder diffraction (XRD, Rigaku/Dmax RB, The Woodlands, TX, USA) and the measurements were performed with graphite monochromated high-density with a θ-θ scan in locked coupled mode, using a Cu k_α radiation source (λ = 0.15406 nm). The absorbance of N-rGO was recorded with an ultraviolet-visible spectrophotometer (UV-Vis, Shimadzu, UV-1800, Roodepoort, South Africa). The GO and N-rGO were first dispersed in absolute ethanol and then sonicated for 30 min before UV-Visible spectrophotometric analysis. Charge recombination analysis of N-rGO was investigated with a PerkinElmer LS 55 spectrofluorometer (Akron, OH, USA) fitted with solid sample accessory. Excitation was performed at 310 nm, and the emission spectrum recorded from 450 to 550 nm with an excitation slit and emission slit at 5 nm and 2 nm, respectively (slid position). Electrical conductivity of N-rGO was determined by four-point probe (Keithley 2400 source-meter, Beaverton, OR, USA) measurements which were carried out on pellets with a thickness of 0.2 mm formed from N-rGO (0.03 g).

3. Results and Discussion

The physicochemical characteristics of N-rGO synthesized with different nitrogen precursors and at different doping temperatures of 600, 700 and 800 °C are presented. Both factors influence the level of nitrogen-doping, morphology, crystallinity, thermal stability, optical, and electrical conductivity properties of N-rGO.

3.1. Morphology

The nitrogen atoms from the different nitrogen precursors were successfully introduced into the GO lattice. N-rGO synthesized from: 4-aminophenol is represented by N-rGO-1N, 4-nitroaniline is represented by N-rGO-2N while 4-nitro-o-phenylenediamine is represented by N-rGO-3N. N-rGO in the SEM images showed a thick and overlapping sheet structure (Supplementary Materials—Figure S1). This was attributed to flake-like structures. A similar observation for SEM images of N-rGO was reported by Jiang et al. [48]. Further details of the structures were evaluated by HR-TEM. For comparison, Figure 2 presents the HR-TEM images and the selected area electron diffraction (SAED) patterns of GO and N-rGO. Both GO and N-rGO exhibited a wrinkled structure (Figure 2a), which increased after doping (Figure 2c). The more wrinkled structure in N-rGO is caused by the stimulation of defects such as pores, holes and cavities which were introduced during the process of doping [49]. The different doping temperatures and nitrogen precursors revealed no effect on the morphology of N-rGO.

The SAED patterns recorded to study the crystalline structures of GO and N-rGO revealed two diffraction rings which were associated with the (002) and (100) planes for both GO (Figure 2b) and N-rGO (Figure 2d). The presence of hexagonal diffraction spots in the electron diffraction pattern observed, indicates that N-rGO has a well-ordered structure while the occurrence of structural distortion after doping as revealed by the ring-like diffraction pattern. The observed disorder might be due to the introduction of functional groups, and the overlapping graphene sheets [50]. The diffraction spots in hexagonal positions in N-rGO are reflective of the preservation of the original honey-comb-like atomic structure of graphene [51]. The GO also shows a ring-like structure (distortion) which is due to ring spacing (Figure 2b).

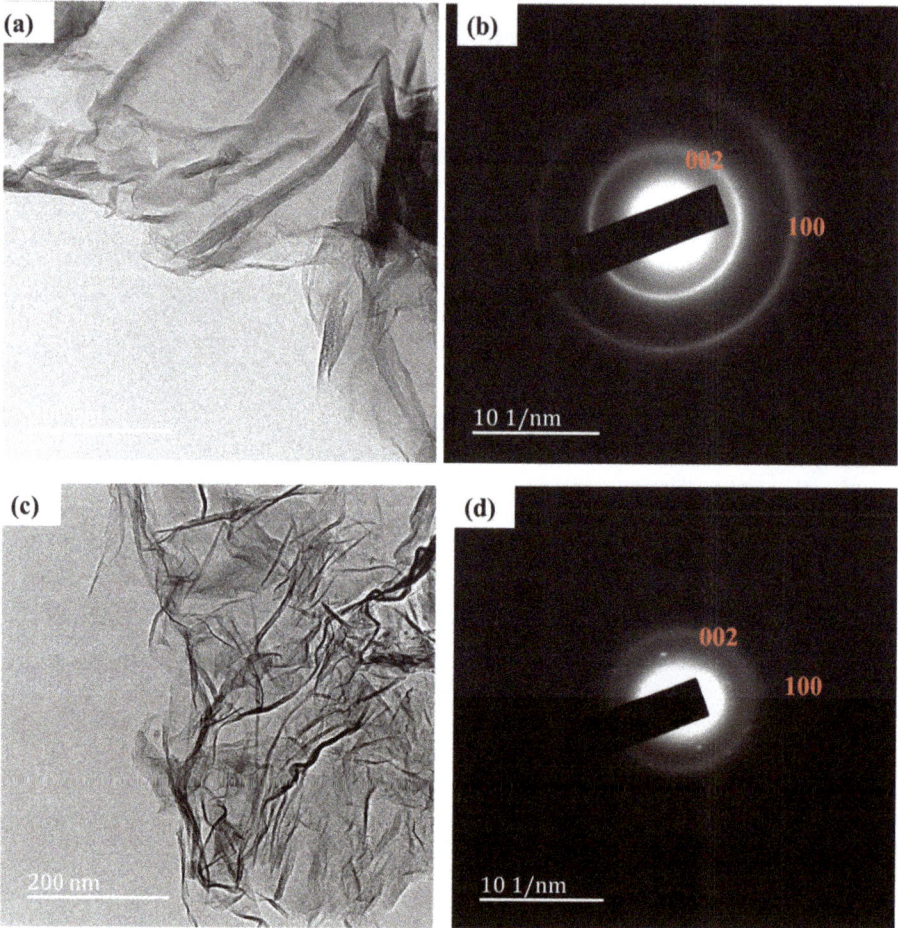

Figure 2. HR-TEM image of (**a**) GO, (**b**) SAED pattern of GO, (**c**) HR-TEM image of N-rGO-1N-600 °C and (**d**) SAED pattern of N-rGO-1N-600 °C.

The different interlayer spacings (d_{002} spacing) on the edge of GO and at the cross-sections of layers of N-rGO as observed in the HR-TEM images (Figure 3a,b) showed that the carbon atom layers were not identical; this is indicative that the N-rGO consists of few-layers of graphene sheets (Table 1). After nitrogen-doping, the d_{002} spacing was found to decrease. For example, the d_{002} spacing of N-rGO-1N-600 °C was found to be 0.37 nm which is smaller than that of GO (0.47 nm). The decrease in d_{002} spacing after nitrogen-doping was attributed to the reduction of oxygen functional groups such as carboxyl, epoxy and hydroxyl groups [52].

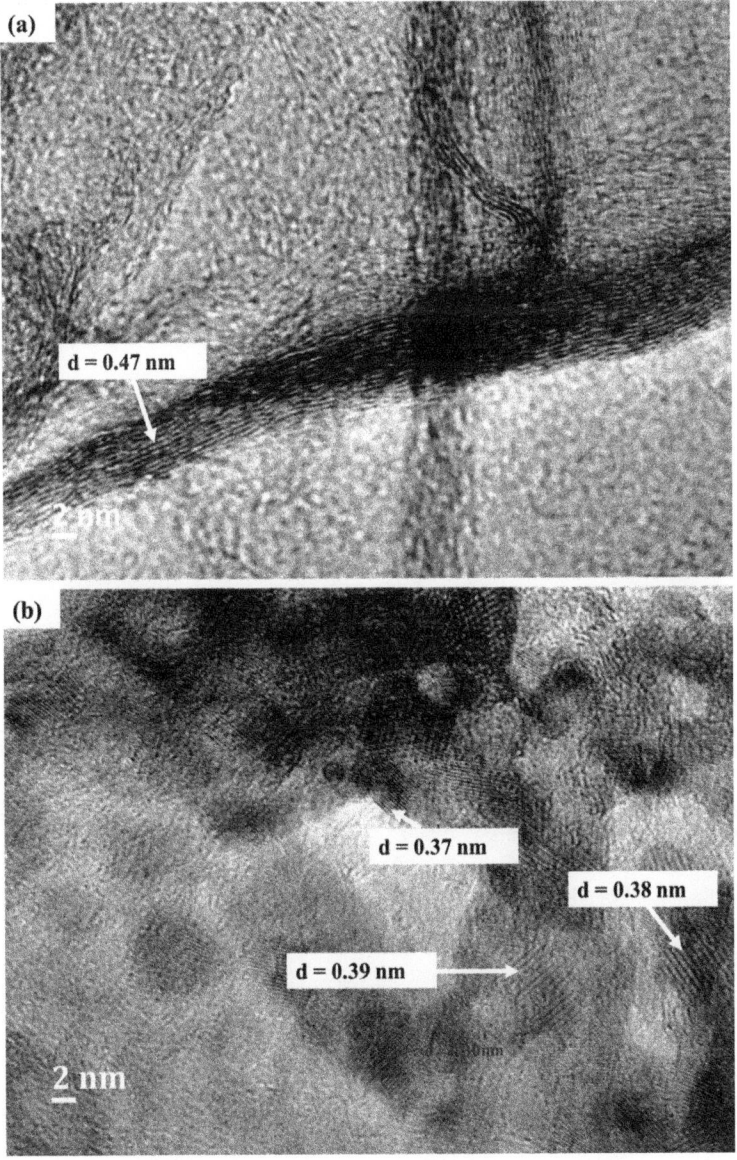

Figure 3. The d_{002} interlayer spacing of (**a**) GO and (**b**) N-rGO-1N-600 °C.

Table 1. Comparison of the d_{002} interlayer spacing of GO and N-rGO synthesized at different temperatures and with different nitrogen precursors.

Sample	Interlayer Spacing/nm
GO	0.47 ± 0.03
N-rGO-1N-600 °C	0.39 ± 0.01
N-rGO-1N-700 °C	0.40 ± 0.02
N-rGO-1N-800 °C	0.44 ± 0.02
N-rGO-2N-600 °C	0.38 ± 0.01
N-rGO-2N-700 °C	0.39 ± 0.01
N-rGO-2N-800 °C	0.40 ± 0.02
N-rGO-3N-600 °C	0.36 ± 0.01
N-rGO-3N-700 °C	0.38 ± 0.01
N-rGO-3N-800 °C	0.39 ± 0.01

The d_{002} spacing of N-rGO varied with different doping temperatures and nitrogen precursors used. The d_{002} spacing was found to increase with an increase in doping temperature. For instance, N-rGO-1N-600 °C, N-rGO-1N-700 °C and N-rGO-1N-800 °C have d_{002} spacings of 0.39, 0.40 and 0.44 nm, respectively. The smaller d_{002} spacing observed at a lower doping temperature (600 °C) is attributed to structural strains and the less crystalline nature of N-rGO. A larger d_{002} spacing was observed for N-rGO-1N synthesized from the nitrogen precursor, 4-aminophenol, at 800 °C. The interlayer spacing increased due to the distortion introduced by the inclusion of nitrogen. Such intercalation structural distortion has been widely reported for different carbon nanomaterials, such as carbon nanotubes [53]. Therefore, Raman spectroscopy was further used to investigate the effect of various temperatures and nitrogen precursors on the graphitic nature (structural properties) of N-rGO.

3.2. Structural Properties

The structural and electronic properties of N-rGO were investigated by Raman spectroscopy. Two major peaks were observed, namely the G band peak (between 1580 and 1606 cm^{-1}) which originates from the Raman E_{2g} mode, and the D-band peak (between 1347 and 1363 cm^{-1}) which is the disorder-induced band. The intensities of the D-band of N-rGO-3N-600 °C and N-rGO-3N-700 °C; and the G-bands of N-rGO-1N-600 °C and N-rGO-1N-700 °C, are of the same value (Table 2). However, their I_D/I_G ratios are different. This indicates that the nitrogen dopant distribution in N-rGO was not homogeneous [54].

Table 2. Crystallinity analysis of N-rGO.

Sample	D-Band/cm^{-1}	G-Band/cm^{-1}	I_D/I_G	I_{2D}/I_G	La/nm
GO	1355 ± 1	1601 ± 1	0.82	0.0589	5.37
N-rGO-1N-600 °C	1354 ± 1	1585 ± 1	1.04	0.0199	4.20
N-rGO-1N-700 °C	1360 ± 1	1585 ± 1	0.88	0.0212	5.00
N-rGO-1N-800 °C	1363 ± 1	1575 ± 1	0.86	0.0287	5.11
N-rGO-2N-600 °C	1357 ± 1	1603 ± 1	1.08	0.0159	4.07
N-rGO-2N-700 °C	1356 ± 1	1602 ± 1	1.02	0.0207	4.31
N-rGO-2N-800 °C	1360 ± 1	1601 ± 1	0.85	0.0253	5.37
N-rGO-3N-600 °C	1355 ± 1	1606 ± 1	1.77	0.0103	2.49
N-rGO-3N-700 °C	1355 ± 1	1605 ± 1	1.40	0.0197	3.14
N-rGO-3N-800 °C	1361 ± 1	1602 ± 1	0.88	0.0212	5.00

The G-bands for all N-rGO samples, showed a slight shift in frequency for all doping temperatures. A shift in the D-band (from 1354 to 1363 cm^{-1}) of N-rGO-1N was observed as the doping temperature increased. This was due to the change in the bond length and symmetry of the C–C and C=C bonds in the graphene lattice and compressive stress on graphene during the annealing process [55]. The asymmetric line shape and shift of the G-band (from 1601 to 1606 cm^{-1}) of GO and N-rGO-3N-600 °C, may be due to the increase in the percentage of nitrogen incorporated. Results showed that the changes in the D- and G-bands are associated with the increase in defects/dopant concentration.

It has been shown that relaxation or the change of lattice constant is highly asymmetric with lattice constant increasing by 0.32% with 2% in boron substitution and decreases very slightly with N substitution [42]. Panchakarla et al. [42] have shown that the inter-planar separation reduces by almost 2.7% in B-doped bilayer reduced graphene oxide while it remains almost unchanged in N-doped bilayers. However, there is a resultant large decrease and a slight increase in frequency shift in G-band with either B or N substitution. Boron affords a homogeneous distribution as such, disorder or the number of possible configurations increases with the concentration of dopant atoms and result in more prominent peaks of the D-band compared to nitrogen. However, G-band stiffens both with boron and nitrogen-doping and the intensity of the D-band is higher with respect to that of the G-band in all the doped samples.

The intensities of the G- and D-bands differ, and this is evident in the I_D/I_G ratio of N-rGO (Table 2). The I_D/I_G ratio is an indication of the degree of disorder and graphitic nature of N-rGO. A broader width of the D-band, narrower width of the G-band and a larger I_D/I_G ratio, suggest that N-rGO possesses many defective sites and different bonding structures (e.g., C-O, C-N) in the graphene lattice [56]. The I_D/I_G ratios of all N-rGO decreased with increase in doping temperature (from 600 to 800 °C) for the same nitrogen precursor. The larger I_D/I_G ratios observed at the lowest doping temperature of 600 °C imply a higher level of disorder (lower crystallinity). The highest doping temperature of 800 °C resulted in highly crystalline N-rGO because more amorphous products in N-rGO were reduced. Similar observations were reported by Capasso et al. [57]. Table 2 also shows the increase in defects upon an increase in nitrogen content (nitrogen precursor) in the graphene oxide lattice. In the case of N-rGO-3N that was obtained from a nitrogen precursor with the largest number of nitrogen atoms, a marked shift in the G-band was observed. This due to the increase in nitrogen atoms introduced in bond formation within the sp^2 carbon lattice of the GO. N-rGO-3N samples were observed to be less crystalline, with I_D/I_G ratios of 1.77, 1.40 and 0.88 at doping temperatures of 600, 700 and 800 °C, respectively. While other N-rGO-1N and N-rGO-2N samples which were synthesized from 4-aminophenol and 4-nitroaniline, respectively, were more crystalline.

Apart from the characteristic D-band and G-band, GO and N-rGO have a third peak; 2D-band. The 2D band represents the second order of the D-band, which is alluded to as an overtone of the D-band. Its occurrence is due to two phonon lattice vibrational processes; however, it is not associated with defects, like D-band. Therefore, the 2D-band is regarded as a strong band in graphene even when there is no presence of the D-band. The observed 2D-band peak of GO had higher intensity compared to N-rGO, 2D-band. The intensity ratios of the G-band and 2D-band (I_{2D}/I_G ratio) have been used to investigate the electron concentration of the N-rGO. Results showed that the I_{2D}/I_G ratio changes as the number of nitrogen atoms in the graphene lattice increases (Table 2). The different nitrogen precursors have different nitrogen atoms insertion capacity into rGO thus, the change in I_{2D}/I_G.

The I_D/I_G ratio of all N-rGO tends to increase as the I_{2D}/I_G ratio decreases. A similar trend was also reported by Zafar et al. [54]. This is because N-rGO consists of extra scattering effect that arises from nitrogen induced electron doping. The 2D band is mostly dependent on the electron/hole scattering rate which is influenced by lattice and charge carrier doping. Therefore, the I_D/I_G ratio would increase the electron-defect elastic scattering rate, while the I_{2D}/I_G ratio would increase the electron-electron inelastic scattering (Coulomb interaction). However, the evaluation of doping level of N-rGO using the I_{2D}/I_G ratio and blue-shifting of G-band is complicated because the G-band and 2D-band features are greatly affected by strains, defects, and number of layers.

The crystallite size (La) of N-rGO, which depends on the I_D/I_G ratio, was calculated with the aid of an equation reported by Mallet-Ladeira et al. [58] (which is an alternative to the Tuinstra–Koenig (TK) law given in Equation (1):

$$HWHM = 71 - 5.2\ La \tag{1}$$

where *HWHM* stands for the half width at half maximum which is the half of the full width at half maximum (FWHM) when the function is symmetric. The crystallite size decreases remarkably with an increase in the I_D/I_G ratio of N-rGO. N-rGO synthesized at the lowest temperature (600 °C) produced N-rGO with smaller crystallite size than that prepared at the highest temperature (800 °C). The crystallite sizes of N-rGO vary with the type of nitrogen precursor used. A smaller crystallite size of (2.49 nm) was observed in N-rGO-3N while N-rGO-2N and N-rGO-1N had a crystallite size of 4.07 and 4.20 nm, respectively at 600 °C. The crystallite size of GO (5.37 nm) was larger than that of N-rGO which indicates that thermal treatment and the type of nitrogen precursor affected the crystallite size of N-rGO. Previous studies have reported a link between the increase in I_D/I_G ratio and smaller crystallite size which is due to the formation of small crystals during reduction [50].

3.3. Thermal Stability

The decomposition behavior of N-rGO was investigated by the thermogravimetric analysis (TGA). The synthesized N-rGO exhibited different thermal stabilities. TGA weight loss curves of N-rGO-1N, N-rGO-2N, and N-rGO-3N are shown in Figure 4a, Figure 4b, and Figure 4c, respectively.

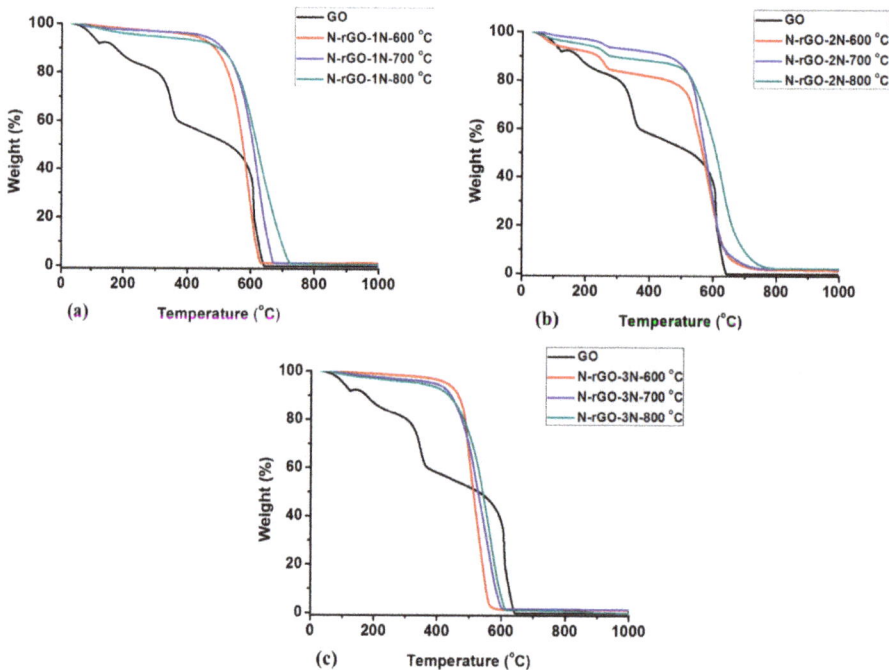

Figure 4. TGA thermograms of (a) N-rGO-1N, (b) N-rGO-2N and (c) N-rGO-3N.

The thermogram of GO showed a sequence of reaction steps because of the different oxygen-containing functional groups present in GO. These include carbonyl (C=O), hydroxyl (C-OH), epoxide (C–O–C) and single-bonded oxygen at the surface (C-O) [59,60]. The oxygen functional groups have different thermal decomposition temperatures. The thermogram of GO revealed that the decomposition occurred in different reaction steps, namely the initial step, second step, and final step. The initial step represents the rapid decline in GO weight, and this occurred before 100 °C and ended at 320 °C. The weight loss around 100 °C is due to the evolution of water (H_2O). The water lost in this step was physically absorbed between the layers of GO. The weight loss above 100 to 320 °C is attributed to the loss of CO_x groups (carbon monoxide and carbon dioxide) [61]. The second step was a slow step where the GO continues to decompose, possibly due to the loss of sp^2 carbon atoms in hexagonal structure that occurs between the decomposition temperatures of 320 and 645 °C. The drastic weight loss between 320 and 645 °C is caused by the loss of labile oxygen-containing functional groups such as hydroxyl and epoxy group due to their bond strength [62]. The final step exhibits a slower reduction in mass that ends at 1000 °C, signaling complete decomposition of GO to char.

All N-rGO samples were found to have different decomposition temperatures. The N-rGO synthesized at the higher doping temperatures (700 and 800 °C) showed no massive mass loss in the decomposition range of 100–300 °C, revealing the efficient removal of oxygen functional groups during the thermal process in the synthesis. All the thermograms (Figure 4) showed that N-rGO synthesized at the highest doping temperature (800 °C) were more thermally stable while samples prepared at the lowest doping temperature (600 °C) displayed the lower thermal stability. The most stable N-rGO were synthesized at doping temperature of 800 °C with decomposition temperatures of 591, 577 and 520 °C for N-rGO-1N, N-rGO-2N, and N-rGO-3N, respectively. At a doping temperature of 600 °C, N-rGO-1N, N-rGO-2N, and N-rGO-3N showed decomposition temperatures of 545, 536 and 488 °C, respectively. The nitrogen precursor 4-aminophenol produced N-rGO-1N that are more structured with fewer defects resulting in a higher thermal stability. On the other hand, the other nitrogen precursors, namely 4-nitroaniline and 4-nitro-*o*-phenylenediamine produced, N-rGO of lower thermal stability which suggests that the samples contain a greater extent of nitrogen-doping. There is a distinct decomposition pattern of N-rGO-2N around 300 to 500 °C which is associated with the removal of stable functionalities. Similar observation was reported by Khandelwal et al. [63]. The thermal stability of N-rGO samples correlates to its lower crystallinity, which is supported with microscopic studies and Raman spectroscopy analysis (Table 2). The thermal stability of N-rGO is also associated with nitrogen bonding configuration (nitrogen functionalities) in a graphene network. Kumar et al. [64] reported that pyridinic-N configuration are mostly dominant at lower doping temperatures. However, at higher doping temperature, graphitic N is more dominant, and this results in more thermal stable N-rGO (with graphitic N). This is evidence of a temperature-dependent nitrogen configuration doping in N-rGO. Hence, it is possible to achieve selective configurative nitrogen-doping, a major breakthrough in tuning physicochemical properties of N-rGO.

3.4. Surface Chemistry

3.4.1. Surface Area and Porosity

The surface areas, pore volumes, and pore size distributions of N-rGO obtained at varying doping temperatures are shown in Table 3. All as-synthesized N-rGO exhibited different surface areas and pore volumes/sizes. GO had a surface area of 59.46 m^2 g^{-1}, but after thermal annealing and doping with nitrogen, the surface area and pore volume increased. Thermal annealing during doping of GO caused additional exfoliation which resulted in increased surface areas and pore volumes in N-rGO due to perforations of the sheets [65].

The effect of doping temperatures and nitrogen precursors on the surface areas, pore volumes, and pore sizes were investigated. It was observed that for all precursors, N-rGO synthesized at the highest doping temperature (800 °C) had a smaller surface area than N-rGO synthesized at the lowest doping temperature (600 °C). However, it has been reported that higher doping temperatures during synthesis of N-rGO create smaller nanocrystalline graphene sheets, porous structures, large surface areas and more defects [66]. In this work, the low surface area is caused by the collapse of the carbon skeleton structure during the annealing process, therefore reducing the surface area of N-rGO.

Apart from doping temperatures, the surface area was also influenced by the nitrogen content in N-rGO. The largest surface area (154.02 m^2 g^{-1}) was observed for the nitrogen precursor 4-nitro-*o*-phenylenediamine (N-rGO-3N-600 °C) while the smaller surface area (65.05 m^2 g^{-1}) was obtained for N-rGO-1N-800 °C which was synthesized from 4-aminophenol. The high surface area in N-rGO-3N and N-rGO-2N may suggest a high percentage of pyridinic-N site in these N-rGO [67].

A good trend of BET surface area and pore sizes was observed in N-rGO samples. This illustrated that the surface area and pore volume increase with an increase in nitrogen content of N-rGO. This is because of the formation of extra pores on the surface of GO after doping, which is associated with extra exfoliation and perforation on the sheets [49,68].

Table 3. A comparison of the surface areas and porosities of N-rGO synthesized at different temperatures and with different nitrogen precursors.

Sample	Surface Area/m² g^{-1}	Pore Volume/cm³ g^{-1}	Pore Size/nm
GO	59.46	0.0564	11.39
N-rGO-1N-600 °C	87.52	0.1876	15.11
N-rGO-1N-700 °C	74.98	0.1652	17.32
N-rGO-1N-800 °C	65.05	0.1083	19.84
N-rGO-2N-600 °C	110.55	0.4324	24.89
N-rGO-2N-700 °C	99.21	0.3003	26.45
N-rGO-2N-800 °C	90.34	0.2537	27.53
N-rGO-3N-600 °C	154.02	0.5029	25.96
N-rGO-3N-700 °C	130.67	0.4986	28.89
N-rGO-3N-800 °C	95.08	0.2835	34.67

The nitrogen adsorption-desorption isotherms (Figure 5) for these materials can be classified as represent a Type IV isotherms [69]. The Type IV isotherms were accompanied by a well-defined H$_3$ hysteresis loops which are associated with capillary condensation. For N-rGO-1N-600 °C, N-rGO-2N-600 °C and N-rGO-3N-600 °C, the H$_3$ type hysteresis loops ranged from 0.49, 0.46 and 0.45 P/P$_o$, respectively, to 1.0 P/P$_o$. This demonstrates the presence of micro- and meso-porous structures within the N-rGO layers with plate-like slit-shaped pores [70].

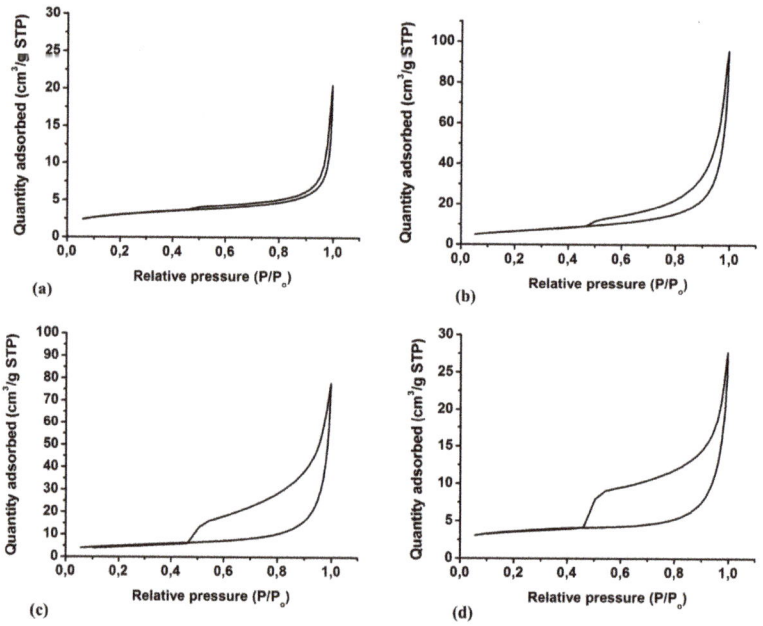

Figure 5. N$_2$ adsorption-desorption isotherms of (**a**) GO, (**b**) N-rGO-1N-600 °C, (**c**) N-rGO-2N-600 °C and (**d**) N-rGO-3N-600 °C.

3.4.2. Functional Groups

N-rGO and GO were characterized by means of FTIR spectroscopy to investigate the effect of doping temperatures and nitrogen precursors on the functional groups present in the N-rGO. The FTIR spectral patterns of all the N-rGO samples were used to identify the presence of different functional groups, by comparison with that of GO (Figure 6).

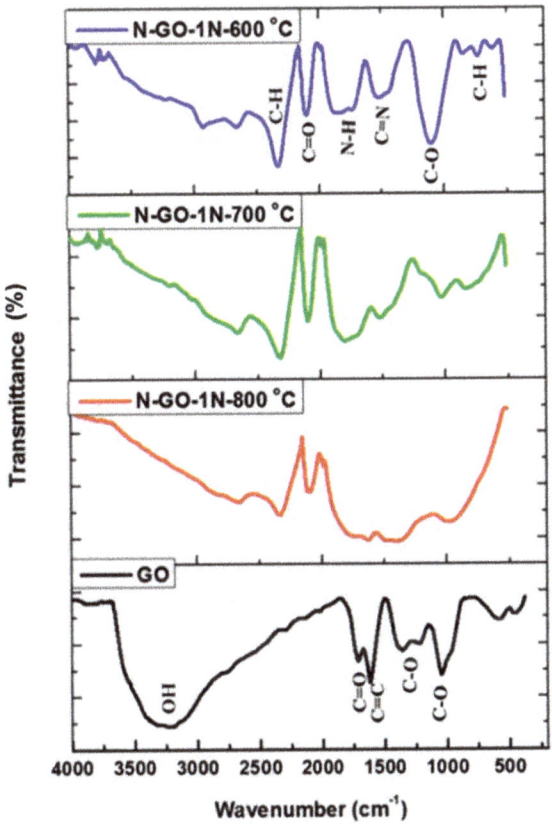

Figure 6. FTIR spectra of GO and N-rGO-1N at doping temperatures of 600, 700 and 800 °C.

The FTIR spectrum of GO showed different functional groups including hydroxyl (O-H), carbonyl (C=O), (C=C) and (C-O), indexed at 3157, 1733, 1614 and 1255/1155 cm^{-1}, respectively, which are similar to those previously reported [71]. After nitrogen-doping of GO, peaks for C=N stretching vibrations and N-H bending vibrations occurred at 1348 and 1660 cm^{-1}, respectively. Higher doping temperatures tend to reduce the peak for the O-H stretching vibration. The C-H stretching vibration peak is observed at 2489 cm^{-1} while the peak at 650 cm^{-1} (fingerprint region) is assigned to the C-H bending vibration (hybridized sp^2 bonding). The FTIR spectra of N-rGO synthesized from other nitrogen precursors (4-nitroaniline and 4-nitro-o-phenylenediamine) revealed the presence of similar functional groups as for N-rGO-1N synthesized from 4-aminophenol.

3.4.3. Nitrogen Contents

Elemental analysis (CHNS/O) of the N-rGO samples prepared was performed to study the relationship between the nitrogen precursors and the elemental composition of N-rGO. All the synthesized N-rGO samples contained different compositions carbon, oxygen and nitrogen (Table 4). The highest doping temperature for each nitrogen precursor produced N-rGO with a lower nitrogen content, while the lowest doping temperature resulted in largest nitrogen content for all nitrogen precursors. Thus, a doping temperature of 600 °C was found to be the best temperature for the nitrogen-doping of GO. A similar trend was reported by Song et al. [56] where GO was doped by hydrothermal treatment with ammonia as the nitrogen precursor at doping temperatures of 160, 190, 220, 250 and 280 °C. Thus, the largest nitrogen content occurred at a doping temperature of 160 °C.

Table 4. Elemental composition of N-rGO.

Sample	Elemental Analysis	
	Nitrogen/%	Oxygen/%
GO	-	42.985 ± 5
N-rGO-1N-600 °C	8.351 ± 5	1.496 ± 5
N-rGO-1N-700 °C	7.053 ± 5	1.255 ± 5
N-rGO-1N-800 °C	6.982 ± 5	0.464 ± 5
N-rGO-2N-600 °C	10.204 ± 5	0.789 ± 5
N-rGO-2N-700 °C	7.697 ± 5	0.613 ± 5
N-rGO-2N-800 °C	7.490 ± 5	0.239 ± 5
N-rGO-3N-600 °C	15.431 ± 5	3.475 ± 5
N-rGO-3N-700 °C	11.981 ± 5	3.029 ± 5
N-rGO-3N-800 °C	9.578 ± 5	1.082 ± 5

Varying the nitrogen precursors was found to influence the level of doping of N-rGO. 4-nitro-o-phenylenediamine which contains the most nitrogen atoms in its structure (3 N atoms per molecule), resulted in a high nitrogen content than for the other nitrogen precursors. The nitrogen content of N-rGO-3N synthesized from 4-nitro-o-phenylenediamine at temperatures of 600, 700 and 800 °C was 15.431, 11.981 and 9.578%, respectively. As the nitrogen content in N-rGO increased, the oxygen content was also found to increase. The decrease in oxygen content from a doping temperature of 600 to 800 °C was attributed to deoxygenation in N-rGO. A higher oxygen content was observed in the N-rGO-3N-600 °C. This is because 4-nitro-o-phenylenediamine contains more oxygen atoms in its structure than the other precursors, therefore, during the doping process, oxygen was also introduced. A correlation was observed between the crystallinity and elemental composition of N-rGO. Less crystalline N-rGO was found to contain a higher nitrogen content. This is exemplified by N-rGO-3N-600 °C with a greater density of defects (high I_D/I_G ratio) and nitrogen content. An increase in nitrogen-doping also resulted in a decrease of crystallite size (Table 2). This is consistent with the findings reported by Zhang et al. [72].

The nitrogen bonding configuration in N-rGO affects the electronic properties [73]. Thus, the incorporation of nitrogen in GO and the C-N bonding configurations were investigated by means of XPS (Figure 7). The trend in nitrogen content observed from the XPS analysis correlates with the results obtained from elemental analysis (Table 4). The C 1s and N 1s peaks in N-rGO appear at about 284 and 400 ± 0.1 eV, respectively. The C 1s spectra (Supplementary Materials–Table S1) of N-rGO synthesized from different nitrogen precursors show a slight shift in the binding energies of the peaks that correspond to C=C, C-N, C-O, carboxylate (O=C-O) and carbonyl (C=O) bonds. Zhang et al. [74] and Sheng et al. [75] reported that the C-O bonding configuration disappears after the doping and annealing process. However, this was not the case here since the C-O bonding peak remained, and this indicates that most oxygen groups in N-rGO were not completely removed. N-rGO-3N-600 °C showed C, O, and N peaks with percentage compositions of 81.5, 9.5 and 8.5%, respectively. N-rGO-3N-600 °C had a higher nitrogen content and lower carbon and oxygen content than N-rGO-1N-600 °C and N-rGO-2N-600 °C.

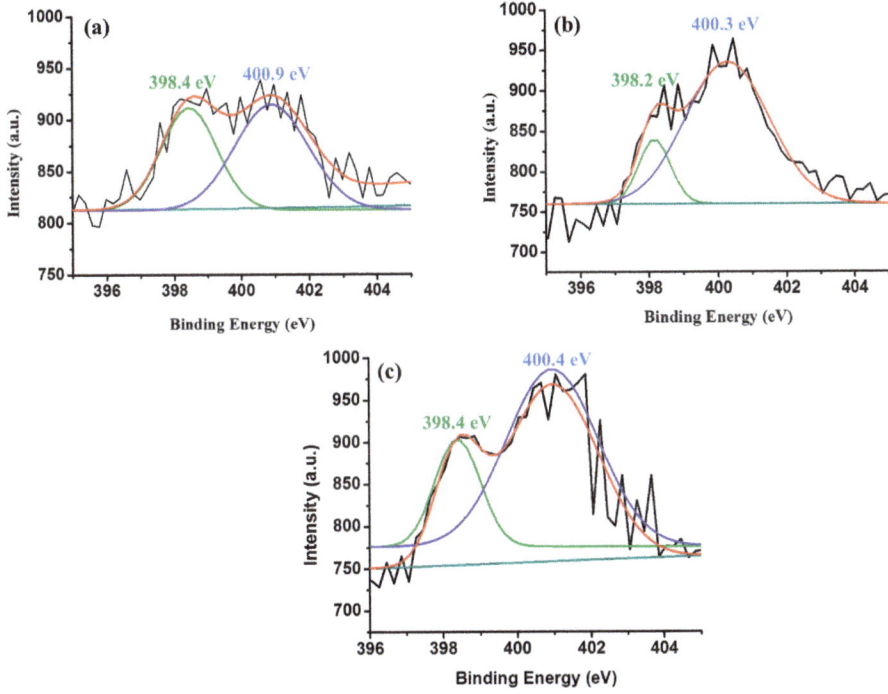

Figure 7. XPS high-resolution N 1s spectra of (**a**) N-rGO-1N-600 °C, (**b**) N-rGO-2N-600 °C and (**c**) N-rGO-3N-600 °C.

Nitrogen-doped reduced graphene oxide is reported to contain three most desired, nitrogen bonding configurations namely pyrrolic-N, pyridinic-N, and graphitic N with different components of N 1s at 399.8–401.2 eV, 398.1–99.3 eV and 401.1–402.7 eV, respectively [76,77]. However, these N 1s positions vary in comparatively wide ranges in different studies. Figure 7 shows a comparison of the different XPS spectra and the presence of various nitrogen (N 1s) species in N-rGO. The N 1s spectra of all N-rGO samples were fitted into two peaks, namely pyrrolic-N and pyridinic-N. The N 1s spectra for N-rGO-1N-600 °C and N-rGO-2N-600 °C are lower in intensity, therefore it was not possible to determine the percentage of each bonding configuration. However, the N 1s spectrum for N-rGO-3N-600 °C was fitted into two N 1s peak, namely pyrrolic-N at 400.4 ± 0.1 eV with a 46% content and pyridinic-N at 398.4 ± 0.1 eV with a 54% constant. These results suggest that N-rGO-3N-600 °C possesses a greater content of pyridinic-N than pyrrolic-N. Increase of pyridinic-N in N-rGO is related with lower thermal stability and higher surface area. This suggests that there are more defects in N-rGO-3N-600 °C. The lower pyrrolic-N content in N-rGO-3N-600 °C is due to the lower stability of pyrrolic-N which occurs in carbon materials that are doped at lower temperatures [78].

The doping temperatures control the type of nitrogen bonding configuration. For example, low doping temperatures are reported to produce N-rGO in which pyrrolic-N and pyridinic-N dominate [56]. Lu et al. [24] also noted that N-rGO synthesized at low temperatures acquired a higher degree of microstructural disorder associated with the higher nitrogen content. Moreover, it was found that the quality of the resulting N-rGO microstructure was directly dependent on the doping temperature.

3.5. Phase Composition

The structure and phase compositions of all N-rGO were investigated by powder-XRD. The X-ray diffractograms of GO, N-rGO-1N-600 °C, N-rGO-2N-600 °C and N-rGO-3N-600 °C are presented in Figure 8. In the case of GO a diffraction peak was observed at 13.8° (2θ). However, all the samples with nitrogen are devoid of this peak at 2θ = 13.8°. This may be due to the deoxidization of oxygen-containing functional groups in the N-rGO structure. The (004) diffraction peak indicates the crystallinity of the synthesized GO and N-rGO. The diffractograms of N-rGO-1N-600 °C, N-rGO-2N-600 °C and N-rGO-3N-600 °C have a broad diffraction peak at 25.1°, 25.5° and 25.6°, respectively, indicating a high graphitic degree. The nitrogen-rich sample, N-rGO-3N-600 °C have a higher shift of the 2θ angle compared with N-rGO-1N-600 °C and N-rGO-2N-600 °C. The 2θ angle shift may be caused by strain, stress, defects, and dislocation induced in the crystal lattice during nitrogen-doping.

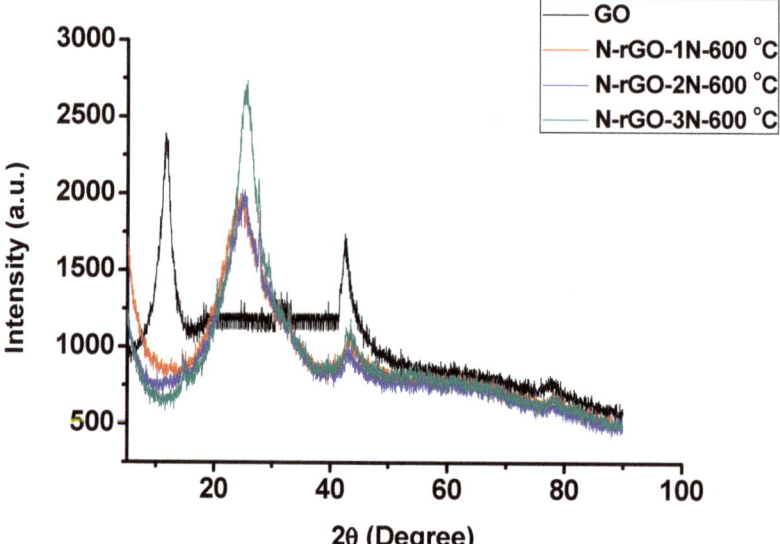

Figure 8. Powder X-ray diffractograms of GO, N-rGO-1N-600 °C, N-rGO-2N-600 °C and N-rGO-3N-600 °C.

Microstructural parameters (lattice dimensions, dislocation density and micro-stain) were determined from the diffraction 2θ angles and the Scherrer equation. Table 5 shows various XRD parameters for N-rGO.

Table 5. Powder-XRD-parameters for N-rGO prepared at a doping temperature of 600 °C.

Sample	2θ/°	FWHM/β_{hkl}	Interlayer Spacing/nm	Crystallite Size/nm
GO	13.8	2.12	0.640	3.93
N-rGO-1N-600 °C	25.1	7.8	0.186	1.09
N-rGO-2N-600 °C	25.5	10.8	0.183	0.79
N-rGO-3N-600 °C	25.6	11.4	0.182	0.75

FWHW = full width at half maximum.

The peak intensity and peak width of 2θ vary significantly depending on the doping of GO. The FWHM of 2θ of N-rGO increased with increased nitrogen content, which in turn depended on the nitrogen precursor. However, the shift in 2θ causes a decrease in the d_{002} spacing which is associated with reduction of epoxy, hydroxyl, and carboxyl functional groups on the GO framework. For instance, the broad (002) peak and decrease in crystallinity in N-rGO-3N-600 °C is due to the increased nitrogen content. Increasing the nitrogen content causes an increase in structural strain of N-rGO-3N-600 °C, thus resulting in enhanced surface defects of the graphite layer which probably led to broadening of the FWHM of the peak. The increase if FWHM observed concurred with the decrease in thermostability from TGA analysis (Figure 4). These changes also correspond to the variations in lattice distortions and d_{002} spacings. For example, the d_{002} spacing of GO (0.640 nm) was larger than those of N-rGO and this correlates with the increase in nitrogen content. N-rGO-3N-600 °C with a higher nitrogen content of 8.5% had a d_{002} spacing of 0.182 nm, while a lower nitrogen content (3.0%) in N-rGO-1N-600 °C resulted in a d_{002} spacing of 0.186 nm. The d_{002} spacings of the N-rGO samples obtained from XRD correspond with those determined from HR-TEM analysis (Table 1). Hence, it can be concluded that nitrogen-doping has an influence on the d_{002} spacing. The crystallite sizes of the synthesized N-rGO also tend to decrease with a higher nitrogen content in N-rGO, which indicates an increase in the number of defects induced. These observations of crystallite size (XRD) correlate with the calculated crystallite sizes from Raman spectroscopy (Table 2). N-rGO-3N-600 °C had a smaller crystallite size (0.75 nm) than either GO or the other N-rGO (N-rGO-1N and N-rGO-2N).

3.6. Optical Properties

The optical properties of the N-rGO were investigated by UV-Vis spectrophotometry. From the work reported by Loryuenyong et al. [79], the maximum absorption wavelength of GO was reported to be around 230–270 nm. In Figure 9, GO exhibited a maximum absorption peak at 234 nm that is associated with π-π* and n-π* transitions of C=C and C=O bonds, respectively [80]. In contrast, N-rGO shows an absorption peak at 262–275 nm. The shift to a longer wavelength indicates the deoxygenation and restoration of the electronic π-conjugation of GO [81]. The peaks between 260–275 nm in the N-rGO spectra are attributed to π-π* transitions of the double bonds. The introduction of more lone electrons creates more n-π* transitions which has a tendency of shifting absorption longer wavelength (since energy is inversely proportional to the wavelength). This shows a characteristic of sp^2 hybridization bands and lone pairs of nitrogen. A significant increase in absorbance was noted which shifted towards the visible light range as the nitrogen content increases. This shift also enables N-rGO to have better capability of light-harvesting compared with GO. This is because freer electrons are easier to excite than bound (π-electrons) and therefore π-π* are fewer than n-π*, hence the shift to longer wavelength, a phenomenon which is required for light-harvesting. Mohamed et al. [82] reported that heteroatom-doped graphene with an absorption frequency ranging from 300–650 nm had limited photocatalytic activity resulting in a lower light-harvesting capability. The nitrogen-rich sample (N-rGO-3N-600 °C) exhibited the slight shift in absorption peak.

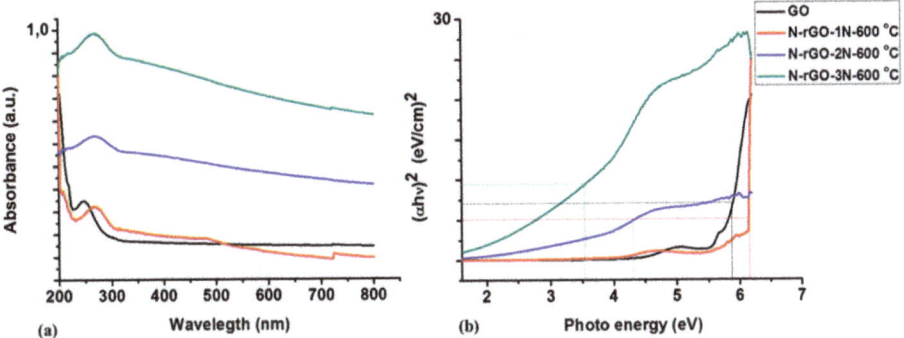

Figure 9. (a) UV-Vis absorption spectra and (b) Tauc plots for GO, N-rGO-1N-600 °C, N-rGO-2N-600 °C and N-rGO-3N-600 °C.

In the Tauc plot for N-rGO and GO (Figure 9b), N-rGO exhibited two absorption edges which correspond to rGO. The optical band gaps obtained from the Tauc plots are: 5.9 eV for GO, 6.2 eV for N-rGO-1N-600 °C, 4.4 eV for N-rGO-2N-600 °C and 3.5 eV for N-rGO-3N-600 °C. The optical band-gap (Table 6) was also recalculated by the Planck's quantum equation to confirm the trend displayed in the Tauc plots. N-rGO-3N-600 °C showed a slight decrease in energy band-gap (4.5 eV). While N-rGO with a lower nitrogen content; i.e., N-rGO-1N-600 °C and N-rGO-2N-600 °C had a band-gap of 4.8 and 4.6 eV, respectively.

Table 6. Energy band-gap of N-rGO from a doping temperature of 600 °C obtained from Planck's quantum equation.

Sample	Wavelength/nm	Band-Gap Energy/eV
GO	234	5.3
N-rGO-1N-600 °C	262	4.8
N-rGO-2N-600 °C	271	4.6
N-rGO-3N-600 °C	275	4.5

The decrease in band-gap energies of N-rGO may be due to compensation of the band-gap states associated with the incorporation of dopant atoms, and this resulted in the Fermi level moving up in the direction of the conduction band edge [83]. This implies that a higher nitrogen content in N-rGO induces a lower rate of electron hole (e^-/h^+) recombination, than for N-rGO with a lower nitrogen content. The large band gaps for N-rGO-1N-600 °C and N-rGO-2N-600 °C indicate that lower nitrogen content may not be ideal for light-harvesting. Smaller band-gap energies can lead to enhancement of visible light trapping than larger band-gap energies. The e^-/h^+ recombination of N-rGO was therefore investigated by photoluminescence spectroscopy. A comparison of the e^-/h^+ recombination dynamics of N-rGO are presented in Figure 10.

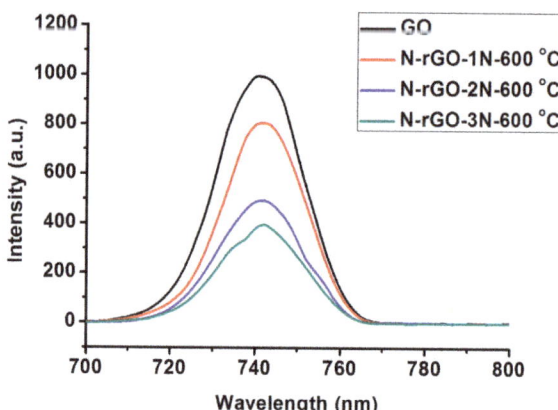

Figure 10. A comparison of the photoluminescence spectra of N-rGO.

The intra-band-gap, which is associated with local defect functions as a trap for free carriers, affects the recombination and electron transport [84]. All N-rGO from different nitrogen precursors luminescence at e^-/h^+ recombination rate of 745 nm as shown in Figure 10. However, their photoluminescence peak intensities are different. The variation of nitrogen precursor showed the enhancement of photoluminescence peak intensity. Van Khai et al. [85] reported that the doping temperatures tend to cause a shift in wavelength, which is due to the presence of quaternary-N, whereas the presence of pyrrolic-N and pyridinic-N was closely related with enhancement of photoluminescence peak intensity. Therefore, the enhanced photoluminescence intensity in N-rGO corresponds to the decrease density of pyrrolic-N. However, the increased density of pyridinic-N may be correlated with the decrease of non-radiative recombination. N-rGO-1N-600 °C with a lower nitrogen content had a higher photoluminescence peak intensity while N-rGO-3N-600 °C with a higher nitrogen content was of lower intensity This further suggests that N-rGO-3N-600 °C has a lower rate of a e⁻/h⁺ recombination.

3.7. Electrical Conductivity Properties

To study the effect of nitrogen content on the electrical conductivity of N-rGO, the current–voltage (I–V) characteristics (Figure 11). All the N-rGO samples exhibited a linear I-V relationship. However, the I-V slope of GO sample is almost close to zero. This is due to high oxygen content (oxygen functional group) which cause GO to behave like an insulating material [86]. Generally, the structure of GO is amorphous because of distortions from the sp^3-oxygen (C–O–C, C–OH, and COOH). Additionally, because of the random dispersion, the sp^2-hybridized benzene rings are isolated by sp^3-hybridized rings, in this way prompting the insulating characteristics. In the case of N-rGO, the I–V slope significantly increased, demonstrating that the electrical conductivity of N-rGO was enhanced. The enhanced electrical conductivity can be attributed to reduction of oxygen functional groups and restoration of sp^2 carbon network.

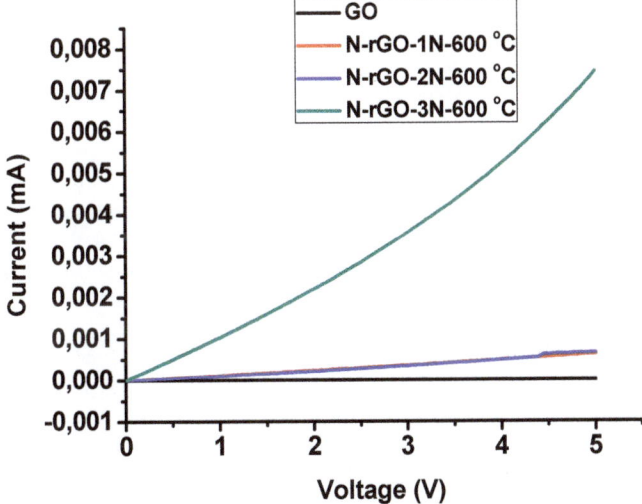

Figure 11. Current–voltage curve of N-rGO at doping temperature of 600 °C.

The electrostatic investigation on the effect of nitrogen-doping on electrical band-gap of GO was carried out by first determining resistivity (ρ), using Equation (2).

$$\rho = \left(\frac{\pi}{\ln 2}\right)\left(\frac{V}{I}\right)t \tag{2}$$

where V is the voltage, I is the current, and t is the sheet thickness. The obtained ρ was then used in the estimation of electrical band-gap (E_g) given Equation (3).

$$E_g = 2k\frac{\ln \rho}{\frac{1}{T}} \tag{3}$$

where k is the Boltzmann constant (0.000086 eV/K) and T is the temperature in Kelvin. It was generally observed that nitrogen-doping has an effect of reducing electrical band-gap of GO structure (Table 7). However, N-rGO-1N-700 °C and N-rGO-2N-800 °C had non-linear relations an indication of two conductive regimes. This can be attributed to dominant N configuration, doping concentration, and the respective temperatures. At 700 °C doping temperature and much lower dopant concentration a pyrrolic-N configuration is dominant a higher conjugation on average is expected and therefore much lower electrical band-gap compared to the pyridinic configuration at the same temperature.

Table 7. Effect of nitrogen-doping temperature on electrical band-gap of N-rGO.

Sample	Resistivity/Ω mm^{-1}	Conductivity/mmS m^{-1}	Band-Gap/eV
GO	7.89	0.127	3.240
N-rGO-1N-600 °C	4.92	0.203	2.500
N-rGO-2N-600 °C	4.63	0.216	2.403
N-rGO-3N-600 °C	5.00	0.200	2.526
N-rGO-1N-700 °C	3.29	0.304	1.866
N-rGO-2N-700 °C	5.04	0.198	2.538
N-rGO-3N-700 °C	4.93	0.203	2.502
N-rGO-1N-800 °C	4.59	0.218	2.392
N-rGO-2N-800 °C	8.03	0.125	3.269
N-rGO-3N-800 °C	4.63	0.216	2.405

The electrical conductivities are shown in Table 8, obtained from four-point probe measurements, this is exhibiting an increase in conductivity with an increase in nitrogen content. N-rGO-3N-600 °C sample exhibited the highest electrical conductivity and excellent ultra-low electrical resistivity. So far, the mechanism of enhancement of electrical conductivity of N-rGO is not yet understood. It is believed that the interaction of N-C and interfacial structure are key variables of controlling the electrical conduction. The high electrical conductivity in N-rGO-3N-600 °C may be attributed the restoration π-electrons conjugated network in graphene, prompting to more formation of percolation pathways within the sp^2 carbon atoms. The dominant nitrogen bonding configuration in N-rGO-3N-600 °C is pyridinic-N, which allows electron within π-graphene structure, lower stone-wall defects, creating high electron percolation pathways and quite conduction gap, hence, enhanced conductivity. Therefore, N-rGO-3N-600 °C that was synthesized have basal pyridinic-N than edge substitution, due to the availability of electrons within conduction space, elevating the density of state near the Dirac point. Thus, creates a specialized band around the Dirac point. The created band gives rise to a finite density of state near the Dirac point and enhance the electrical conductivity. Hence, it can be suggested that the electrical conductivity of N-rGO might be dependent on the nitrogen content which is incorporated into the structure of GO. Consequently, N-rGO-3N-600 °C serves as a promising material for various applications such as electronic and opto-electronic devices.

Table 8. Electrical conductivity of the N-rGO at doping temperature of 600 °C.

Sample	Sheet Resistance/Ω sq^{-1}	Bulk Resistivity/Ω cm	Electrical Conductivity/S cm^{-1}
GO	9147399.8	82894.2	1.21×10^{-5}
N-rGO-1N-600 °C	8764.1	794.6	0.00126
N-rGO-2N-600 °C	8768.8	554.9	0.00182
N-rGO-3N-600 °C	830.5	7.5	0.133

4. Conclusions

In conclusion, N-rGO has been successfully synthesized from solid nitrogen precursors (4-aminophenol, 4-nitroaniline and 4-nitro-o-phenylenediamine). The incorporation of N atom into the GO lattice at various doping temperatures caused a significant effect on the physicochemical properties such as surface morphology, surface chemistry, surface area, and porosity. Microscopic studies showed a more wrinkled-like structure on N-rGO than for GO due to the presence of nitrogen atoms in the GO framework. By lowering the doping temperature, a higher nitrogen content was incorporated into the GO lattice. The nitrogen content of N-rGO varied for different nitrogen precursors. N-rGO exhibited lower thermal stability as the level of nitrogen-doping increased, due to more defects and distortions experienced in the N-rGO structure. The enhancement of surface area and high degree of disorder on N-rGO were attributed to the removal of oxygen-containing functional groups.

N-rGO-1N-600 °C, N-rGO-2N-600 °C and N-rGO-3N-600 °C had a nitrogen content of 3.0, 3.7 and 8.5%, respectively. The nitrogen-rich precursor, 4-nitro-*o*-phenylenediamine, lead to higher doping of N-rGO. N-rGO-3N-600 °C was found to have the highest nitrogen content of 8.5% and a high surface area of 154.02 m^2 g^{-1} though it was less crystalline and manifested low thermal stability. The peak fitting of N 1s in all N-rGO samples produced two major components of pyridinic-N and pyrrolic-N with different nitrogen content. N-rGO showed good absorption and luminescence in the near UV region. The photoluminescence peak intensity and band-gap values were highly dependent on nitrogen content. A higher nitrogen content in N-rGO exhibited a smaller optical band-gap of 4.5 eV with lower photoluminescence peak intensity. N-rGO-3N-600 °C exhibited higher electrical conductivity of 0.133 S cm^{-1}.

Supplementary Materials: The following are available online at http://www.mdpi.com/1996-1944/12/20/3376/s1, Figure S1: SEM images of (a) GO, (b) N-rGO-1N-600 °C, (c) N-rGO-1N-700 °C, (d) N-rGO-1N-800 °C, (e) N-rGO-2N-600 °C, (f) N-rGO-2N-700 °C, (g) N-rGO-2N-800 °C, (h) N-rGO-3N-600 °C, (i) N-rGO-3N-700 °C and (j) N-rGO-3N-800 °C, Table S1: Atomic percentage (%) of N 1s and C 1s peak binding energy (eV).

Author Contributions: Conceptualization, N.P.D.N., M.A.O. and V.O.N.; methodology, N.P.D.N., M.A.O. and V.O.N.; software, N.P.D.N., M.A.O. and V.O.N.; validation, N.P.D.N., M.A.O. and V.O.N.; formal analysis, N.P.D.N., M.A.O. and V.O.N.; investigation, N.P.D.N., M.A.O. and V.O.N.; resources, V.O.N.; data curation, N.P.D.N., M.A.O. and V.O.N.; writing—original draft preparation, N.P.D.N., M.A.O. and V.O.N.; writing—review and editing, N.P.D.N.; visualization, N.P.D.N., M.A.O. and V.O.N.; supervision, V.O.N.; project administration, V.O.N.; funding acquisition, V.O.N.

Funding: The authors would like to thank the National Research Foundation (NRF, Grant numbers—101357 and 103979), Eskom Tertiary Education Support Programme (TESP), University of KwaZulu-Natal (UKZN) and UKZN Nanotechnology Platform for financial support and facilities.

Acknowledgments: The authors wish to sincerely thank Bice Martincigh, Olatunde Olatunji, and Nicholas Rono for their input and proof-reading the manuscript.

Conflicts of Interest: The authors declare that they have no conflict of interest.

References

1. Ye, M.; Zhang, Z.; Zhao, Y.; Qu, L. Graphene platforms for smart energy generation and storage. *Joule* **2018**, *2*, 245–268. [CrossRef]
2. Lee, S.D.; Lee, H.S.; Kim, J.Y.; Jeong, J.; Kahng, Y.H. A systematic optimization for graphene-based supercapacitors. *Mater. Res. Express* **2017**, *4*, 085601–085610. [CrossRef]
3. Yao, Y.; Ping, J. Recent advances in graphene-based freestanding paper-like materials for sensing applications. *Trends Anal. Chem.* **2018**, *105*, 75–88. [CrossRef]
4. Bhatt, K.; Rani, C.; Vaid, M.; Kapoor, A.; Kumar, P.; Kumar, S.; Shriwastawa, S.; Sharma, S.; Singh, R.; Tripathi, C. A comparative study of graphene and graphite-based field effect transistor on flexible substrate. *Pramana-J. Phys.* **2018**, *90*, 71–76. [CrossRef]
5. Priyadarsini, S.; Mohanty, S.; Mukherjee, S.; Basu, S.; Mishra, M. Graphene and graphene oxide as nanomaterials for medicine and biology application. *J. Nanostructure Chem.* **2018**, *8*, 123–137. [CrossRef]
6. Lu, G.; Yu, K.; Wen, Z.; Chen, J. Semiconducting graphene: Converting graphene from semimetal to semiconductor. *Nanoscale* **2013**, *5*, 1353–1368. [CrossRef] [PubMed]
7. Czerniak-Reczulska, M.; Niedzielska, A.; Jędrzejczak, A. Graphene as a material for solar cells applications. *Adv. Mater. Sci.* **2015**, *15*, 67–81. [CrossRef]
8. Liu, H.; Liu, Y.; Zhu, D. Chemical doping of graphene. *J. Mater. Chem.* **2011**, *21*, 3335–3345. [CrossRef]
9. Whitby, R.L.D. Chemical control of graphene architecture: Tailoring shape and properties. *ACS Nano* **2014**, *8*, 9733–9754. [CrossRef]

10. Chen, D.; Tang, L.; Li, J. Graphene-based materials in electrochemistry. *Chem. Soc. Rev.* **2010**, *39*, 3157–3180. [CrossRef]
11. Thirumal, V.; Pandurangan, A.; Jayavel, R.; Ilangovan, R. Synthesis and characterization of boron doped graphene nanosheets for supercapacitor applications. *Synth. Met.* **2016**, *220*, 524–532. [CrossRef]
12. Li, S.; Wang, Z.; Jiang, H.; Zhang, L.; Ren, J.; Zheng, M.; Dong, L.; Sun, L. Plasma-induced highly efficient synthesis of boron doped reduced graphene oxide for supercapacitors. *Chem. Commun.* **2016**, *52*, 10988–10991. [CrossRef] [PubMed]
13. Usachov, D.Y.; Fedorov, A.V.; Vilkov, O.Y.; Petukhov, A.E.; Rybkin, A.G.; Ernst, A.; Otrokov, M.M.; Chulkov, E.V.; Ogorodnikov, I.I.; Kuznetsov, M.V. Large-scale sublattice asymmetry in pure and boron-doped graphene. *Nano Lett.* **2016**, *16*, 4535–4543. [CrossRef] [PubMed]
14. Megawati, M.; Chua, C.K.; Sofer, Z.; Klimova, K.; Pumera, M. Nitrogen-doped graphene: Effect of graphite oxide precursors and nitrogen content on the electrochemical sensing properties. *Phys. Chem. Chem. Phys.* **2017**, *19*, 15914–15923. [CrossRef] [PubMed]
15. Xing, Z.; Ju, Z.; Zhao, Y.; Wan, J.; Zhu, Y.; Qiang, Y.; Qian, Y. One-pot hydrothermal synthesis of nitrogen-doped graphene as high-performance anode materials for lithium ion batteries. *Sci. Rep.* **2016**, *6*, 26141–26150. [CrossRef] [PubMed]
16. Cai, W.; Wang, C.; Fang, X.; Yang, L.; Chen, X. Synthesis and characterization of nitrogen-doped graphene films using C_5NCl_5. *Appl. Phys. Lett.* **2015**, *106*, 253101–253105. [CrossRef]
17. Wang, X.; Sun, G.; Routh, P.; Kim, D.-H.; Huang, W.; Chen, P. Heteroatom-doped graphene materials: Syntheses, properties and applications. *Chem. Soc. Rev.* **2014**, *43*, 7067–7098. [CrossRef]
18. Rao, C.N.R.; Gopalakrishnan, K.; Govindaraj, A. Synthesis, properties and applications of graphene doped with boron, nitrogen and other elements. *Nano Today* **2014**, *9*, 324–343. [CrossRef]
19. Usachov, D.; Vilkov, O.; Gruneis, A.; Haberer, D.; Fedorov, A.; Adamchuk, V.; Preobrajenski, A.; Dudin, P.; Barinov, A.; Oehzelt, M. Nitrogen-doped graphene: Efficient growth, structure, and electronic properties. *Nano Lett.* **2011**, *11*, 5401–5407. [CrossRef]
20. Kumar, M.P.; Kesavan, T.; Kalita, G.; Ragupathy, P.; Narayanan, T.N.; Pattanayak, D.K. On the large capacitance of nitrogen doped graphene derived by a facile route. *RSC Adv.* **2014**, *4*, 38689–38697. [CrossRef]
21. Ambrosi, A.; Chua, C.K.; Latiff, N.M.; Loo, A.H.; Wong, C.H.A.; Eng, A.Y.S.; Bonanni, A.; Pumera, M. Graphene and its electrochemistry–an update. *Chem. Soc. Rev.* **2016**, *45*, 2458–2493. [CrossRef] [PubMed]
22. Lin, T.; Huang, F.; Liang, J.; Wang, Y. A facile preparation route for boron-doped graphene, and its CdTe solar cell application. *Energy Environ. Sci.* **2011**, *4*, 862–865. [CrossRef]
23. Poh, H.L.; Šimek, P.; Sofer, Z.; Tomandl, I.; Pumera, M. Boron and nitrogen doping of graphene via thermal exfoliation of graphite oxide in a BF_3 or NH_3 atmosphere: Contrasting properties. *J. Mater. Chem. A* **2013**, *1*, 13146–13153. [CrossRef]
24. Lu, Y.-F.; Lo, S.-T.; Lin, J.-C.; Zhang, W.; Lu, J.-Y.; Liu, F.-H.; Tseng, C.-M.; Lee, Y.-H.; Liang, C.-T.; Li, L.-J. Nitrogen-doped graphene sheets grown by chemical vapor deposition: Synthesis and influence of nitrogen impurities on carrier transport. *ACS Nano* **2013**, *7*, 6522–6532. [CrossRef] [PubMed]
25. Wang, T.; Wang, L.-X.; Wu, D.-L.; Xia, W.; Jia, D.-Z. Interaction between nitrogen and sulfur in co-doped graphene and synergetic effect in supercapacitor. *Sci. Rep.* **2015**, *5*, 9591–9599. [CrossRef] [PubMed]
26. Luo, Z.; Lim, S.; Tian, Z.; Shang, J.; Lai, L.; MacDonald, B.; Fu, C.; Shen, Z.; Yu, T.; Lin, J. Pyridinic-N doped graphene: Synthesis, electronic structure, and electrocatalytic property. *J. Mater. Chem.* **2011**, *21*, 8038–8044. [CrossRef]
27. Zhang, S.; Tsuzuki, S.; Ueno, K.; Dokko, K.; Watanabe, M. Upper limit of nitrogen content in carbon materials. *Angew. Chem. Int. Ed.* **2015**, *54*, 1302–1306. [CrossRef]
28. He, W.; Jiang, C.; Wang, J.; Lu, L. High-rate oxygen electroreduction over graphitic-N species exposed on 3D hierarchically porous nitrogen-doped carbons. *Angew. Chem. Int. Ed.* **2014**, *53*, 9503–9507. [CrossRef]
29. Park, S.; Hu, Y.; Hwang, J.O.; Lee, E.-S.; Casabianca, L.B.; Cai, W.; Potts, J.R.; Ha, H.-W.; Chen, S.; Oh, J. Chemical structures of hydrazine-treated graphene oxide and generation of aromatic nitrogen doping. *Nat. Commun.* **2011**, *3*, 638. [CrossRef]
30. Zhao, L.; He, R.; Rim, K.T.; Schiros, T.; Kim, K.S.; Zhou, H.; Gutiérrez, C.; Chockalingam, S.; Arguello, C.J.; Pálová, L. Visualizing individual nitrogen dopants in monolayer graphene. *Science* **2011**, *333*, 999–1003. [CrossRef]

31. Li, N.; Wang, Z.; Zhao, K.; Shi, Z.; Gu, Z.; Xu, S. Large scale synthesis of N-doped multi-layered graphene sheets by simple arc-discharge method. *Carbon* **2010**, *48*, 255–259. [CrossRef]
32. Zhang, X.; Hsu, A.; Wang, H.; Song, Y.; Kong, J.; Dresselhaus, M.S.; Palacios, T. Impact of chlorine functionalization on high-mobility chemical vapor deposition grown graphene. *ACS Nano* **2013**, *7*, 7262–7270. [CrossRef] [PubMed]
33. Li, X.-J.; Yu, X.-X.; Liu, J.-Y.; Fan, X.-D.; Zhang, K.; Cai, H.-B.; Pan, N.; Wang, X.-P. Synthesis of nitrogen-doped graphene via thermal annealing graphene with urea. *Chin. J. Chem. Phys.* **2012**, *25*, 321–326. [CrossRef]
34. Wang, H.; Zhou, Y.; Wu, D.; Liao, L.; Zhao, S.; Peng, H.; Liu, Z. Synthesis of boron-doped graphene monolayers using the sole solid feedstock by chemical vapor deposition. *Small* **2013**, *9*, 1316–1320. [CrossRef] [PubMed]
35. Wu, T.; Shen, H.; Sun, L.; Cheng, B.; Liu, B.; Shen, J. Nitrogen and boron doped monolayer graphene by chemical vapor deposition using polystyrene, urea and boric acid. *New J. Chem.* **2012**, *36*, 1385–1391. [CrossRef]
36. Zhou, S.; Liu, N.; Wang, Z.; Zhao, J. Nitrogen-doped graphene on transition metal substrates as efficient bifunctional catalysts for oxygen reduction and oxygen evolution reactions. *ACS Appl. Mater. Interfaces* **2017**, *9*, 22578–22587. [CrossRef]
37. Guo, N.; Xi, Y.; Liu, S.; Zhang, C. Greatly enhancing catalytic activity of graphene by doping the underlying metal substrate. *Sci. Rep.* **2015**, *5*, 12051–12058. [CrossRef]
38. Du, D.; Li, P.; Ouyang, J. Nitrogen-doped reduced graphene oxide prepared by simultaneous thermal reduction and nitrogen doping of graphene oxide in air and its application as an electrocatalyst. *ACS Appl. Mater. Interfaces* **2015**, *7*, 26952–26958. [CrossRef]
39. Zabet-Khosousi, A.; Zhao, L.; Pálová, L.; Hybertsen, M.S.; Reichman, D.R.; Pasupathy, A.N.; Flynn, G.W. Segregation of sublattice domains in nitrogen-doped graphene. *J. Am. Chem. Soc.* **2014**, *136*, 1391–1397. [CrossRef]
40. Wang, H.; Maiyalagan, T.; Wang, X. Review on recent progress in nitrogen-doped graphene: Synthesis, characterization, and its potential applications. *ACS Catal.* **2012**, *2*, 781–794. [CrossRef]
41. Van Nang, L.; Van Duy, N.; Hoa, N.D.; Van Hieu, N. Nitrogen-doped graphene synthesized from a single liquid precursor for a field effect transistor. *J. Electron. Mater.* **2016**, *45*, 839–845. [CrossRef]
42. Panchakarla, L.; Subrahmanyam, K.; Saha, S.; Govindaraj, A.; Krishnamurthy, H.; Waghmare, U.; Rao, C. Synthesis, structure, and properties of boron-and nitrogen-doped graphene. *Adv. Mater.* **2009**, *21*, 4726–4730. [CrossRef]
43. Bao, J.F.; Kishi, N.; Soga, T. Synthesis of nitrogen-doped graphene by the thermal chemical vapor deposition method from a single liquid precursor. *Mater. Lett.* **2014**, *117*, 199–203. [CrossRef]
44. Zhang, C.; Lin, W.; Zhao, Z.; Zhuang, P.; Zhan, L.; Zhou, Y.; Cai, W. CVD synthesis of nitrogen-doped graphene using urea. *Sci. China Phys. Mech.* **2015**, *58*, 107801–107805. [CrossRef]
45. Vishwakarma, R.; Kalita, G.; Shinde, S.M.; Yaakob, Y.; Takahashi, C.; Tanemura, M. Structure of nitrogen-doped graphene synthesized by combination of imidazole and melamine solid precursors. *Mater. Lett.* **2016**, *177*, 89–93. [CrossRef]
46. Wang, Z.; Li, P.; Chen, Y.; Liu, J.; Tian, H.; Zhou, J.; Zhang, W.; Li, Y. Synthesis of nitrogen-doped graphene by chemical vapour deposition using melamine as the sole solid source of carbon and nitrogen. *J. Mater. Chem. C* **2014**, *2*, 7396–7401. [CrossRef]
47. Hummers, W.S., Jr.; Offeman, R.E. Preparation of graphitic oxide. *J. Am. Chem. Soc.* **1958**, *80*, 1339. [CrossRef]
48. Jiang, M.-H.; Cai, D.; Tan, N. Nitrogen-doped graphene sheets prepared from different graphene-based precursors as high capacity anode materials for lithium-ion batteries. *Int. J. Electrochem. Sci.* **2017**, *12*, 7154–7165. [CrossRef]
49. Yokwana, K.; Ray, S.C.; Khenfouch, M.; Kuvarega, A.T.; Mamba, B.B.; Mhlanga, S.D.; Nxumalo, E.N. Facile synthesis of nitrogen doped graphene oxide from graphite flakes and powders: A comparison of their surface chemistry. *J. Nanosci. Nanotechnol.* **2018**, *18*, 5470–5484. [CrossRef]
50. Kumar, N.A.; Nolan, H.; McEvoy, N.; Rezvani, E.; Doyle, R.L.; Lyons, M.E.; Duesberg, G.S. Plasma-assisted simultaneous reduction and nitrogen doping of graphene oxide nanosheets. *J. Mater. Chem. A* **2013**, *1*, 4431–4435. [CrossRef]
51. Pham, P. A library of doped-graphene images via transmission electron microscopy. *J. Carbon Res.* **2018**, *4*, 34. [CrossRef]

52. Yen, M.-Y.; Hsieh, C.-K.; Teng, C.-C.; Hsiao, M.-C.; Liu, P.-I.; Ma, C.-C.M.; Tsai, M.-C.; Tsai, C.-H.; Lin, Y.-R.; Chou, T.-Y. Metal-free, nitrogen-doped graphene used as a novel catalyst for dye-sensitized solar cell counter electrodes. *RSC Adv.* **2012**, *2*, 2725–2728. [CrossRef]
53. Ayala, P.; Arenal, R.; Rümmeli, M.; Rubio, A.; Pichler, T. The doping of carbon nanotubes with nitrogen and their potential applications. *Carbon* **2010**, *48*, 575–586. [CrossRef]
54. Zafar, Z.; Ni, Z.H.; Wu, X.; Shi, Z.X.; Nan, H.Y.; Bai, J.; Sun, L.T. Evolution of Raman spectra in nitrogen-doped graphene. *Carbon* **2013**, *61*, 57–62. [CrossRef]
55. Matsoso, B.J.; Ranganathan, K.; Mutuma, B.K.; Lerotholi, T.; Jones, G.; Coville, N.J. Time-dependent evolution of the nitrogen configurations in N-doped graphene films. *RSC Adv.* **2016**, *6*, 106914–106920. [CrossRef]
56. Song, J.-h.; Kim, C.-M.; Yang, E.; Ham, M.-H.; Kim, I. The effect of doping temperature on the nitrogen-bonding configuration of nitrogen-doped graphene by hydrothermal treatment. *RSC Adv.* **2017**, *7*, 20738–20741. [CrossRef]
57. Capasso, A.; Dikonimos, T.; Sarto, F.; Tamburrano, A.; De Bellis, G.; Sarto, M.S.; Faggio, G.; Malara, A.; Messina, G.; Lisi, N. Nitrogen-doped graphene films from chemical vapor deposition of pyridine: Influence of process parameters on the electrical and optical properties. *Beilstein J. Nanotechnol.* **2015**, *6*, 2028–2038. [CrossRef]
58. Mallet-Ladeira, P.; Puech, P.; Toulouse, C.; Cazayous, M.; Ratel-Ramond, N.; Weisbecker, P.; Vignoles, G.L.; Monthioux, M. A Raman study to obtain crystallite size of carbon materials: A better alternative to the Tuinstra–Koenig law. *Carbon* **2014**, *80*, 629–639. [CrossRef]
59. Pei, S.; Cheng, H.-M. The reduction of graphene oxide. *Carbon* **2012**, *50*, 3210–3228. [CrossRef]
60. Mowry, M.; Palaniuk, D.; Luhrs, C.C.; Osswald, S. In situ Raman spectroscopy and thermal analysis of the formation of nitrogen-doped graphene from urea and graphite oxide. *RSC Adv.* **2013**, *3*, 21763–21775. [CrossRef]
61. Justh, N.; Berke, B.; László, K.; Szilágyi, I.M. Thermal analysis of the improved Hummers' synthesis of graphene oxide. *J. Therm. Anal. Calorim.* **2018**, *131*, 2267–2272. [CrossRef]
62. Chang, B.Y.S.; Huang, N.M.; An'amt, M.N.; Marlinda, A.R.; Norazriena, Y.; Muhamad, M.R.; Harrison, I.; Lim, H.N.; Chia, C.H. Facile hydrothermal preparation of titanium dioxide decorated reduced graphene oxide nanocomposite. *Int. J. Nanomedicine* **2012**, *7*, 3379–3387. [PubMed]
63. Khandelwal, M.; Kumar, A. One-pot environmentally friendly amino acid mediated synthesis of N-doped graphene–silver nanocomposites with an enhanced multifunctional behavior. *Dalton Transactions* **2016**, *45*, 5180–5195. [CrossRef] [PubMed]
64. Kumar, A.; Ganguly, A.; Papakonstantinou, P. Thermal stability study of nitrogen functionalities in a graphene network. *J. Phys. Condens. Matter* **2012**, *24*, 235501–235506. [CrossRef] [PubMed]
65. Youn, H.C.; Bak, S.M.; Kim, M.S.; Jaye, C.; Fischer, D.A.; Lee, C.W.; Yang, X.Q.; Roh, K.C.; Kim, K.B. High-Surface-Area Nitrogen-Doped Reduced Graphene Oxide for Electric Double-Layer Capacitors. *ChemSusChem* **2015**, *8*, 1875–1884. [CrossRef]
66. Liu, S.; Peng, W.; Sun, H.; Wang, S. Physical and chemical activation of reduced graphene oxide for enhanced adsorption and catalytic oxidation. *Nanoscale* **2014**, *6*, 766–771. [CrossRef]
67. Yang, S.-Y.; Chang, K.-H.; Huang, Y.-L.; Lee, Y.-F.; Tien, H.-W.; Li, S.-M.; Lee, Y.-H.; Liu, C.-H.; Ma, C.-C.M.; Hu, C.-C. A powerful approach to fabricate nitrogen-doped graphene sheets with high specific surface area. *Electrochem. Commun.* **2012**, *14*, 39–42. [CrossRef]
68. Fu, C.; Song, C.; Liu, L.; Xie, X.; Zhao, W. Synthesis and properties of nitrogen-doped graphene as anode materials for lithium-ion batteries. *Int. J. Electrochem. Sci.* **2016**, *11*, 3876–3886. [CrossRef]
69. Sing, K.S. Reporting physisorption data for gas/solid systems with special reference to the determination of surface area and porosity (Recommendations 1984). *Pure Appl. Chem.* **1985**, *57*, 603–619. [CrossRef]
70. Qiao, X.; Liao, S.; You, C.; Chen, R. Phosphorus and nitrogen dual doped and simultaneously reduced graphene oxide with high surface area as efficient metal-free electrocatalyst for oxygen reduction. *Catalysts* **2015**, *5*, 981–991. [CrossRef]
71. Hanifah, M.F.R.; Jaafar, J.; Aziz, M.; Ismail, A.F.; Rahman, M.A.; Othman, M.H.D. Synthesis of graphene oxide nanosheets via modified hummers' method and its physicochemical properties. *J. Teknol.* **2015**, *74*, 189–192. [CrossRef]
72. Zhang, C.; Fu, L.; Liu, N.; Liu, M.; Wang, Y.; Liu, Z. Synthesis of nitrogen-doped graphene using embedded carbon and nitrogen sources. *Adv. Mater.* **2011**, *23*, 1020–1024. [CrossRef]

73. Zhang, J.; Zhao, C.; Liu, N.; Zhang, H.; Liu, J.; Fu, Y.Q.; Guo, B.; Wang, Z.; Lei, S.; Hu, P. Tunable electronic properties of graphene through controlling bonding configurations of doped nitrogen atoms. *Sci. Rep.* **2016**, *6*, 28330–28340. [CrossRef]
74. Zhang, W.; Lin, C.-T.; Liu, K.-K.; Tite, T.; Su, C.-Y.; Chang, C.-H.; Lee, Y.-H.; Chu, C.-W.; Wei, K.-H.; Kuo, J.-L.; et al. Opening an electrical band gap of bilayer graphene with molecular doping. *ACS Nano* **2011**, *5*, 7517–7524. [CrossRef]
75. Sheng, Z.-H.; Shao, L.; Chen, J.-J.; Bao, W.-J.; Wang, F.-B.; Xia, X.-H. Catalyst-free synthesis of nitrogen-doped graphene via thermal annealing graphite oxide with melamine and its excellent electrocatalysis. *ACS Nano* **2011**, *5*, 4350–4358. [CrossRef]
76. Baldovino, F.; Quitain, A.; Dugos, N.P.; Roces, S.A.; Koinuma, M.; Yuasa, M.; Kida, T. Synthesis and characterization of nitrogen-functionalized graphene oxide in high-temperature and high-pressure ammonia. *RSC Adv.* **2016**, *6*, 113924–113932. [CrossRef]
77. Chen, F.; Guo, L.; Zhang, X.; Leong, Z.Y.; Yang, S.; Yang, H.Y. Nitrogen-doped graphene oxide for effectively removing boron ions from seawater. *Nanoscale* **2017**, *9*, 326–333. [CrossRef]
78. Chen, M.; Shao, L.-L.; Guo, Y.-X.; Cao, X.-Q. Nitrogen and phosphorus co-doped carbon nanosheets as efficient counter electrodes of dye-sensitized solar cells. *Chem. Eng. J.* **2016**, *304*, 303–312. [CrossRef]
79. Loryuenyong, V.; Totepvimarn, K.; Eimburanapravat, P.; Boonchompoo, W.; Buasri, A. Preparation and characterization of reduced graphene oxide sheets via water-based exfoliation and reduction methods. *Adv. Mater. Sci. Eng.* **2013**, *2013*, 1–5. [CrossRef]
80. Jamil, A.; Mustafa, F.; Aslam, S.; Arshad, U.; Ahmad, M.A. Structural and optical properties of thermally reduced graphene oxide for energy devices. *Chin. Phys. B* **2017**, *26*, 086501–086508. [CrossRef]
81. Vinoth, R.; Ganesh Babu, S.; Bahnemann, D.; Neppolian, B. Nitrogen-doped reduced graphene oxide hybrid metal free catalyst for effective reduction of 4-nitrophenol. *Sci. Adv. Mater.* **2015**, *7*, 1–7. [CrossRef]
82. Mokhtar Mohamed, M.; Mousa, M.A.; Khairy, M.; Amer, A.A. Nitrogen graphene: A new and exciting generation of visible light driven photocatalyst and energy storage application. *ACS Omega* **2018**, *3*, 1801–1814. [CrossRef]
83. Marschall, R.; Wang, L. Non-metal doping of transition metal oxides for visible-light photocatalysis. *Catal. Today* **2014**, *225*, 111–135. [CrossRef]
84. Chuang, C.H.; Wang, Y.F.; Shao, Y.C.; Yeh, Y.C.; Wang, D.Y.; Chen, C.W.; Chiou, J.W.; Ray, S.C.; Pong, W.F.; Zhang, L.; et al. The effect of thermal reduction on the photoluminescence and electronic structures of graphene oxides. *Sci. Rep.* **2014**, *4*, 4521–4527. [CrossRef]
85. Van Khai, T.; Na, H.G.; Kwak, D.S.; Kwon, Y.J.; Ham, H.; Shim, K.B.; Kim, H.W. Influence of N-doping on the structural and photoluminescence properties of graphene oxide films. *Carbon* **2012**, *50*, 3799–3806. [CrossRef]
86. Van Khai, T.; Na, H.G.; Kwak, D.S.; Kwon, Y.J.; Ham, H.; Shim, K.B.; Kim, H.W. Significant enhancement of blue emission and electrical conductivity of N-doped graphene. *J. Mater. Chem.* **2012**, *22*, 17992–18003. [CrossRef]

© 2019 by the authors. Licensee MDPI, Basel, Switzerland. This article is an open access article distributed under the terms and conditions of the Creative Commons Attribution (CC BY) license (http://creativecommons.org/licenses/by/4.0/).

Article

Thermoresistive Properties of Graphite Platelet Films Supported by Different Substrates

Mariano Palomba [1], Gianfranco Carotenuto [1], Angela Longo [1,*], Andrea Sorrentino [1], Antonio Di Bartolomeo [2,3,*], Laura Iemmo [2,3], Francesca Urban [2,3], Filippo Giubileo [3], Gianni Barucca [4], Massimo Rovere [5], Alberto Tagliaferro [5], Giuseppina Ambrosone [6] and Ubaldo Coscia [6,7]

[1] Institute for Polymers, Composites and Biomaterials—National Research Council (IPCB-CNR). SS Napoli/Portici, Piazzale E. Fermi, 1-80055 Portici (NA), Italy; mariano.palomba@cnr.it (M.P.); giancaro@unina.it (G.C.); andrea.sorrentino@cnr.it (A.S.)
[2] Department of Physics 'E.R.Caianello', University of Salerno, Via Giovanni Paolo II, 132—84084 Fisciano (SA), Italy; liemmo@unisa.it (L.I.); furban@unisa.it (F.U.)
[3] Superconducting and Other Innovative Materials and Devices Institute—National Research Council (SPIN-CNR), Via Giovanni Paolo II, 132—84084 Fisciano (SA), Italy; filippo.giubileo@spin.cnr.it
[4] Department SIMAU, Polytechnic University of Marche, Via Brecce Bianche, I-60131 Ancona, Italy; g.barucca@staff.univpm.it
[5] Department of Applied Science and Technology, Politecnico di Torino. Corso Duca degli Abruzzi, 24, 10129 Torino, Italy; massimo.rovere@polito.it (M.R.); alberto.tagliaferro@polito.it (A.T.)
[6] Department of Physics 'Ettore Pancini', University of Naples 'Federico II', Via Cintia, I-80126 Napoli, Italy; giuseppina.ambrosone@unina.it (G.A.); coscia@fisica.unina.it (U.C.)
[7] CNISM, Naples Unit, Via Cintia, I-80126 Napoli, Italy
* Correspondence: angela.longo@cnr.it (A.L.); adibartolomeo@unisa.it (A.D.B.)

Received: 6 October 2019; Accepted: 2 November 2019; Published: 5 November 2019

Abstract: Large-area graphitic films, produced by an advantageous technique based on spraying a graphite lacquer on glass and low-density polyethylene (LDPE) substrates were studied for their thermoresistive applications. The spray technique uniformly covered the surface of the substrate by graphite platelet (GP) unities, which have a tendency to align parallel to the interfacial plane. Transmission electron microscopy analysis showed that the deposited films were composed of overlapped graphite platelets of different thickness, ranging from a few tens to hundreds of graphene layers, and Raman measurements provided evidence for a good graphitic quality of the material. The GP films deposited on glass and LDPE substrates exhibited different thermoresistive properties during cooling–heating cycles in the −40 to +40 °C range. Indeed, negative values of the temperature coefficient of resistance, ranging from -4×10^{-4} to -7×10^{-4} °C^{-1} have been observed on glass substrates, while positive values varying between 4×10^{-3} and 8×10^{-3} °C^{-1} were measured when the films were supported by LDPE. These behaviors were attributed to the different thermal expansion coefficients of the substrates. The appreciable thermoresistive properties of the graphite platelet films on LDPE could be useful for plastic electronic applications.

Keywords: graphite platelet coatings; LDPE; thermal expansion coefficient; thermoresistive properties

1. Introduction

Plastic electronics is an emerging technological field with a remarkable potential in the areas of robotics, solar energy, sensors, health care, industrial automation, etc. [1–5]. Plastic electronic devices offer the unique characteristics of stretchability, flexibility, transparency, lightweightness, etc., which can be exploited for future industrial applications. Additionally, the processing technologies applied for the fabrication of plastic electronic devices (e.g., contact printing, roll-to-roll, ink-jet, spraying, etc.)

are inexpensive and powerful, compared to the traditional approaches available for silicon-based electronics. However, all these technologies require further optimization to allow the production of these materials on a large scale.

In this field, plastics are useful both for fabricating printed circuit boards and for making the active and passive electronic components of a circuit. These components can be easily achieved by incorporating functional organic materials (e.g., chromophores, fluorophores, conductive or magnetic fillers, etc.) into an adequate polymer matrix [6–8]. In particular, graphite platelets, carbon nanotubes, fullerene, graphene, and other carbonaceous materials have been extensively studied and utilized to obtain conductive, thermoresistive, and semiconductive polymeric nanocomposites [9–17]. Furthermore, the surfaces of polymers such as poly (methyl methacrylate), polyethylene terephthalate and low-density polyethylene (LDPE) have been made conductive by depositing graphite or graphene layers onto them for the fabrication of printed radio frequency devices [18], electrically conductive paths [19], piezoresistive sensors [20], and strain gauges [9]. These layers can be deposited by chemical vapor deposition [21], casting and drying inks [18], micromechanical techniques based on spreading an alcoholic suspension of graphite nanoplatelets [22,23] and spraying conductive composites [9]. In particular, this last technique is easy, inexpensive, and industrially scalable for the fabrication of large area films.

In this study, the properties of graphite platelet (GP) films, obtained by spraying a commercial lacquer on different substrates (LDPE and glass), were investigated. The deposited coatings were morphologically and structurally characterized by scanning electron microscopy (SEM), transmission electron microscopy (TEM), Fourier-transform infrared spectroscopy (FT-IR), and Raman spectroscopy. The thermal properties of the commercial lacquer, pure LDPE and graphite platelets deposited on LDPE were also explored by thermogravimetric analysis (TGA). The thermal expansion coefficients of the LDPE substrates coated by GP films were determined by dynamic–mechanical thermal analysis (DMTA). Thermoresistive measurements of graphite platelets films on glass and LDPE substrates, i.e., the measurements of the electrical resistance as a function of the temperature, were carried out during the cooling–heating cycles between −40 °C and +40 °C. Owing to the thermal expansion coefficient mismatch, the substrate could dramatically affect the temperature coefficient of resistance (TCR) of the GP film, which exhibited a negative TCR on glass and a positive TCR on the LDPE substrate.

2. Materials and Methods

Large area thin films of graphite-based material were deposited on glass and LDPE substrates by spray technology, using a commercial lacquer, Graphit 33 (from Kontakt Chemie, Zele, Belgium), which is commonly used in optical and electrical fields [24]. In order to produce a full cone jet spot, the spray nozzle was horizontally directed, taking it at a distance of 20 cm from the substrate surface. After spraying, the coated substrates were dried in air at room temperature, for 4 h.

Scanning electron microscopy analysis of the sample surface was performed using a FEI Quanta 200 FEG (FEI, Hillsboro, Oregon, USA) microscope. Due to the conductive nature of the graphite-based material, samples were observed without any preparation except for the electrical grounding of the surface. The inner structure of the deposited material was investigated through transmission electron microscopy measurements carried out by a Philips CM200 (Philips, Amsterdam, The Netherlands) microscope, operating at 200 kV and equipped with a LaB_6 filament. For TEM observations, two kinds of samples were prepared. In one case, the Graphit 33 was sprayed in acetone, the solid phase was isolated by centrifugation and deposited on a TEM copper grid covered with a thin carbon film. In another case, in order to see the inner structure of the deposited layer the Graphit 33 was sprayed on a substrate and prepared in a cross-section by the conventional thinning procedure, consisting of mechanical polishing by grinding papers, diamond pastes, and a dimple grinder; final thinning was carried out by an ion beam system (Gatan PIPS), using Ar ions at 5 kV.

Fourier-transform infrared spectra of dry Graphit 33 were made by a MIR/FIR Spectrometer (Frontier, PerkinElmer, Milan, Italy). The samples preparation was performed by mixing powered

Graphit 33 with a crystalline KBr powder in an adequate ratio (1% by weight) and this mixture was cold pressed under vacuum at 8 tons, for 10 min, to obtain a transparent pellet.

The thermal gravimetric analysis, in the 40–600 °C range was used to establish the thermal stability of the coating, the pure LDPE, and the LDPE coated by the GP film. Such analyses were performed by a TA-Instrument (Q500, Milan, Italy), operating in flowing nitrogen, with a constant heating rate of 10 °C/min.

Raman spectra of GP films on glass and LDPE substrates were performed by a Raman spectrometer InviaH-Renishaw (InviaH, Renishaw, plc, New Mils, Wotton-under Edge, Glowcester Shire, GL128JR, UK). A green argon laser with a 514.5 nm wavelength and a beam size approximately 2 µm in diameter was selected for the analysis. A microscope with a 50X magnification power was used with an exposure time of 10 s. Extended scans from 100 cm^{-1} to 3500 cm^{-1} was performed using a laser power of ca. 5 mW (5% of available laser power). An in-house MATLAB software (Matlab 9.7.0.1190202 R2019b, Mathworks Inc., Natick, MA, USA) was used to correct the quantum efficiency of the detector, conduct the baseline subtraction, and carry out the data processing.

Thermal expansion and contraction tests of an LDPE sample coated by GP (18.7 × 8.9 mm and ca. 90 µm thick) was carried out by means of cooling–heating cycles in the −40 to +40 °C range, at a rate of 5 °C/min, using a thermal mechanical analyzer (TMA2940, TA-Instruments, New Castle, USA).

Electrical measurements were executed under vacuum (~2 mbar) in a coplanar configuration by silver paint contacts (1-cm long and 1-mm spaced) spread on their surfaces. Vacuum was adopted as a precaution to avoid ice formation during the low temperature cycle, as well as to prevent possible effects of moisture or other adsorbates. Current–voltage (I–V) characteristics were taken in a Janis Research ST-500 probe station equipped (Janis Research, Woburn, MA, USA) with 4 micromanipulators connected to a Source-Measurement Unit (SMU) Keithley 4200-SCS (Tektronix, Inc., Beaverton, OR, USA). From each I–V and from monitoring the resistance during a period of 60 s it was estimated that the mean resistance of the samples at different temperatures T during cooling–heating cycles from −40 °C to +40 °C performed at a rate of about 5 °C/min.

3. Results and Discussion

3.1. Characterization of Graphit 33 Lacquer

Large-area thin films of graphite-based material were deposited onto the glass and LDPE substrates through spray technology, using a commercial lacquer such as Graphit 33 (see Figure 1).

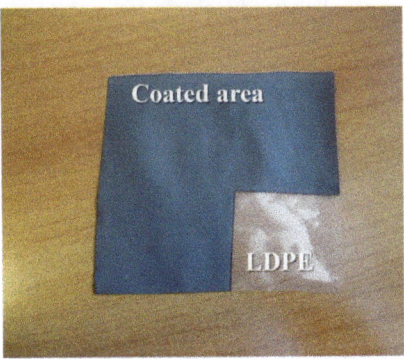

Figure 1. Large-area film after spraying Graphit 33 lacquer onto the low-density polyethylene (LDPE) film.

Thermo-gravimetric analysis and infrared spectroscopy were carried out to identify the composition and concentration of the lacquer. The TGA–thermogram of a dry Graphit 33 sample is shown in Figure 2. The sample is characterized by a weight loss in the 250–400 °C range, with a maximum degradation rate at ca. 330 °C, as can be seen from the derivative curve (red line in Figure 2). This weight loss could be attributed to the thermal degradation of the polymeric binder contained in the product, and it was estimated as ca. 18% by weight. As evidenced by the thermogram, the absence of weight loss at low temperature confirmed the absence of a volatile solvent after drying.

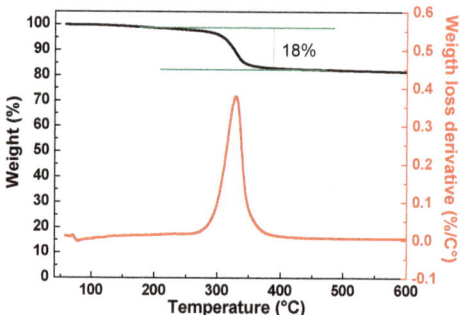

Figure 2. TGA–thermogram and derivative thermogravimetric plot profile of a typical dried Graphit 33 sample.

The FT–IR spectrum of the dry Graphit 33 in KBr is shown in Figure 3. The main absorption bands were centered at 3440 cm^{-1} (–OH group stretching), 2927 cm^{-1} (C–H group stretching), 1631 cm^{-1} (C=C group stretching), and 1092 cm^{-1} (C–O group stretching).

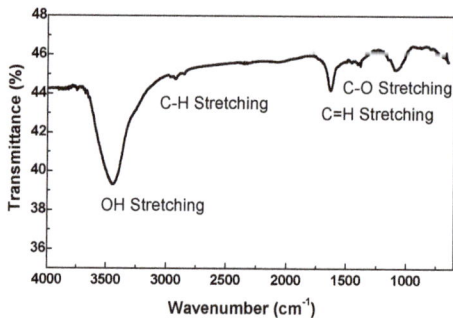

Figure 3. FT–IR spectrum of a typical dried Graphit 33 sample in KBr.

TEM measurements were carried out to investigate the inner structure of the graphite phase. In particular, Figure 4A,B show the solid phase extracted by the Graphit 33 by using acetone to remove the polymeric binder. It was evident that the filler was made of platelets that in the images appear to be largely superimposed. The platelets were hundreds of nanometers large, and their thicknesses were quite small, considering the low contrast visible in the images. Selected area electron diffraction (SAED) measurements were used to investigate the phase of the filler. Figure 4C shows a typical SAED pattern. All the diffraction rings could be associated with the pure graphite phase (International Centre for Diffraction Data, ICDD, card n°. 41–1487) confirming that the only crystalline phase inside the Graphit 33 was graphite.

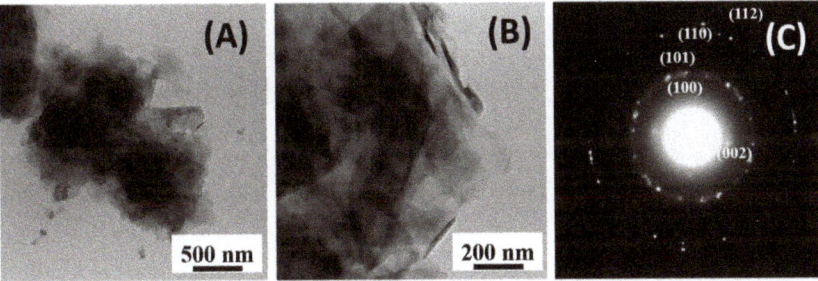

Figure 4. Graphite platelets extracted by the Graphit 33 lacquer: (**A**,**B**) bright field TEM images taken at different magnifications; and (**C**) corresponding SAED pattern in which the diffraction rings can be associated with the families of graphite atomic planes.

All experimental observations shown so far reveal that the Graphit 33 lacquer is composed mainly of graphite platelets. The presence of the binder allows us to deposit a continuous graphite coating on several types of substrates (glass, silicon, LDPE, etc.) using spray technology.

3.2. Morphological and Structural Characterizations of the GP Coatings

Coating morphology and structure, after spraying Graphit 33 on different substrates, were investigated by SEM and TEM techniques. In particular, Figure 5A,B show the SEM images of the deposited coatings on LDPE and glass substrates, respectively. The coatings on both substrates were quite rough, porous, and were made of small graphite platelets, which covered all substrates without discontinuities. TEM micrographs of a cross-sectioned GP coating are displayed in Figure 5C,D. The coating is rather wrinkled, as shown in Figure 5C, with a thickness ranging from 2.3 to 3.6 µm and a mean value of about 2.5 µm was measured on the large areas. TEM and SEM analyses suggest a surface roughness of 500–1000 nm. The current application of this study deals with macroscopic thermoresistors, therefore, the surface roughness of the platelet films did not present any issue. Indeed, the metal contacts could have an arbitrary size and could be formed by a silver paste coating or by metal sputtering (Au), using a shadow mask. Clearly, in the case of a microscopic device, the roughness of the film could hamper the fabrication of micro-nano patterns, through an advanced lithographic process.

By increasing the magnification (Figure 5D), it is possible to observe that coating is made of overlapping graphite platelets of different thickness, ranging from tens to hundreds of graphene layers. Among the platelets, the polymeric binder visibly generates a lighter contrast typical of amorphous material (see arrows). Although, the platelets assume all orientations locally in the selected area electron diffraction (SAED) pattern of a large part of the coating (shown in the inset of Figure 5C) it could be observed that the ring corresponding to the (0001) atomic planes had a non-uniform intensity that could be correlated to the platelet's tendency to align, in average, parallel to the interfacial plain.

Further structural characterizations were performed by Raman spectroscopy. The following spectra are related to GP films deposited on glass (Figure 6A) and LDPE (Figure 6B).

Spectra analysis of GP films deposited both on glass and LDPE substrates indicated that the coatings present similar characteristics, with a narrow D peak and a sharp and narrow G peak, right shouldered (D*) due to the presence of defects [25–27]. The mean value of the ratio between the intensities of the two peaks, I_d/I_g, over the sampled points, was about 0.55 in the case of GP on glass and about 0.80 for GP on LDPE, revealing that the films were made of a good quality graphitic material with a lower presence of defects in the film deposited on glass. Three other peaks could be detected in both spectra at higher wavenumbers (2300 to 3400 cm^{-1})–an intense left shouldered 2D peak (indicating a multilayered structure), a D + G peak, and a 2G peak normal in width and intensity for a graphitic material. No substrate signal was detected because the thickness of the GP film was larger than the penetration length of laser radiation used in the Raman apparatus.

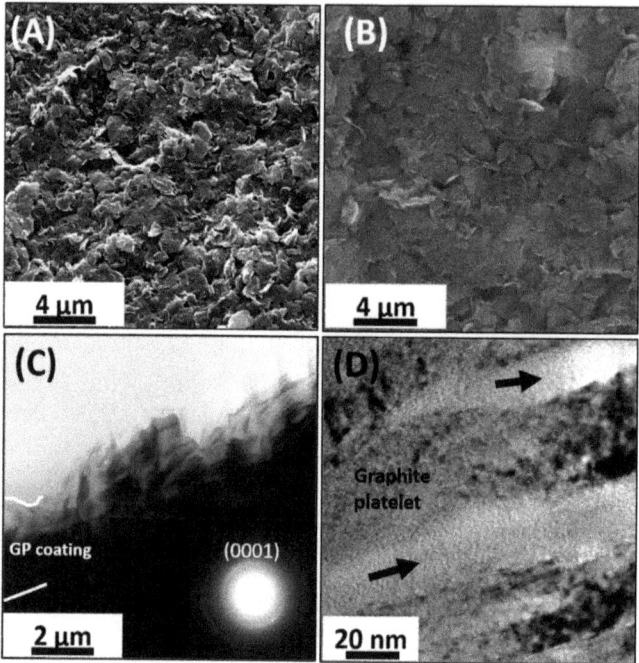

Figure 5. SEM-micrographs of surface topography of GP films deposited on different substrates: low-density polyethylene (LDPE) (**A**) and glass (**B**). Bright field TEM images of cross-sectioned GP films at different magnifications, (**C,D**). The inset in (**C**) is the selected area electron diffraction (SAED) pattern of the coating shown in (**C**). Dark arrows in (**D**) evidence the presence of amorphous materials among graphite platelets.

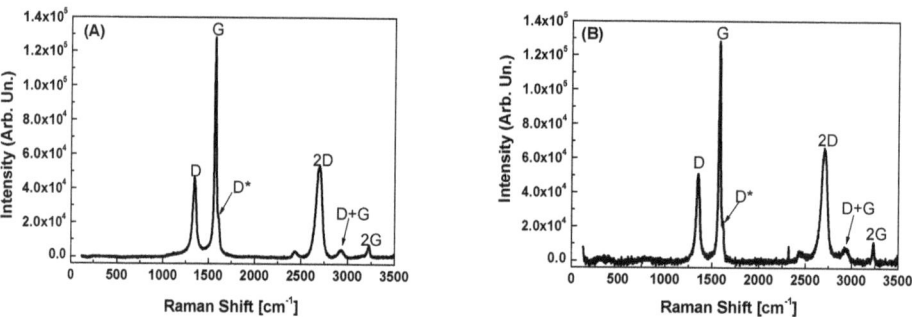

Figure 6. Raman spectra of graphite platelet (GP) films deposited on glass (**A**) and LDPE (**B**).

3.3. Thermal Properties of Pure LDPE and LDPE Coated by GP Films

The thermal stability of the pure LDPE and LDPE coated by GP was evaluated by TGA measurements. According to the TGA–thermograms, up to a temperature of ca. 130 °C, both samples were stable since they did not show any weight loss. Furthermore, by comparing the residual mass at 600 °C, which had TGA curves in the 40–600 °C range, it could be seen that the average amount of coating was ca. 1% by weight of the full LDPE/GP system. The graphite coating had the effect of increasing the pure LDPE substrate thermal stability by ca. 22 °C (see Figure 7).

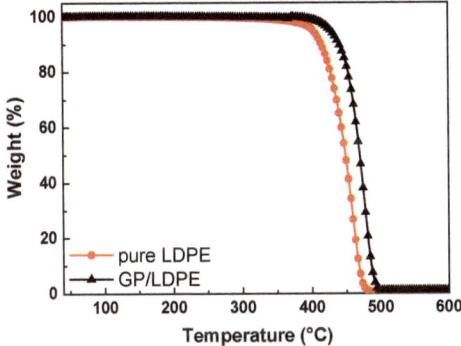

Figure 7. TGA–thermograms of pure LDPE and LDPE coated by GP.

In order to correlate the effects of the thermal expansion of the LDPE substrate with the electrical properties of graphite platelet films, tests of thermal expansion and contraction were carried out on a sample of LDPE coated by GP. In particular, this characterization was carried out between −40 °C and 40 °C, because in this temperature range, LDPE did not show any phase transition such as crystallisation, melting, or glass transition, and the stress–strain response induced by temperature variations was quite reversible [28].

The length of the sample, L, was recorded applying a constant force of 0.01 N and varying the temperature in the above range at the rate of 5 °C/min. The strain of the sample, defined as $\varepsilon = (L - L_0)/L_0$, where L_0 is the initial sample length at 20 °C, is plotted in Figure 8, for two consecutive cooling–heating cycles. A small hysteresis was evident between the heating and cooling curves. It could be due to both a thermal relaxation of the LDPE molecules and a small temperature gradient normally present in the furnace during cooling. The coefficient of linear thermal expansion, CTE, of the LDPE coated by the GP sample was calculated as $\varepsilon/\Delta T$, where ΔT is the temperature change during the test. CTE value results to be about 1.7×10^{-4} °C^{-1} on the investigated temperature range. This value was in good agreement with those reported in the literature for LDPE films ($1-2 \times 10^{-4}$ °C^{-1}) [29].

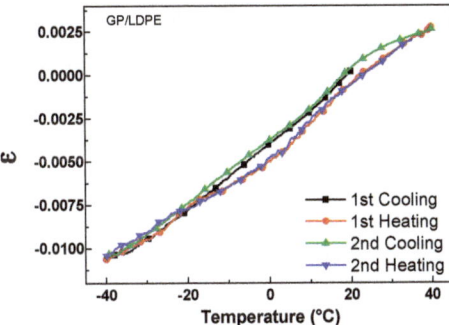

Figure 8. Strain, ε, vs. temperature of the LDPE coated by GP for two consecutive cooling–heating cycles.

3.4. Thermoresistive Characterizations of Graphite Platelet Films on Glass and LDPE

All electrical measurements were carried out under vacuum in two probe configuration. The I–V characteristics of the GP films deposited on glass and LDPE were linear, indicating ohmic contacts, as shown in Figure 9.

The resistance values, R_0, of the samples at the temperature of 20 °C were determined by the fit of the plotted experimental data.

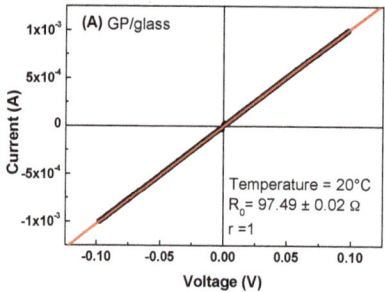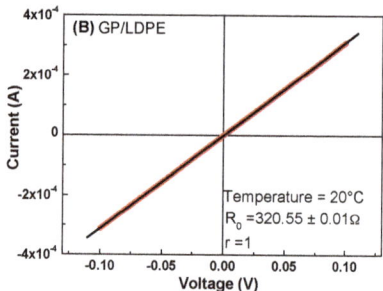

Figure 9. I–V characteristics of the GP films deposited on glass (**A**) and LDPE (**B**) substrates.

The thermoresistive properties of the GP films were investigated by recording the resistance values, R, of the samples starting from 20 °C and performing the cooling–heating cycles in the −40 to 40 °C range. The high resistance value of GP on LDPE could be due to the larger surface roughness of the polymer substrate and its different chemical nature, compared to the glass ones that determined a greater degree of inhomogeneity, as evidenced by the increased I_d/I_g ratio obtained by Raman analysis.

In Figure 10, the R/R_0 ratios are plotted against temperature for representative samples of GP films deposited on glass and LDPE, respectively. Clearly, different thermoresistive behaviors could be observed in the examined range. Indeed, the resistance of GP film on glass slowly decreases with increasing T, similar to that of graphite [30], while GP on LDPE shows an increase in resistance in the whole range. Thus, thermoresistive properties of GP films strongly depend on the substrates. In fact, TCR defined as:

$$TCR = \frac{1}{R}\frac{dR}{dT} \quad (1)$$

is negative for GP on glass and positive for GP on LDPE.

Taking into account the cooling–heating cycles of Figure 10A,B, TCR varies in the -4×10^{-4} to -7×10^{-4} °C^{-1} and 4×10^{-3} to 8×10^{-3} °C^{-1} ranges for GP on glass and LDPE, respectively. Furthermore, a greater reproducibility of the resistance–temperature characteristics in the case of GP film deposited on the glass was observed. On the other hand, the appreciable thermoresistive sensitivity of GP on LDPE made these materials useful for flexible electronic applications, although more work has to be done to reduce the hysteresis during the thermal cycles and obtain a GP material with a more reversible thermoresistive response.

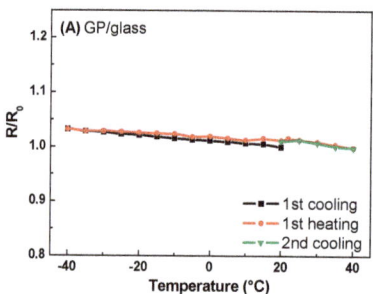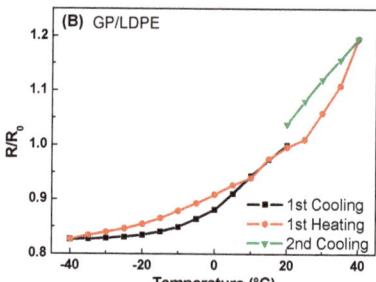

Figure 10. R/R_0 vs. temperature during the cooling–heating cycles for GP films deposited on glass (**A**) and LDPE (**B**) substrates.

The observed behaviors could be attributed to the different CTE of the substrates. Glass has a CTE (6–9 × 10^{-6} °C^{-1}) close to that of graphite (4–8 × 10^{-6} °C^{-1}), and therefore, the resistance of the GP film as compared to T, decreases, as in the case of graphite [29]. On the other hand, as reported in

Section 3.3, the CTE of the coated LDPE (1.7×10^{-4} °C^{-1}) is more than one order of magnitude greater than that of graphite, thus, the thermal expansion (contraction) of the polymer substrate could induce strains in the GP film, which tend to increase (decrease) its resistance. For example, in the case of the investigated sample, by comparing the data in Figures 8 and 10B, the fractional change of the electrical resistance, $(R - R_0)/R_0$, of the GP film could be correlated to the strain, ε, of the coated LDPE substrate, as shown in Figure 11.

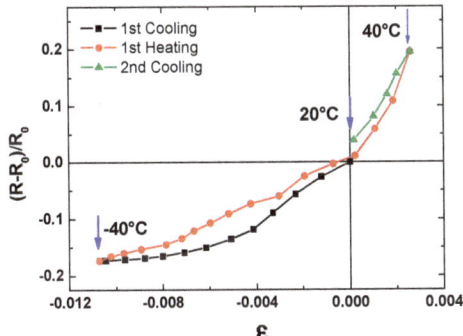

Figure 11. The fractional change of the electrical resistance, $(R - R_0)/R_0$, vs. the strain, ε, of the GP film on the LDPE substrate.

As can be seen, the slopes of the curves in the −40 to 20 °C range were lower than those in the 20 to 40 °C range. Indeed, below the sample deposition temperature (20 °C), a greater compaction of the platelets enhanced the decrease (increase) in the resistance during heating (cooling), as in the case of graphite. This effect tends to counterbalance the increase (decrease) in resistance due to the expansion (contraction) of the substrate and, therefore, the resulting resistance increases (decreases) more slowly in this temperature range.

Additionally, the larger resistance hysteresis occurring in the GP film on LDPE during the cooling–heating cycles (Figure 10B) could be attributed to the greater CTE of the LDPE, compared to that of glass. Indeed, the platelets deposited on LDPE were subjected to a greater mobility due to the strains of this polymer substrate and the occurrence of possible nano/micro fractures in the films that could cause a different assembly of the platelets when the sample passed again for the same temperature during a thermal cycle, leading to a different value of its initial resistance.

4. Conclusions

It was observed that large-area conductive thin films could be produced by spraying Graphit 33 lacquer on glass and LDPE substrates. Raman spectra analysis revealed that the graphitic material deposited on both substrates was of good quality. According to the morphological and structural investigations by SEM and TEM, the films consist of overlapped graphite platelets that cover the surfaces of the substrates, mostly in a coplanar manner. It was found that the resistance of the film as a function of temperature in the −40 to 40 °C range decreased if the substrate was glass and increased in the case of the LDPE substrate. Therefore, the temperature coefficient of resistance changed from negative to positive values, respectively. It was demonstrated that the different thermoresistive properties of the GP films depend on the thermal expansion characteristics of the substrates on which they have been deposited. The appreciable thermoresistive sensitivity of GP films on LDPE makes these structures promising for applications in plastic electronics, however, more work has to be carried out to reduce the hysteresis observed during the thermal cycles to obtain a GP material with more reproducible thermoresistive properties.

Author Contributions: Conceptualization, U.C., G.C. and A.D.B; methodology, F.G.; software, F.G. and M.R.; validation, M.P., A.L., G.C., L.I., F.U. and G.A.; formal analysis, L.I., F.U., F.G., A.L. and M.P.; investigation, L.I., F.U., G.B., M.R, A.T., A.L., A.S. and M.P.; resources, A.D.B; G.C., U.C., A.T., A.S. and G.B; data curation, F.U., L.I., G.B., M.R, A.T., A.L., A.S. and M.P.; writing—original draft preparation, G.C., M.P., A.L., A.D.B., G.B., A.T., M.R., A.S. and U.C.; writing—review and editing A.D.B., G.B., A.T., G.A. and U.C.; supervision U.C., G.A.,G.C., G.B. and A.D.B.

Funding: This research received no external funding.

Acknowledgments: The authors are grateful to Maria Cristina Del Barone of LAMEST laboratory (IPCB-CNR) for TEM cross-section preparation and to Maria Rosaria Marcedula of Thermo-analysis Laboratory (IPCB-CNR) for FT-IR and TGA tests. The authors are also grateful to A. Vanzanella and the electronic workshop of INFN Napoli for the fruitful discussions and their support during the assembling of the experimental setup. The authors acknowledge the funding support by the project Pico & Pro, MIUR Project ARS01_01061, 2018–2021.

Conflicts of Interest: The authors declare no competing financial interest.

References

1. Singh, T.B.; Sariciftci, N.S. Progress in plastic electronics devices. *Annu. Rev. Mater. Res.* **2006**, *36*, 199–230. [CrossRef]
2. Stoppa, M.; Chiolerio, A. Wearable Electronics and Smart Textiles: A Critical Review. *Sensors* **2014**, *14*, 11957–11992. [CrossRef] [PubMed]
3. Wang, X.; Dong, L.; Zhang, H.; Yu, R.; Pan, C.; Wang, Z.L. Recent Progress in Electronic Skin. *Adv. Sci.* **2015**, *2*, 1500169. [CrossRef] [PubMed]
4. Lu, N.; Kim, D.-H. Flexible and Stretchable Electronics Paving the Way for Soft Robotics. *Soft Robot.* **2014**, *1*, 53–62. [CrossRef]
5. Rogers, J.A.; Someya, T.; Huang, Y. Materials and Mechanics for Stretchable Electronics. *Science* **2010**, *327*, 1603–1607. [CrossRef]
6. Kukhta, A.V.; Kolesnik, E.E.; Lesnikovich, A.I.; Nichik, M.N.; Kudlash, A.N.; Vorobyova, S.A. Organic-Inorganic Nanocomposites: Optical and Electrophysical Properties. *Lumin. Nanocompos.* **2007**, *37*, 333–339. [CrossRef]
7. Khan, S.; Lorenzelli, L. Recent advances of conductive nanocomposites in printed and flexible electronics. *Smart Mater. Struct.* **2017**, *26*, 083001. [CrossRef]
8. Raj, P.M.; Muthana, P.; Xiao, T.D.; Wan, L.; Balaraman, D.; Abothu, I.R.; Bhattacharya, S.; Swaminathan, M.; Tummala, R. Magnetic nanocomposites for organic compatible miniaturized antennas and inductors. *Conf. Pap. IEEE* **2005**, 272–275. [CrossRef]
9. Kondratov, A.P.; Zueva, A.M.; Varakin, R.S.; Taranec, I.P.; Savenkova, I.A. Polymer film strain gauges for measuring large elongations. *IOP Conf. Ser. Mater. Sci. Eng.* **2018**, *312*, 012013. [CrossRef]
10. Tripathi, S.N.; Rao, G.S.S.; Mathur, A.B.; Jasra, R. Polyolefin/graphene nanocomposites: A review. *RSC Adv.* **2017**, *7*, 23615–23632. [CrossRef]
11. Khare, R.; Bose, S. Carbon Nanotube Based Composites—A Review. *J. Miner. Mater. Charact. Eng.* **2005**, *4*, 31–46. [CrossRef]
12. Turkani, V.S.; Maddipatla, D.; Narakathu, B.B.; Bazuin, B.J.; Atashbar, M.Z. A carbon nanotube based NTC thermistor using additive print manufacturing processes. *Sens. Actuators Phys.* **2018**, *279*, 1–9. [CrossRef]
13. Wang, S.; Kowalik, D.P.; Chung, D.D.L. Self-sensing attained in carbon-fiber–polymer-matrix structural composites by using the interlaminar interface as a sensor. *Smart Mater. Struct.* **2004**, *13*, 570–592. [CrossRef]
14. Hirotani, J.; Amano, J.; Ikuta, T.; Nishiyama, T.; Takahashi, K. Carbon nanotube thermal probe for quantitative temperature sensing. *Sens. Actuators Phys.* **2013**, *199*, 1–8. [CrossRef]
15. Dong, Q.; Guo, Y.; Sun, X.; Jia, Y. Coupled electrical-thermal-pyrolytic analysis of carbon fiber/epoxy composites subjected to lightning strike. *Polymer* **2015**, *56*, 385–394. [CrossRef]
16. Sibinski, M.; Jakubowska, M.; Sloma, M. Flexible Temperature Sensors on Fibers. *Sensors* **2010**, *10*, 7934–7946. [CrossRef]
17. Dinh, T.; Phan, H.-P.; Qamar, A.; Woodfield, P.; Nguyen, N.-T.; Dao, D.V. Thermoresistive Effect for Advanced Thermal Sensors: Fundamentals, Design Considerations, and Applications. *J. Microelectromech. Syst.* **2017**, *26*, 966–986. [CrossRef]

18. Huang, X.; Leng, T.; Zhang, X.; Chen, J.C.; Chang, K.H.; Geim, A.K.; Novoselov, K.S.; Hu, Z. Binder-free highly conductive graphene laminate for low cost printed radio frequency applications. *Appl. Phys. Lett.* **2015**, *106*, 203105. [CrossRef]
19. Longo, A.; Verucchi, R.; Aversa, L.; Tatti, R.; Ambrosio, A.; Orabona, E.; Coscia, U.; Carotenuto, G.; Maddalena, P. Graphene oxide prepared by graphene nanoplatelets and reduced by laser treatment. *Nanotechnology* **2017**, *28*, 224002. [CrossRef]
20. Bonavolontà, C.; Camerlingo, C.; Carotenuto, G.; De Nicola, S.; Longo, A.; Meola, C.; Boccardi, S.; Palomba, M.; Pepe, G.P.; Valentino, M. Characterization of piezoresistive properties of graphene-supported polymer coating for strain sensor applications. *Sens. Actuators Phys.* **2016**, *252*, 26–34. [CrossRef]
21. De Castro, R.K.; Araujo, J.R.; Valaski, R.; Costa, L.O.O.; Archanjo, B.S.; Fragneaud, B.; Cremona, M.; Achete, C.A. New transfer method of CVD-grown graphene using a flexible, transparent and conductive polyaniline-rubber thin film for organic electronic applications. *Chem. Eng. J.* **2015**, *273*, 509–518. [CrossRef]
22. Palomba, M.; Longo, A.; Carotenuto, G.; Coscia, U.; Ambrosone, G.; Rusciano, G.; Nenna, G.; Barucca, G.; Longobardo, L. Optical and electrical characterizations of graphene nanoplatelet coatings on low density polyethylene. *J. Vac. Sci. Technol. B Nanotechnol. Microelectron. Mater. Process. Meas. Phenom.* **2018**, *36*, 01A104. [CrossRef]
23. Coscia, U.; Palomba, M.; Ambrosone, G.; Barucca, G.; Cabibbo, M.; Mengucci, P.; de Asmundis, R.; Carotenuto, G. A new micromechanical approach for the preparation of graphene nanoplatelets deposited on polyethylene. *Nanotechnology* **2017**, *28*, 194001. [CrossRef] [PubMed]
24. Lim, K.-H.; Kim, S.-K.; Chung, M.-K. Improvement of the thermal diffusivity measurement of thin samples by the flash method. *Thermochim. Acta* **2009**, *494*, 71–79. [CrossRef]
25. Ferrari, A.C.; Robertson, J. Interpretation of Raman spectra of disordered and amorphous carbon. *Phys. Rev. B* **2000**, *61*, 14095. [CrossRef]
26. Abdelkader, A.M.; Patten, H.V.; Li, Z.; Chen, Y.; Kinloch, I.A. Electrochemical exfoliation of graphite in quaternary ammonium-based deep eutectic solvents: a route for the mass production of graphane. *Nanoscale* **2015**, *7*, 11386–11392. [CrossRef]
27. Pimenta, M.A.; Dresselhaus, G.; Dresselhaus, M.S.; Canc, L.G.; Jorioa, A.; Saitoe, R. Studying disorder in graphite-based systems by Raman spectroscopy. *Phys. Chem. Chem. Phys.* **2007**, *9*, 1276–1291. [CrossRef]
28. Peacock, A. *Handbook of Polyethylene. Structures: Properties, and Applications*; Marcel Dekker, INC.: New York, NY, USA, 2000.
29. Mark, D.H. Thermal characterization of polymeric materials, Edith A. Turi, Ed., Academic, New York, 1981, 972 pp. Price: $98.00. *J. Polym. Sci. Polym. Lett. Ed.* **1982**, *20*, 281–282, Book review. [CrossRef]
30. Iwashita, H.; Imagawa, H.; Nishiumi, W. Variation of temperature dependence of electrical resistivity with crystal structure of artificial products. *Carbon* **2013**, *61*, 602–608. [CrossRef]

© 2019 by the authors. Licensee MDPI, Basel, Switzerland. This article is an open access article distributed under the terms and conditions of the Creative Commons Attribution (CC BY) license (http://creativecommons.org/licenses/by/4.0/).

Article

Effect of the Organic Functional Group on the Grafting Ability of Trialkoxysilanes onto Graphene Oxide: A Combined NMR, XRD, and ESR Study

Massimo Calovi [1], Emanuela Callone [2,*], Riccardo Ceccato [1], Flavio Deflorian [1], Stefano Rossi [1] and Sandra Dirè [2,*]

[1] Department of Industrial Engineering, University of Trento, 38123 Trento, Italy; massimo.calovi@unitn.it (M.C.); riccardo.ceccato@unitn.it (R.C.); flavio.deflorian@unitn.it (F.D.); stefano.rossi@unitn.it (S.R.)
[2] "Klaus Müller" Magnetic Resonance Laboratory, Department of Industrial Engineering, University of Trento, 38123 Trento, Italy
* Correspondence: emanuela.callone@unitn.it (E.C.); sandra.dire@unitn.it (S.D.); Tel.: +39-04-6428-2463 (E.C.); +39-04-6128-2456 (S.D.)

Received: 31 October 2019; Accepted: 19 November 2019; Published: 21 November 2019

Abstract: The functional properties displayed by graphene oxide (GO)-polymer nanocomposites are strongly affected by the dispersion ability of GO sheets in the polymeric matrix, which can be largely improved by functionalization with organosilanes. The grafting to GO of organosilanes with the general formula $RSi(OCH_3)_3$ is generally explained by the condensation reactions of silanols with GO reactive groups. In this study, the influence of the organic group on the $RSi(OCH_3)_3$ grafting ability was analyzed in depth, taking into account the interactions of the R end chain group with GO oxidized groups. Model systems composed of commercial graphene oxide reacted with 3-aminopropyltrimethoxysilane (APTMS), 3-mercaptopropyltrimethoxysilane (MPTMS), and 3-methacryloxypropyltrimethoxysilane, (MaPTMS), respectively, were characterized by natural abundance ^{13}C, ^{15}N and ^{29}Si solid state nuclear magnetic resonance (NMR), x-ray diffraction (XRD), and electron spin resonance (ESR). The silane organic tail significantly impacts the grafting, both in terms of the degree of functionalization and direct interaction with GO reactive sites. Both the NMR and XRD proved that this is particularly relevant for APTMS and to a lower extent for MPTMS. Moreover, the epoxy functional groups on the GO sheets appeared to be the preferential anchoring sites for the silane condensation reaction. The characterization approach was applied to the GO samples prepared by the nitric acid etching of graphene and functionalized with the same organosilanes, which were used as a filler in acrylic coatings obtained by cataphoresis, making it possible to correlate the structural properties and the corrosion protection ability of the layers.

Keywords: graphene oxide; organosilanes; grafting; solid state NMR; XRD; ESR

1. Introduction

Recently, large interest has been devoted to the use of graphene as a nanofiller in polymeric matrices thanks to its peculiar features [1,2]. To ensure the effective improvement of polymer properties by graphene addition, additives are required in order to obtain a homogeneous distribution within the matrix. In this regard, graphene flakes are usually oxidized to obtain graphene oxide (GO) [3,4], which is subsequently functionalized to increase chemical affinity with the polymer matrix [5].

Among the different functionalizing agents, organoalkoxysilanes have been widely used with success, exploiting both their grafting ability [6,7] and the possibility to ad hoc select the end chain organic functionality. Organoalkoxysilanes with the general formula $RSi(OR')_3$ are characterized by two different functional groups: -OR' are hydrolyzable groups forming reactive silanols suitable for condensation reactions, while R is a non-hydrolyzable organic group that imparts the desired features [8]. Generally, organoalkoxysilane grafting to GO is considered to take place by condensation reactions among silanols and the functional groups present on the basal plane of the GO flakes, producing an increase in the interplanar distance of the lamellae that depends on the silane organic chain, with the consequence of improving the dispersion in a polymer matrix [9].

The recent literature reports several studies on GO functionalization with silanes for a wide range of applications. For example, through the reaction of vinyltrimethoxysilane (VTMS) and GO in an aqueous medium, Abass et al. obtained VTMS-reduced graphene oxide (rGO) nanospheres and pointed out the role of silane in both the successful exfoliation of rGO layers and the enhancement of the electrical conductivity [9]. With the final aim to prepare high-performance thermosetting resins, Xu et al. modified GO with aminopropyltrimethoxysilane (APTMS); after reduction with hydrazine, the thermal resistance of functionalized GO appeared improved as did its dispersion ability in organic solvent, allowing the use of APTMS-rGO as filler for polybenzoxazine resins [10]. Moreover, Chen et al. studied the APTMS-grafted GO as filler for polybenzoxazole (PBO) fibers and obtained a new hierarchical reinforcement. They proved that not only the surface roughness and wettability of the PBO fibers increased after grafting GO with APTMS, but the atomic oxygen erosion resistance of PBO fibers and its composites was also improved, making these materials suitable for aerospace applications [11].

In previous reports [12,13], we showed that the addition of silane-functionalized GO flakes to an acrylic cataphoretic bath allowed for an increase in the corrosion resistance properties of the composite coatings. GO was produced by nitric acid etching of commercial graphene, obtaining flakes with a low degree of oxidation suitable for preserving the electrical properties of the pristine material. The graphene oxide flakes were functionalized with trialkoxysilanes with R groups characterized by different end chain functions, obtaining different results both in terms of the dispersion of the lamellae into the polymer matrix and the properties of the protective layers. Interestingly, aside from the proof of successful grafting and good dispersion ability of silane-modified GO flakes in the cataphoretic bath, the results showed that not only the filler load but, above all, the choice of the silane, had a key role in obtaining the desired properties in the final coating.

In order to improve the design of composite coatings with functionalized GO, avoiding trial–error processes and reducing the production costs of the final product, a fundamental step is the understanding of the structure-properties relationships of the material. The different reactivity toward GO of the used organoalkoxysilanes pointed out that the organic functional group R plays a fundamental role in the grafting process, which needs to be clarified. Unfortunately, the obtained silanized GO flakes [12] presented a low degree of functionalization that was sufficient to increase the dispersion ability and compatibility with the acrylic resin, but was not suitable for a deep structural characterization of the materials. Therefore, a model GO sample with a high degree of oxidation (i.e., a high amount of functional groups available as anchoring sites for trialkoxysilanes) must be selected in order to exploit the information that can be obtained through the combination of different spectroscopic techniques. Accordingly, the commercial product Graphenea (Ga), was functionalized with the organoalkoxysilanes bearing different end chain groups, following the procedure adopted in Calovi et al. [13], and the samples were characterized by solid state nuclear magnetic resonance (NMR) analysis and x-ray powder diffraction (XRD) to study the degree of functionalization, the preferential anchoring sites, and the type of interaction between silane and GO. Moreover, electron spin resonance (ESR) was used to evaluate the type and amount of GO conductive defects in correlation with the different degree of functionalization obtainable with the three silanes.

2. Materials and Methods

3-aminopropyltrimethoxysilane (APTMS Merck KGaA, Darmstadt, Germany), 3-mercaptopropyltrimethoxysilane (MPTMS Merck KGaA, Darmstadt, Germany), and 3-methacryloxypropyltrimethoxysilane (MaPTMS, Merck KGaA, Darmstadt, Germany), nitric acid

(Merck KGaA, Darmstadt, Germany), toluene (Merck KGaA, Darmstadt, Germany), ethanol (Merck KGaA, Darmstadt, Germany), and acetone (Merck KGaA, Darmstadt, Germany) were used as received. Graphene powder (G, COMETOX s.r.l. (Milan, Italy)) with an average diameter of 25 µm was provided by COMETOX s.r.l. The graphene oxide aqueous dispersion (0.4 wt.% concentration, Graphenea Inc., Cambridge, MA, USA) was supplied by Graphenea (Donostia, Gipuzkoa, Spain).

Graphene oxide powders (GO) were obtained by the reaction of G with nitric acid, according to [12]. GO powders were reacted in a 1:0.1 molar ratio with 3-aminopropyltrimethoxysilane, 3-mercaptopropyltrimethoxysilane, and 3-methacryloxypropyltrimethoxysilane [12], and the functionalized samples were labeled GO-N, GO-S, and GO-M, respectively.

The Ga sample was obtained by drying the graphene oxide aqueous dispersion at 60 °C for 24 h and grinding the solid residues with a mortar. Ga powders were then subjected to the functionalization process with APTMS, MPTMS, and MaPTMS in a 1:0.1 molar ratio under the same conditions employed for the GO powders. The functionalized samples were labeled Ga-N, Ga-S, and Ga-M, respectively. Scheme 1 shows the structures of both Ga and trialkoxysilanes used for the functionalization.

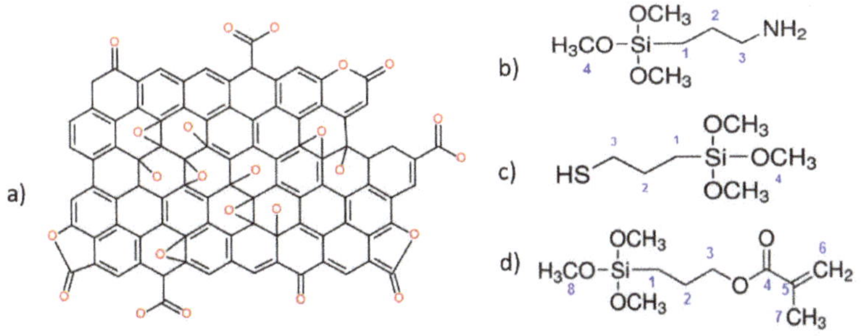

Scheme 1. Molecular structure of (a) Graphenea, (b) 3-aminopropyltrimethoxysilane (APTMS), (c) 3-mercaptopropyltrimethoxysilane (MPTMS), and (d) 3-methacryloxypropyltrimethoxysilane (MaPTMS). The numbering of carbon atoms used for the peak assignment in the nuclear magnetic resonance (NMR) spectra is also shown.

Solid state NMR analyses were carried out with a Bruker 400WB spectrometer (Bruker, Billerica, MA, US) operating at a proton frequency of 400.13 MHz. The magic angle spinning (MAS) NMR spectra were acquired with cross-polarization (CP) and noise dephasing single pulse (SP) pulse sequences under the following conditions: ^{13}C frequency, 100.48 MHz; $\pi/2$ pulse 3.5 µs; decoupling length 5.9 µs; 7 k scans and recycle delay 15 s. For CP: recycle delay 5 s and 20 k scans; contact time 0.5 ms; proton decoupled pulse $\pi/4$; pulse 2.5 µs; recycle delay 10; and 128 scans. ^{29}Si frequency, 79.49 MHz; $\pi/2$ pulse 4.1 µs; contact time 5 ms; decoupling length 5.9 µs; 10 k scans; and recycle delay 10 s. The samples, diluted with KBr in order to avoid skin depth effect (RF penetration) and probe tuning problems [14], were packed in 4 mm zirconia rotor and spun at 10 kHz under air flow. ^{15}N frequency 40.54 MHz; $\pi/2$ pulse 2.2 µs; contact time 2 ms; decoupling length 5.9 µs; 80 k scans; and recycle delay 3 s. Adamantane, Q_8M_8, and glycine were used as external secondary references. The silicon sites were labeled according to the usual T^n notation, where T represents the trifunctional SiCO$_3$ unit and n (n = 0 ÷ 3) is the number of bridging oxygen atoms. The lineshape analysis was performed using Bruker TopSpin software and the fitting was considered acceptable with a confidence level of 90%.

The ESR (Bruker, Billerica, MA, US) spectra were acquired at room temperature with a Bruker EMX cw spectrometer equipped with a rectangular cavity working in the X band at 9.77 GHz microwave frequency with a modulation frequency of 100 kHz. The intensity of the signal was normalized with respect to the weight of the powder sample. The magnetic field and g-value were calibrated with the DPPH powder sample (diphenyl picrylhydrazyl free radical, g = 2.0036).

Powder x-ray diffraction spectra were collected by means of a Rigaku D-Max III-D powder diffractometer (Rigaku, Tokyo, Japan) using Cu-Kα radiation (λ = 0.154056 nm) and a graphite monochromator in the diffracted beam. A θ–2θ Bragg-Brentano configuration was adopted with the following scan conditions: scan range 3–80° (in 2θ); and a sampling interval and counting time of 0.05° and 5 s, respectively. Jade8® software (MDI, Livermore, CA, USA) was used for the fitting procedure of the experimental peaks in order to evaluate the peak position and full width at half maximum (FWHM) values, after the background correction.

3. Results and Discussion

A requirement to properly characterize the interaction between functionalizing agents and graphene oxide is the availability of a valuable number of reactive sites on the sheets. Three organotrimethoxysilanes $(H_3CO)_3Si-CH_2-CH_2-CH_2-X$ were used with the objective of assessing the effect of the end chain group X (X = –NH$_2$, –SH and –(OOC)–CH$_2$=CH–CH$_3$) on the reactivity toward graphene oxide functionalization and possibly elucidate the preferential organosilane-GO interactions. A commercial high oxidation degree graphene oxide (Ga) was employed for the combined spectroscopic study, aiming to possibly state a correlation with the properties imparted by different silane-functionalized graphene oxide powders to protective layers, as reported in our previous study [12].

^{13}C solid state NMR is suitable to explore the structural features of the graphene oxide flakes and the extent of silane grafting. Accordingly, both the ^{13}C proton decoupled MAS (Figure 1a) and CPMAS (Figure 1b) experiments were run on pristine and functionalized Ga. The spectra resolution was good enough to point out the presence of different functional groups on the graphene layers.

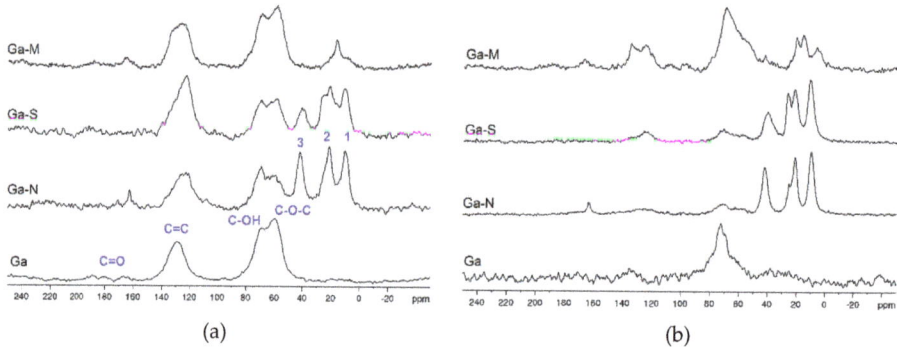

Figure 1. ^{13}C solid state NMR spectra of the samples Ga, Ga-N, Ga-S, and Ga-M. (a) Decoupled MAS and (b) CPMAS.

The two main resonances in the Ga ^{13}C MAS NMR spectrum (Figure 1a) belonged to sp^2 aromatic carbons (about 130 ppm) and alcoholic and epoxide groups (about 70 and 59 ppm, respectively), respectively. Two weak and broad bands were also detectable in the range of 200 ÷ 160 ppm, which can be attributed to small amounts of ketones and edge carboxyls [15]. The main identified resonances in the spectrum of Ga are summarized in Table 1 [16].

The amount of the different functional groups was evaluated from the profile fitting analysis of the quantitative Ga spectrum (Table 1, Figure S1). With respect to a pure graphene sheet that should give rise to a single resonance at about 100–130 ppm, and taking into account all the identified carbon groups, the sp^2 carbons were about 30% of the total, therefore assessing the high degree of oxidation of Ga sample.

Table 1. Assignment and amount of the main identified peaks [15,16] calculated from the profile fitting of the ^{13}C MAS spectrum of Ga (Figure S1, Electronic Supplementary Materials).

δ(iso) ppm	Functional Group	Amount %
190	C=O	4.4
180	C=O	0.9
164	O=C–O	2.0
129	C=C	30.2
95	O–C–O (lactol)	3.0
69	C–OH	30.6
58	C–O–C (epoxy)	28.9

Figure 1a also shows the spectra of the functionalized Ga samples. In the Ga-N and Ga-S spectra, the peaks of the α-, β-, and γ-methylene groups of the propyl chains (labeled 1, 2, and 3 in Figure 1a according to Scheme 1) of APTMS and MPTMS, respectively, are clearly detected in the range of 0–50 ppm. In contrast, unresolved broad resonances both in the methylene region and carbonyl range (167 ppm) characterize the Ga-M spectrum. Interestingly, in the carbonyl region of the Ga-N spectrum, a sharp peak was detected at 163.2 ppm, whose features appeared different with respect to those of the signals found in pristine Ga; its not negligible presence both in the MAS and CPMAS spectra (Figure 1a,b) of Ga-N deserves a more in-depth description, presented further on. In Ga-N and Ga-S, the aromatic resonances (150–100 ppm) showed an upfield shift of about 5 ppm and the signal lineshape changed as a consequence of the functionalization, particularly for Ga-S. Instead, the Ga-M sample did not show remarkable differences for the aromatic peak, maintaining the same lineshape with an upfield shift less than 2 ppm.

It is worth noting that in Ga-N and Ga-S, the silane functionalization led to a modification of the epoxy resonance, which appeared broader and reduced in intensity with respect to the one in pristine Ga. Conversely, the signal appeared almost unchanged in the Ga-M sample.

Table 2 reports, for both the pristine and functionalized Ga samples, the total peak area of the oxidized groups (C–OH, C–O–C, and C=O) were normalized considering the aromatic signal (in the range of 134–125 ppm). Considering the low number of carbonyl groups (Table 1), the Ga functionalization degree can be estimated by comparing the relative intensity of alcoholic and epoxy signals and those due to α-, β-, and γ-methylene carbons of the organosilanes. The most effective functionalization was found in Ga-N, whereas the degree of grafting was almost negligible in the case of Ga-M, as also confirmed by the Fourier transform infrared (FTIR) analysis (Figure S2).

Table 2. Comparison of the number of oxidized C sites, extent of functionalization, and preferential anchoring sites in the different samples. Data were obtained through the profile fitting analysis of the ^{13}C MAS spectra.

Sample	(C–OH + C–O–C(epoxy) + C=O) Normalized Area [1] (70–60, 180 ppm)	[COH + COC (epoxy)]/(R'O)$_3$-Si-R Ratio [2]	COH/COC (epoxy) Ratio
Ga	2.1		1.1
Ga-N	1.3	1.8	1.8
Ga-S	1.0	2.3	1.3
Ga-M	1.8	10 [3]	1.0

[1] Normalized with respect to the aromatic peak in the range of 134–125. [2] Ratio between COH + COC integrals and the integration of the region at 40–0 ppm, containing the three propyl methylene groups belonging to silane divided by three. According to propagation error theory, the reported value is a lower limit. [3] For sample Ga-M, the integration was done in the region 25–0 ppm, containing three peaks belonging to MaPTMS and the resulted integral is divided by three; this amount is subtracted from the 70–60 resonance that also convolutes the C-3 of the MaPTMS for comparison with the Ga-N and Ga-S results.

The profile fitting analysis of the spectra permits the calculation of the amount of alcoholic and epoxide functions in the different samples to appreciate the different consumption of the Ga reactive sites with changing the employed silane in the grafting process. The COH/COC (epoxy) ratio (Table 2) suggests that the epoxide groups are the preferential anchoring sites in Ga-S and even more so in Ga-N, whereas the epoxide involvement appeared negligible in Ga-M. This indicates that the reactive organic groups linked to the silane propyl chain have a relevant influence on the grafting ability onto the graphene oxide surface, which probably overcomes the well-known electronic inductive effect on the condensation ability of the alkoxide group [17]. In particular, the preferential consumption of the epoxide functions in Ga-N suggests that the amine terminal group could play a role in the grafting, in addition to the expected anchoring obtained through methoxy group condensation. As a matter of fact, the epoxide opening should generate both C–OH and C–NH groups in the GO sheet, whose ^{13}C signals should be at about 70 and 60 ppm, justifying the noticed increase in the C–OH peak. The direct interaction of the end sites is expected to be less relevant with the thiol group and negligible for the methacrylate one.

It must be mentioned that in the literature the –COOH group has been reported as the preferential anchoring site, leading to the possible formation of amide and thioester derivatives [18]. Unfortunately, the –COOH amount in our samples was very low and too broad to be evaluated quantitatively for a detailed discussion. Qualitatively, it can be seen that the –COOH related peaks were not anymore detectable in Ga-N (where a small peak at about 164 ppm suggests the formation of an amide group) and in the Ga-S spectra (no thioester signals at about 200 ppm are visible); in contrary, the –COOH broad resonances seemed unchanged in Ga-M, for which the overall grafting was quite poor.

To obtain more insight into the interactions between Ga and organosilanes, the ^{13}C CPMAS spectra have been analyzed. The improved resolution achievable for sp^3 carbons can be appreciated in particular from the signal to noise (S/N) ratio in the Ga-N and Ga-S spectra (Figure 1b).

By studying in detail the resonances of the methylene carbons of the APTMS propyl chain in the Ga-N spectrum, the following information were obtained: (i) the upfield position of Cα at about 10 ppm, and the absence of –OMe peaks indicate a good condensation degree; (ii) the Cβ peak splits in two resonances at 25.4 and 21.0 ppm, suggesting the presence of amino groups with different protonation degrees, since it is well known that the second carbon close to the terminal nitrogen is sensitive to its changes [19,20]; and (iii) the peak at 42 ppm belongs to Cγ and results were quite insensitive to the structural rearrangements.

The two main broad peaks observable in the spectra of functionalized samples (Figure 1b) belong to Ga. The one centered at 125 ppm is due to the aromatic carbons and the two peaks at 71.2 and 59.7 ppm were assigned to alcoholic and epoxide functions, which were also detected in the CPMAS spectrum of pristine Ga (Figure 1b). Moreover, as remarked above for the MAS spectrum, in the carbonyl region of Ga-N, a sharp resonance was detected at 163.2 ppm, which can be tentatively attributed to edge carboxyl carbons [15] or amide groups [21–23].

To shed light on the peculiarities shown by the Ga-N sample, and further investigate the interaction of the aminopropyl chains with the Ga sheets, a sample with Ga:APTMS = 1:1 molar ratio was prepared. Due to the high APTMS amount, it is likely that both the grafting process and the silane self-condensation reaction take place. Focusing on the Cβ resonance (^{13}C CPMAS spectrum, Figure S3), the intensity of the downfield component (25 ppm) increased when compared to the same resonance in the Ga-N spectrum of Figure 1b, reaching roughly the 1:1 ratio with the upfield component (21 ppm). Moreover, the intensity of the signal at 163 ppm also increased with respect to the Ga-N spectrum.

These results prompted us to record a long ^{15}N CPMAS experiment at nitrogen natural abundance. Excitingly, the ^{15}N spectra of both the Ga-N 1:01 and Ga 1:1 samples showed two resonances at 33 and 89 ppm (Figure 2), which can be attributed to the primary amine and amide functional groups, respectively [21,23].e The two peaks presented an intensity ratio 1:0.25 for the Ga-N 1:0.1 sample, reaching a ratio of 1:1 in the Ga with a higher APTMS load. This clearly proves that the unexpected carbon resonance at about 163 ppm, present only in the Ga-N spectra, can be attributed to the reaction between edge carboxylic and amino groups, probably belonging to two subsequent graphene layers.

Accordingly, the possibility of APTMS anchoring graphene oxide layers both through Si–O–C and O=C–NH– bond formation explains its highest ability among the chosen silane series.

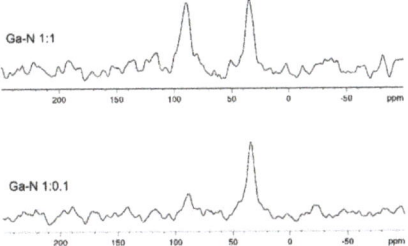

Figure 2. ^{15}N CPMAS NMR spectra of the sample Ga-N prepared with the Ga:APTMS ratio of 1:1 and 1:0.1, respectively.

In the ^{13}C CPMAS spectrum of Ga-S (Figure 1b), the peaks due to the methoxy groups were absent, representing a complete hydrolysis. The spectrum shows the Ga resonances, and the peaks due to the mercaptopropyl chain were found at 10.8 (Cα), 21.3 and 26.3 (Cβ), and 40.2 ppm (Cγ), respectively. Interestingly, in this case, the Cβ resonance appeared to split into two components and the occurrence of chain folding with different –SH interaction could be adduced. It is known that upon MPTMS condensation, the detection in the spectrum of two signals at about 11 (Cα) and 28 ppm (Cβ and Cγ) indicate a free propyl-SH chain [20,24]. The detection of the resonances at 40 and 21.3 ppm suggests the formation of disulfide bonds that can be created only in the case of the close proximity of the MPTMS molecules grafted onto the graphene sheets. Kao et al. [24] proposed that –S-S– bond formation is favored by the reactive sites of graphene oxide, similarly to that found with carbon nanotubes (CNT) [25].

Finally in the case of the Ga-M sample, in addition to the Ga peaks, metacryloxy resonances could be distinguished at 7.1 (C-1), 16.2 (C-7), 21.1 (C-2), 135.1 and 125.4 (C-5 and C-6, sharp peaks overlapped with the aromatic band), and 166.9 ppm (C-4), respectively, whereas C-3 was hidden by the C-OH Ga resonance. The chemical shift of C-4 proved that the metacryloxy ending group was preserved [26].

The reduced ability toward the condensation of MPTMS was confirmed by the ^{29}Si CPMAS spectrum of Ga-M (Figure S4), which is characterized by a large number of T^2 units (−58 ppm). T^3 units (−66 ppm) increased from Ga-M to Ga-S and finally to Ga-N, in accordance with the increase in the condensation degree (Table S1).

Both pure and functionalized Ga samples were analyzed using the XRD technique (Figure 3) in order to confirm previous findings on functionalized GO [12] and correlate them with the NMR results. The XRD pattern of the starting Ga powders, obtained by drying the commercial aqueous solution, was characterized by an intense 001 basal peak located at 2θ = 11.16° (d = 0.792 nm); its correlation length, evaluated from Scherrer's equation, is 10.2 nm. According to Lerf et al. [27], d-spacing values of about 0.79 nm can be attributed to a fully hydrated graphite oxide structure, with a monomolecular layer of water intercalated between the graphene oxide sheets. Moreover, from the evaluated correlation length, it is possible to estimate the number of stacked graphene oxide sheets oriented along the perpendicular direction to the diffracting (001) plane [28]; in this case, the value is around 12–13 layers. By comparing the XRD patterns of the functionalized samples with the Ga spectrum, the decrease in the intensity of the basal peak of graphene oxide was detected for the functionalized samples; the presence of the siloxane lattice was evidenced by the broad peak at about 2θ = 10° and the band at about 20–22°, both typical of the amorphous silsesquioxane structure [29]. In the Ga-N pattern, a signal at 6.43° (d = 1.37 nm) was clearly visible; analogously to the NMR study, the Ga:APTMS = 1:1 sample was analyzed (Figure S5) to strengthen the assignment of this peak to the graphene oxide counterpart. In Ga-N, the 001 peak shifted toward lower angles in relation with the increase of the interplanar distance

d. The decrease in intensity and the shift of the basal peak of graphene oxide, together with the signal broadening evidenced by an increase in the FWHM value (from 0.79° in Ga to 1.35° in Ga-N), point out the presence of a strong interaction between silane amino groups and graphene oxide sheets, leading to a loss of the long-range order along the perpendicular direction to the diffracting plane; in fact, for the Ga-N sample, a stack of 4–5 sheets could be evaluated.

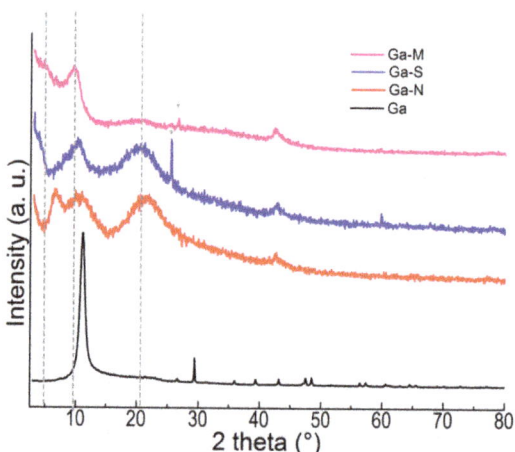

Figure 3. XRD diffractograms of the pristine and functionalized Ga samples.

The progressive increase in the structural disorder of graphene oxide sheets can also be found for the Ga-S and Ga-M samples; in fact, the basal peak changed in a less intense shoulder located at around 3.5° for both samples. This effect can be related to the increase in the interplanar distances among the sheets, together with the loss of long-range order in the z-direction, as revealed by FWHM values higher than that of the Ga-N sample. At the same time, Figure 3 highlights the shift toward low angles of the siloxane band with the trend Ga-N (d = 0.810 nm) > Ga-S (d = 0.883 nm) > Ga-M (d = 0.910 nm), which may be due to the different length and arrangement of the silane organic chains. Additionally, the lineshape of the siloxane peak changed, the highest FWHM value (6.4°) was displayed by the Ga-N sample, attributable to a higher degree of interaction with GO sheets, while Ga-M (FWHM = 2.15°) showed the minimum interaction. Another common feature to arise from the analysis of the spectra was that all the other peaks present in the Ga spectrum vanished after the functionalization, except for a small peak located at d = 0.210–0.213 nm, due to the (100) direction of graphene oxide; this effect could be attributed to the maintenance of the structural order along the direction parallel to the diffracting planes [30]. Nevertheless, from the correlation lengths evaluated from the FWHM values, a lowering of the domain dimensions was recorded from Ga (>30 nm) to Ga-N (3.2 nm) and Ga-S (3.4 nm); again, the scarce interaction in the Ga-M sample also led to a higher degree of order in this direction, with an evaluated correlation length of 7.0 nm.

Finally, the presence of a small peak at 2θ = 26.6° in the Ga-M spectrum, attributable to the presence of graphite, could be interpreted as a consequence of a partial reduction reaction occurring in the graphene oxide layers as well as in the Ga-S sample that presented a very narrow signal at 25.5°.

The experimental approach used to study the silanized Ga samples was also applied to the functionalized graphene oxide samples (GO-N, GO-S, and GO-M) prepared from the nitric acid etching of G powders [12]. Figure 4 shows the ^{13}C CPMAS spectra of the samples GO-N, GO-S, and GO-M. The spectrum of GO (not reported) was flat due to the conductive character of the GO, which interferes with the polarization transfer of the experiment.

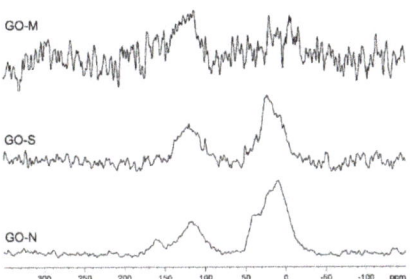

Figure 4. ^{13}C CPMAS spectra of samples GO-N, GO-S, and GO-M.

Theoretically, the CPMAS experiment permitted us to obtain good quality spectra for the organic component of the silane. Unfortunately, the low quantity of the grafted silane combined with the effect of the conductive graphene layers also caused both large widening and low S/N ratio of the peaks for the functionalized GO samples. However, it is worth noting that the broad resonance of graphene aromatic carbons, which is usually not detectable with CP experiments, can be observed in the spectra of Figure 3. This suggests that the functionalization allows partial cross-polarization between the protons of the silane propyl chain and the graphene neighboring C atoms.

Despite the low oxidation and functionalization degrees of the GO powders obtained by nitric acid etching with respect to Ga, the GO-N sample showed similarities to the Ga-N one. The band in the range of 50–0 ppm was given by the convolution of three peaks and there was also a weak C=O signal around 160 ppm, which has been previously assigned to amide bond formation. The GO-S spectrum presents similar features, but with a large decrease in the S/N ratio. For the GO-M sample, on the other hand, it is impossible to observe the peaks previously described, which suggests that the interaction between the silane and GO is very limited.

These results are comparable with those of the Graphenea series samples, as shown in Figure 1b, thus proving the suitability of the approach selected for the structural characterization of GO-organosilane interactions. Moreover, the conclusions of the NMR study on GO functionalized samples were in agreement with the evidence from both the FTIR and XRD study on the same samples reported in a previous paper [12]. Despite the low quality of the recorded XRD patterns and the graphitic structure of the GO samples prepared by nitric acid etching, the functionalization with the organosilanes led the graphitic and silsesquioxane counterparts to interact differently in relation with the different organic groups linked to silicon [12].

It is worth recalling that the properties of the composite layers based on acrylic resin loaded with 1% silanized GO powders made via cataphoresis were found to depend on the organosilane nature [12]. The trend in improving the layers' corrosion resistance was in the order GO-N > GO-S > GO-M, which can be directly related with both the grafting ability and effect on the GO structural order of the different organosilanes evaluated through NMR and XRD.

Finally, another important feature of materials like the graphene oxide powders studied in this work is the presence of defects, which can significantly affect their chemical-physical properties. Most of the defects are related to the presence of unpaired electron spins, thus paramagnetic. Accordingly, they could be easily evaluated through ESR spectroscopy.

Figure 5a,b show the ESR spectra of Ga, GO, and the corresponding functionalized samples. The ESR spectrum of Ga (Figure 5a, black) shows both a broad signal with a sextet hyperfine pattern and a small narrow peak (F) with a g value close to 2. The former signal, typical of diluted Mn^{2+} impurities ($S = 5/2$, $I = 5/2$) in polycrystalline samples, appears most likely as a consequence of the oxidation process with $KMnO_4$ [31,32]. The latter is due to unpaired spins in both the C-related dangling bond and oxygen-based functional groups, such as carboxylic, alcoholic, epoxy and hydroxyl groups, whose presence has been already proved through ^{13}C solid state NMR. It should be noticed that the presence of variable amounts of Mn^{2+} ions can affect the conductivity properties of the flakes as well as modify the barrier effect exerted by the GO fillers in polymeric matrices, leading to non-reproducible results.

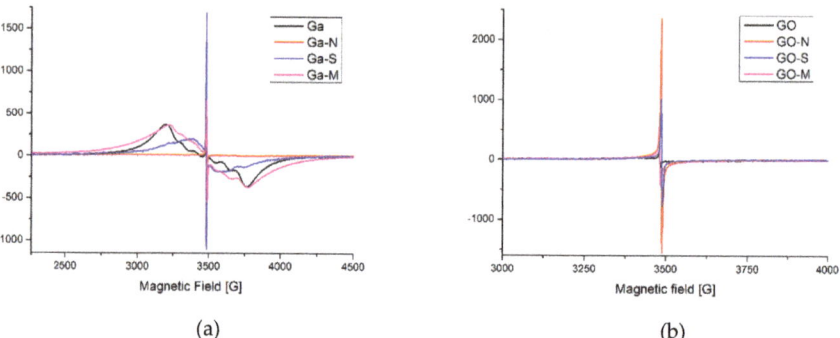

Figure 5. First derivative X-band cwESR spectra of the pristine and functionalized (**a**) Ga and (**b**) GO samples at room temperature.

The GO spectrum (Figure 5b, black) is characterized by a single sharp line centered at the g-value of 2.004. Since the starting graphene material (G) shows a relevant amount of C-based defects, according to the small broad peak with a g-value close to 2 and a high g-value signal probably due to the presence of spurious metals (Figure S6), it is likely that oxidation with nitric acid remarkably reduces the number of paramagnetic defects (vacancies, dangling bonds). Therefore, GO samples, partially oxidized without using KMnO$_4$, offer better guarantees of conductivity and reproducibility with respect to the sample Ga in the preparation of the cataphoretic coatings [33].

The Ga functionalization with APTMS (Ga-N, Figure 5a) significantly reduced both the Mn impurities and the number of dangling bonds (C-related defects), whereas after reaction with MaPTMS (Ga-M) and, overall, MPTMS (Ga-S) the spectra displayed residual amounts of Mn^{2+} and the sharp signal due to the carbon radicals.

The spectra recorded on the GO functionalized samples (Figure 5b) showed only the sharp peaks related to the free radical species with a g-value of about 2.003–2.002.

Table 3 shows the intensity values of the sharp peak (F) attributed to defects/free electrons for a direct comparison of the sample properties, whereas the total area takes into account all the ESR signals, thus including all the other possible sources of spin density such as paramagnetic and metallic impurities, represented in the G and Ga-series by the broad ESR component. Interestingly, both Ga and GO functionalized samples showed the same trend with an increasing presence of defects from Ga-N to Ga-S to Ga-M and from GO-N to GO-S to GO-M. These results, therefore, strengthen the already observed similarity between the two series of samples and the possibility of generalizing the conclusions drawn for the model sample.

It can be noticed that the functionalization with the APTMS of both commercial Ga and GO strongly reduces the spin density. As reported above, functionalization with APTMS also reduced the metallic impurities of Ga. This is in agreement with the reported use of APTMS as fast and effective metal scavengers [34], thanks to the presence of the amino group.

According to [35], it is interesting to compare the ESR signal area that reflects all the unpaired electron spins with the area of the sp^2 carbons in the ^{13}C MAS NMR spectrum that could be related only to the unpaired spins coupled to ^{13}C. Thomas et al. [35] found a correlation between the two types of data in their produced graphene oxide samples. In the present case, effective functionalization (samples Ga-S and Ga-N) seemed to reduce both the whole amount of unpaired spins (Table 3 column 3) and the ratio among the oxidized-C and sp^2 C (Table 2, column 3).

Table 3. ESR results of the modified graphene samples.

Sample	I_F (peak-to-peak) [1]	Total Area (*10^6) [2]
G	7362.1	630
GO	275.8	4.69
GO-N	389.9	0.22
GO-S	1775.7	1.16
GO-M	9927.3	5.54
Ga	136.1	30.6
Ga-N	506.6	2.57
Ga-S	2790.6	19.6
Ga-M	1216.5	55.6

[1] I_F is the intensity of the sharp peak. [2] The total area refers to the overall spectrum integral.

4. Conclusions

The present paper describes the investigation of both the type and extent of interactions between a model commercial graphene oxide and three trialkoxysilanes bearing a different end group of the organic tail. The results of the Ga samples were compared with the ones obtained on GO prepared by nitric acid etching and subjected to the same functionalization, which was previously employed as a filler in protective coatings prepared by cataphoresis [12]. This approach allows for both the generalization of the results obtained on a model sample and for a relationship to be established among the structural and functional features of the GO-polymer nanocomposite layers for corrosion protection.

The multinuclear NMR study of the functionalized Ga samples highlights that the epoxy group is the preferential anchoring site for silane condensation, through its opening upon reaction with the Si–OH groups. The reaction with (X-propyl)trimethoxysilanes with amino-, mercapto-, and methacryloxy-X end groups, respectively, leads to a functionalization degree in the order $-NH_2$ > $-SH$ >> $-O_2CCCH_2CH_3$. The methacryloxy function seems to hinder the grafting, whereas both aminopropyl and mercaptopropyl chains let the silane fill the space between the graphene sheets, most probably substituting the water molecules present in the pristine Ga, according to the XRD results. Moreover, the carbon chemical shifts of the mercaptopropyl chain indicate the formation of S–S bonds. Finally, according to the ^{13}C and ^{15}N results, the APTMS amino groups directly react with the edge carboxylic groups, leading to the formation of amide bonds. This evidence clearly explains the better performance imparted by the GO filler functionalized with APTMS to the acrylic layers deposited by cataphoresis onto metal surfaces [12]. Moreover, ESR spectroscopy proved that the presence of amino groups is beneficial for the removal of Mn impurities in commercial GO, acting as a metal scavenger and reducing the number of defects.

Supplementary Materials: The following are available online at http://www.mdpi.com/1996-1944/12/23/3828/s1, Figure S1: Profile fitting analysis of Ga ^{13}C MAS spectrum; Figure S2: FTIR spectra of samples Ga, Ga-N, Ga-S, and Ga-M; Figure S3: ^{13}C CPMAS NMR spectrum of the sample Ga-N 1:1; Figure S4: ^{29}Si CPMAS spectra of the samples Ga-N, Ga-S, and Ga-M; Figure S5: XRD spectrum of the Ga:APTMS 1:1 sample; Figure S6: First derivative cwESR spectrum of the graphene sample G; Table S1: Semi-quantitative profile fitting of T resonances based on CPMAS spectra.

Author Contributions: Conceptualization, M.C., and S.D.; Methodology, M.C., E.C., and R.C.; Project administration, S.D.; Validation, E.C., M.C., and R.C.; Formal analysis, M.C., E.C., and R.C.; Investigation, E.C., R.C., and M.C.; Resources, S.D, S.R., R.C., and F.D.; Data curation, M.C., E.C., and R.C.; Writing—original draft preparation, M.C., E.C., and R.C.; Writing—review and editing, S.D., E.C., and R.C.; Visualization, E.C, and M.C.; Supervision, S.D, S.R., R.C., and F.D.; Project administration, S.D.

Funding: This research received no external funding

Conflicts of Interest: The authors declare no conflicts of interest.

References

1. Papageorgiou, D.G.; Kinloch, I.A.; Young, R.J. Mechanical properties of graphene and graphene-based nanocomposites. *Prog. Mater. Sci.* **2017**, *90*, 75–127. [CrossRef]
2. Atif, R.; Shyha, I.; Inam, F. Mechanical, thermal, and electrical properties of graphene-epoxy nanocomposites—A review. *Polymers* **2010**, *8*, 281–318. [CrossRef] [PubMed]
3. Romero, A.; Lavin-Lopez, M.P.; Sanchez-Silva, L.; Valverde, J.L.; Paton-Carrero, A. Comparative study of different scalable routes to synthesize graphene oxide and reduced graphene oxide. *Mater. Chem. Phys.* **2018**, *203*, 284–292. [CrossRef]
4. Yao, H.; Hawkins, S.A.; Sue, H.J. Preparation of epoxy nanocomposites containing well-dispersed graphene nanosheets. *Compos. Sci. Technol.* **2017**, *146*, 161–168. [CrossRef]
5. Jing, Q.; Liu, W.; Pan, Y.; Silberschmidt, V.V.; Li, L.; Dong, Z.L. Chemical functionalization of graphene oxide for improving mechanical and thermal properties of polyurethane composites. *Mater. Des.* **2015**, *85*, 808–814. [CrossRef]
6. Lee, C.Y.; Bae, J.H.; Kim, T.Y.; Chang, S.H.; Kim, S.Y. Using silane-functionalized graphene oxides for enhancing the interfacial bonding strength of carbon/epoxy composites. *Compos. Part A Appl. Sci. Manuf.* **2015**, *75*, 11–17. [CrossRef]
7. Pourhashem, S.; Rashidi, A.; Vaezi, M.R.; Bagherzadeh, M.R. Excellent corrosion protection performance of epoxy composite coatings filled with amino-silane functionalized graphene oxide. *Surf. Coat. Technol.* **2017**, *317*, 1–9. [CrossRef]
8. Mittal, K.L. *Silanes and Other Coupling Agents*, 1st ed.; Brill: Leiden, The Netherlands, 2009; Volume 5.
9. Abbas, S.S.; Rees, G.J.; Kelly, N.L.; Dancer, C.E.J.; Hanna, J.V.; McNally, T. Facile silane functionalization of graphene oxide. *Nanoscale* **2018**, *10*, 16231–16242. [CrossRef]
10. Xu, P.; Yan, X.; Cong, P.; Zhu, X.; Li, D. Silane coupling agent grafted graphene oxide and its modification on polybenzoxazine resin. *Compos. Interfaces* **2017**, *24*, 635–648. [CrossRef]
11. Chen, L.; Wei, F.; Liu, L.; Cheng, W.; Hu, Z.; Wu, G.; Du, Y.; Zhang, C.; Huang, Y. Grafting of silane and graphene oxide onto PBO fibers: Multifunctional interphase for fiber/polymer matrix composites with simultaneously improved interfacial and atomic oxygen resistant properties. *Compos. Sci. Technol.* **2015**, *106*, 32–38. [CrossRef]
12. Calovi, M.; Dirè, S.; Ceccato, R.; Deflorian, F.; Rossi, S. Corrosion protection properties of functionalized grapheme-acrylate coatings produced by cataphoretic deposition. *Prog. Org. Coat.* **2019**, *136*, 105261–105272. [CrossRef]
13. Calovi, M.; Rossi, S.; Deflorian, F.; Dirè, S.; Ceccato, R. Effect of functionalized graphene oxide concentration on the corrosion resistance properties provided by cataphoretic acrylic coatings. *Mater. Chem. Phys.* **2020**, *239*, 121984–121996. [CrossRef]
14. Freitas, J.; Cipriano, D.F.; Zucolotto, C.G.; Cunha, A.C.; Emmerich, F.G. Solid-State 13C NMR Spectroscopy Applied to the Study of Carbon Blacks and Carbon Deposits Obtained by Plasma Pyrolysis of Natural Gas. *J. Spectrosc.* **2016**, *2016*, 1543273. [CrossRef]
15. Lu, N.; Huang, Y.; Li, H.; Li, Z.; Yang, J. First Principles NMR Signatures of Graphene Oxide. *J. Chem. Phys.* **2010**, *133*, 034502. [CrossRef]
16. Vieira, M.A.; Goncalves, G.R.; Cipriano, D.F.; Schettino, M.A.; Filho, E.A.S.; Cunha, A.G.; Emmerich, F.G.; Freitas, J.C.C. Synthesis of graphite oxide from milled graphite studied by solid-state ^{13}C nuclear magnetic resonance. *Carbon* **2016**, *98*, 496–503. [CrossRef]
17. Brinker, C.J.; Scherer, G.W. *Sol-Gel Science: The Physics and Chemistry of Sol-Gel Processing*; Academic Press: Cambridge, MA, USA, 1990; ISBN 9780121349707.
18. Bonanni, A.; Chua, C.K.; Pumera, M. Rational design of carboxyl groups perpendicularly attached to a graphene sheet. *Chem. Eur. J.* **2014**, *20*, 217–222. [CrossRef]
19. Caravajal, G.S.; Leyden, D.E.; Quinting, G.R.; Maciel, G.E. Structural characterization of (3-aminopropyl)triethoxysilane-modified silicas by Si-29 and C-13 nuclear magnetic resonance. *Anal. Chem.* **1988**, *60*, 1776–1786. [CrossRef]
20. Zub, Y.L.; Melnyk, I.V.; White, M.G.; Alonso, B. Structural Features of Surface Layers of Bifunctional Polysiloxane Xerogels Containing 3-Aminopropyl Groups and 3-Mercaptopropyl Groups. *Adsorpt. Sci. Technol.* **2008**, *26*, 119–132. [CrossRef]

21. Maeda, S.; Oumae, S.; Kaneko, S.; Kunimoto, K.K. Formation of carbamates and cross-linking of microbial poly(ε-L-lysine) studied by ^{13}C and ^{15}N solid-state NMR. *Polym. Bull.* **2012**, *68*, 745–754. [CrossRef]
22. Daniel, R.; Dreyer, D.R.; Park, S.; Bielawski, C.W.; Ruoff, R.S. The chemistry of graphene oxide. *Chem. Soc. Rev.* **2010**, *39*, 228–240. [CrossRef]
23. Ganesan, K.; Heyer, M.; Ratke, L.; Milow, B. Facile Preparation of Nanofibrillar Networks of "Ureido-Chitin" Containing Ureido and Amine as Chelating Functional Groups. *Chem. Eur. J.* **2018**, *24*, 19332–19340. [CrossRef] [PubMed]
24. Kao, H.M.; Chiu, P.J.; Jheng, G.L.; Kao, C.C.; Tsai, C.T.; Yau, S.L.; Gavin Tsai, H.H.; Chou, Y.K. Oxidative transformation of thiol groups to disulfide bonds in mesoporous silicas: A diagnostic reaction for probing distribution of organic functional groups. *New J. Chem.* **2009**, *33*, 2199–2203. [CrossRef]
25. Hirano, A.; Kameda, T.; Sakuraba, S.; Wada, M.; Tanaka, T.; Kataura, H. Disulfide bond formation of thiols by using carbon nanotubes. *Nanoscale* **2017**, *9*, 5389–5393. [CrossRef] [PubMed]
26. Di Maggio, R.; Callone, E.; Girardi, F.; Dirè, S. Structure-related behavior of hybrid organic–inorganic materials prepared in different synthesis conditions from Zr-based NBBs and 3-methacryloxypropyl trimethoxysilane. *J. Appl. Pol. Sci.* **2012**, *125*, 1713–1723. [CrossRef]
27. Lerf, A.; Buchsteiner, A.; Pieper, J.; Schöttl, S.; Dekany, I.; Szabo, T.; Boehm, H.P. Hydration behavior and dynamics of water molecules in graphite oxide. *J. Phys. Chem. Solids* **2006**, *67*, 1106–1110. [CrossRef]
28. Dikin, D.A.; Stankovich, S.; Zimney, E.J.; Piner, R.D.; Dommett, G.H.B.; Evmenenko, G.; Nguyen, S.B.T.; Ruoff, R.S. Preparation and characterization of graphene oxide paper. *Nat. Lett.* **2007**, *448*, 457–460. [CrossRef] [PubMed]
29. Dirè, S.; Borovin, E.; Ribot, F. Architecture of Silsesquioxanes. In *Handbook of Sol-Gel Science and Technology: Processing, Characterization and Applications*; Klein, L., Aparicio, M., Jitianu, A., Eds.; Springer International Publishing: Basel, Switzerland, 2018; pp. 3119–3151. ISBN 978 3 319 32099 1.
30. Mauro, M.; Maggio, M.; Antonelli, A.; Acocella, M.R.; Guerra, G. Intercalation and Exfoliation Compounds of Graphite Oxide with Quaternary Phosphonium Ions. *Chem. Mater.* **2015**, *27*, 1590–1596. [CrossRef]
31. Panich, M.; Shames, A.I.; Sergeev, N.A. Paramagnetic Impurities in Graphene Oxide. *Appl. Magn. Reson.* **2013**, *44*, 107–116. [CrossRef]
32. Pham, C.V.; Krueger, M.; Eck, M.; Weber, S.; Erdem, E. Comparative electron paramagnetic resonance investigation of reduced graphene oxide and carbon nanotubes with different chemical functionalities for quantum dot attachment. *Appl. Phys. Lett.* **2014**, *104*, 132102. [CrossRef]
33. Rossi, S.; Calovi, M. Addition of graphene oxide plates in cataphoretic deposited organic coatings. *Prog. Org. Coat.* **2018**, *424*, 2017–2018. [CrossRef]
34. Liu, A.M.; Hidajat, K.; Kawi, S.; Zhao, D.Y. A new class of hybrid mesoporous materials with functionalized organic monolayers for selective adsorption of heavy metal ions. *Chem. Commun.* **2000**, *13*, 1145–1146. [CrossRef]
35. Thomas, H.R.; Day, S.P.; Woodruff, W.E.; Vallés, C.; Young, R.J.; Kinloch, I.A.; Morley, G.W.; Hanna, J.V.; Wilson, N.R.; Rourke, J.P. Deoxygenation of Graphene Oxide: Reduction or Cleaning? *Chem. Mater.* **2013**, *25*, 3580–3588. [CrossRef]

© 2019 by the authors. Licensee MDPI, Basel, Switzerland. This article is an open access article distributed under the terms and conditions of the Creative Commons Attribution (CC BY) license (http://creativecommons.org/licenses/by/4.0/).

Article

Chain Model for Carbon Nanotube Bundle under Plane Strain Conditions

Elena A. Korznikova [1], Leysan Kh. Rysaeva [1], Alexander V. Savin [2], Elvira G. Soboleva [3], Evgenii G. Ekomasov [4], Marat A. Ilgamov [5] and Sergey V. Dmitriev [1,6,*]

1. Institute for Metals Superplasticity Problems, Russian Academy of Sciences, Khalturin St., 39, 450001 Ufa, Russia
2. Institute of Physical Chemistry of RAS, Kosygin St., 4, 119991 Moscow, Russia
3. Yurga Institute of Technology (Branch), National Research Tomsk Polytechnic University, 652050 Yurga, Russia
4. South Ural State University (National Research University), Lenin Ave., 76, 454080 Chelyabinsk, Russia
5. Institute of Mechanics, Ufa Federal Research Center, Russian Academy of Sciences, Oktyabrya Ave., 71, 450054 Ufa, Russia
6. National Research Tomsk State University, Lenin Ave., 36, 634050 Tomsk, Russia
* Correspondence: dmitriev.sergey.v@gmail.com

Received: 9 November 2019; Accepted: 22 November 2019; Published: 28 November 2019

Abstract: Carbon nanotubes (CNTs) have record high tensile strength and Young's modulus, which makes them ideal for making super strong yarns, ropes, fillers for composites, solid lubricants, etc. The mechanical properties of CNT bundles have been addressed in a number of experimental and theoretical studies. The development of efficient computational methods for solving this problem is an important step in the design of new CNT-based materials. In the present study, an atomistic chain model is proposed to analyze the mechanical response of CNT bundles under plane strain conditions. The model takes into account the tensile and bending rigidity of the CNT wall, as well as the van der Waals interactions between walls. Due to the discrete character of the model, it is able to describe large curvature of the CNT wall and the fracture of the walls at very high pressures, where both of these problems are difficult to address in frame of continuum mechanics models. As an example, equilibrium structures of CNT crystal under biaxial, strain controlled loading are obtained and their thermal stability is analyzed. The obtained results agree well with previously reported data. In addition, a new equilibrium structure with four SNTs in a translational cell is reported. The model offered here can be applied with great efficiency to the analysis of the mechanical properties of CNT bundles composed of single-walled or multi-walled CNTs under plane strain conditions due to considerable reduction in the number of degrees of freedom.

Keywords: carbon nanotube bundle; plane strain conditions; lateral compression; equilibrium structure; thermal stability; chain model

PACS: 61.48.De

---

1. Introduction

There exist a huge number of carbon polymorphs, including a wide class of sp² structures such as fullerenes, carbon nanotubes (CNT), and graphene. Due to the action of relatively weak van der Waals forces, a great variety of secondary structures can be formed, and some of them can have a long-range order, for example, fullerite crystal composed of fullerenes [1–3], graphite made of graphene layers [4,5], and crystals made of CNTs [6–8]. Such crystalline structures are of great interest since they have properties not exhibited by isolated structural elements [1–8]. Here, we focus on mechanical response of CNT bundles.

Various experimental techniques have been developed to produce CNT forests [9–12]. Mechanical applications of CNTs include ropes [7,13], fibers [14–18], polymer-matrix and metal-matrix composites [19–21], solid lubricants [21,22], etc. In all these applications, superior mechanical properties of CNTs such as tensile strength in the range from 11 to 63 GPa, tensile Young's modulus of the order of 1.0 to 1.3 TPa, and high deformability up to ultimate fracture strain of about 10% are used [23–26]. In addition, they are lightweight, flexible, have high thermal and electrical conductivity, and these properties are useful in a number of applications [27–30].

Not only tension [7,13–18] and compression of vertically aligned CNT brushes and forests [31–38], but also lateral compression of isolated CNT or CNT bundles [39–43] is of interest, and the latter loading scheme has been studied less thoroughly than the former ones. Drawing, winding, micromechanical rolling, and shear pressing were used to produce horizontally aligned CNT bundles from vertically aligned CNT arrays [44–47]. Experimental and computational approaches used for evaluation of mechanical properties of CNTs have been outlined in the review [48]. Carbon nanotube bundles are linear elastic under hydrostatic pressure up to 1.5 GPa at room temperature; the volume compressibility, measured by in situ synchrotron x-ray diffraction, is 0.024 GPa^{-1}; the deformation of the trigonal nanotube lattice under hydrostatic pressure is reversible up to 4 GPa [49]. Using X-ray diffraction and Raman scattering techniques, it has been shown that CNT bundles under non-hydrostatic pressure are not reversible for pressures beyond 5 GPa [50].

Indentation experiments are widely used to assess mechanical properties of vertically aligned CNT forests and brushes [32–38]. In particular, the Young's modulus of 200 nm thick brush is about 17 GPa and the critical buckling stress can be estimated as 0.3 GPa at a load of 0.02 mN [32]. Carbon nanotube forests composed of CNTs of 2.2 mm height and 50 nm diameter have shown an elastic compressive modulus of 2.1 MPa as measured at initial loading condition and 20.8 MPa as measured after the plateau region [33].

Computational studies on the mechanical properties of nanomaterials become increasingly important because they speed up and reduce the cost of research and design. On the other hand, it is a tremendous challenge to predict nonlinear mechanical behaviors of nanomaterials by full atomic molecular dynamics method due to the huge computational cost, particularly for CNT bundles. Development of new computational approaches capable of modeling mechanical response at different scales is a very important task.

Mesoscopic modeling of phase transformations and mechanical deformation mechanism of CNT forest has been addressed in [51,52]. It has been shown that under compression along the tubes a low-density phase composed of vertically aligned CNT bundles transforms into a dense phase with horizontal alignment of CNTs. Carbon nanotubes subject to large deformations obtain different morphological patterns that can be simulated using a continuum shell model [53]. Transverse mechanical properties of CNTs have been studied in [8]. Nonlocal beam, plate, and shell theories employed in modeling of the mechanical properties of nanoscale structures are described in the review [54]. The applicability of the continuum beam model in the mechanics of CNT has been discussed in [55]. It has been shown that the rigidity of CNT crystal does not decrease with increasing tube diameter [6]. Isolated CNT of diameter above a threshold value can have two stable configurations—circular and collapsed [56–58]. In a recent experimental and molecular dynamics study, the irreversible transformation of triple-wall carbon nanotube bundles have been analyzed at pressure up 72 GPa and temperature up to 2400 K [39]. The irreversible transformation threshold pressure has been found to be in between 60 GPa and 72 GPa. Nonlinear coarse-grained stretching and bending potentials for CNTs have been developed to enable simulation of the mechanical behaviors and failure mechanism of the CNT bundles [59].

In spite of the fact that a number of computational methods have been developed for the analysis of mechanical properties of CNT forests, there is always a need to increase the efficiency and accuracy of simulation methods. Continuum mechanics is a powerful and effective tool to successfully describe macroscopic parameters of CNT bundles, but it has some limitations. For example, thermal fluctuations of CNTs and their fracture can be more adequately modelled in frame of atomistic models. On the other hand, full atomic models, as mentioned above, are very demanding on computational resources. One possible compromise is to use atomistic models for particular deformation modes, when the number of degrees of freedom can be substantially reduced.

In this study, in order to reduce the number of degrees of freedom, a full atomic model of CNT bundles under plane strain conditions is substituted by the chain model developed in the work [60] and successfully used to study structure and properties of secondary structures such as folds and scrolls of carbon nanoribbons [60–64] and dynamics of surface ripplocations on a graphite substrate [65]. For simplicity, here we will only consider the case of a bundle composed of single-walled CNTs of equal diameter, but the model can be applied to the cases of CNTs of different diameter, multi-walled CNTs, and include graphene scrolls and cylindrically crumpled graphene.

2. Materials and Methods

The computational model employed in this study is schematically shown in Figure 1. The nanotube bundle is aligned along the z-axis and CNTs of equal diameter create in cross-section a triangular lattice; they are numbered by the indices $i = 1, ..., I$ and $j = 1, ..., J$ (the case of $I = J = 2$ is shown). Only zigzag CNTs are considered for simplicity. The carbon atoms move on the (x, y) plane and each atom represents a rigid row of atoms oriented normal to the (x, y) plane. Within each CNT, carbon atoms are numbered by the index $n = 1, ..., N$, anti-clockwise, starting from the atom with maximal x-coordinate. Thus, total number of atoms in the computational cell is $I \times J \times N$. Atomic positions are defined by the radius-vectors $\mathbf{r}_{ijn} = (x_{ijn}, y_{ijn})$. Periodic boundary conditions are imposed.

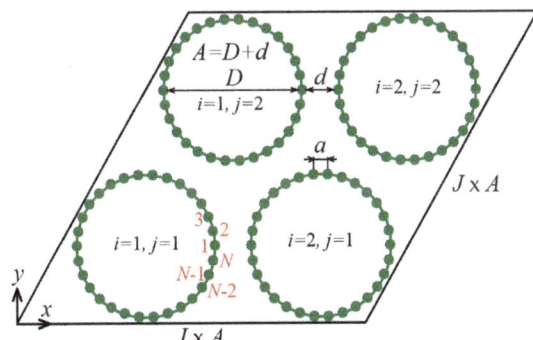

Figure 1. Schematic of the computational cell that includes $I \times J$ carbon nanotubes (CNTs) ($I = J = 2$ in this case) numbered by the indices $i = 1, ..., I$ and $j = 1, ..., J$. Carbon nanotubes in cross-section create a triangular lattice. Within each CNT, carbon atoms are numbered by the index $n = 1, ..., N$ anti-clockwise, starting from the atom with maximal x-coordinate. Atoms move on the (x, y) plane. Each atom represents a row of atoms oriented normal to the (x, y) plane, which moves as a rigid body. The computational cell has the shape of a parallelogram with the sides $I \times A$ and $J \times A$, where A is the distance between centers of neighboring CNTs. Periodic boundary conditions are used.

Let us describe the model geometry. The interatomic distance in graphene is equal to $\rho = 1.418$ Å. The distance between neighboring atomic rows oriented along the armchair direction in graphene is then $a = \rho\sqrt{3}/2 = 1.228$ Å, and this is the distance between atoms in the chain model (see Figure 1). Carbon nanotube diameter is $D = a/\sin(\pi/N)$. Let d be the shortest distance between CNT walls, then the distance between centers of neighboring CNTs is $A = D + d$. The sides of the computational cell in the form of parallelogram are $I \times A$ and $J \times A$. In our simulations, we consider CNTs with $N = 30$ having diameter $D = 11.75$ Å and equilibrium value of $d = 3.088$ Å, which can be compared to the interplanar distance of graphite equal to 3.3 Å.

Carbon nanotube bundle under uniform lateral compression can be efficiently described by the Hamiltonian of the chain model [60]

$$H = K + U_B + U_A + U_{vdW}, \quad (1)$$

which includes kinetic energy

$$K = \frac{M}{2} \sum_{i=1}^{I} \sum_{j=1}^{J} \sum_{n=1}^{N} (\dot{x}_{ijn}^2 + \dot{y}_{ijn}^2), \quad (2)$$

energy of valence bonds

$$U_B = \sum_{i=1}^{I} \sum_{j=1}^{J} \sum_{n=1}^{N} V(|\mathbf{r}_{ijn+1} - \mathbf{r}_{ijn}|), \quad \text{where} \quad V(r) = \frac{k}{2}(r-a)^2, \quad (3)$$

energy of valence angles

$$U_A = \sum_{i=1}^{I} \sum_{j=1}^{J} \sum_{n=1}^{N} P(\theta_{ijn}), \quad \text{where} \quad P(\theta) = \epsilon[\cos(\theta) + 1], \quad (4)$$

and energy of van der Waals interactions

$$U_{vdW} = \sum_{i=1}^{I} \sum_{j=1}^{J} \sum_{n=1}^{N} \sum_{i'=1}^{I} \sum_{j'=1}^{J} \sum_{n'=1}^{N} W(|\mathbf{r}_{ijn} - \mathbf{r}_{i'j'n'}|), \quad \text{where} \quad |n' - n| > 3 \quad \text{when} \quad i = i', j = j'. \quad (5)$$

In Equation (2) M is the carbon atom mass, which is 12 amu. In our simulations, time is measured in picoseconds, energy in eV, and distance in angstrom. In these units $M = 12 \times 1.0364 \times 10^{-4}$. As it can be seen from Equation (3), harmonic potential with stiffness k is used to model deformation of the valence bonds. In order to reproduce the longitudinal stiffness of graphene sheet one should take $k = 405$ N/m [60], which in the units adopted here gives $k = 25.279$.

In Equation (4), cosine of the angle between two valence bonds, $\mathbf{r}_{ijn} - \mathbf{r}_{ijn-1}$ and $\mathbf{r}_{ijn+1} - \mathbf{r}_{ijn}$, is calculated as

$$\cos(\theta_{ijn}) = \frac{(\mathbf{r}_{ijn} - \mathbf{r}_{ijn-1}, \mathbf{r}_{ijn+1} - \mathbf{r}_{ijn})}{|\mathbf{r}_{ijn} - \mathbf{r}_{ijn-1}||\mathbf{r}_{ijn+1} - \mathbf{r}_{ijn}|}. \quad (6)$$

Bending rigidity of graphene sheet is well reproduced with the value of the potential parameter $\epsilon = 3.50$ eV [60].

The van der Waals interactions in Equation (5) are given by the Lennard–Jones potential (5,11) [65]

$$W(r) = \frac{\epsilon}{6}\left[5\left(\frac{\sigma}{r}\right)^{11} - 11\left(\frac{\sigma}{r}\right)^5\right], \quad (7)$$

with the interaction energy $\epsilon = 0.00166$ eV and the equilibrium bond length $\sigma = 3.61$ Å.

Further information on the chain model and on the procedure of fitting its parameters can be found in [60,65].

As it has been mentioned, a CNT of sufficiently large diameter can have either cylindrical or collapsed equilibrium configuration. In the present study we consider CNTs of relatively small diameter ($N = 30$, $D = 11.75$ Å) with only circular stable state when unloaded.

The aim of this study is to evaluate mechanical response of CNT bundle to lateral biaxial compression under plane strain condition with $\varepsilon_{xx} = \varepsilon_{yy} \leq 0$ and $\varepsilon_{xy} = 0$. Firstly, equilibrium configurations are found at zero temperature and then, their stability at room temperature ($T = 300$ K) is analyzed.

Perturbation–relaxation molecular dynamics simulations are done at zero temperature in order to find equilibrium structures at different values of applied strain and $T = 0$ K. The simulation protocol is as follows. The compressive strain is applied by increments $\Delta\varepsilon_{xx} = \Delta\varepsilon_{yy} = -0.0025$ starting from zero strain. After each increment, the positions of atoms are perturbed by adding small random displacements to their x- and y-coordinates. The displacements are uniformly distributed in the range from -10^{-6} to 10^{-6} Å. Then the equilibrium structure is obtained by minimizing potential energy of the system with the help of the gradient method. Energy minimization stops when the absolute value of the maximal force acting on atoms becomes smaller than 10^{-10} eV/Å.

Different computational cell sizes were considered. Calculations with $24 \times 24 = 576$ CNTs have revealed that structures with period doubling are formed as a result of instability at particular value of compressive strain. Such structures can be analyzed with smaller cell size and most of the results reported here are for the cell that includes $6 \times 6 = 36$ CNTs.

Classical molecular dynamics was used to assess stability of equilibrium structures with respect to thermal fluctuations at $T = 300$ K. Temperature in our simulations is defined as

$$T = \frac{M}{2IJNk_B(t_2 - t_1)} \int_{t_1}^{t_2} \sum_{i=1}^{I} \sum_{j=1}^{J} \sum_{n=1}^{N} (\dot{r}_{ijn}, \dot{r}_{ijn}) d\tau, \qquad (8)$$

where $k_B = 8.617 \times 10^{-5}$ eV·K^{-1} is the Boltzmann constant and the averaging time is $t_2 - t_1 = 10$ ps. For a given temperature T, the initial velocities of atoms are assigned according to the Maxwellian distribution. Random initial displacements of atoms are assigned in a way to increase the potential energy of the system by the amount equal to the kinetic energy.

Equations of atomic motion that stem from the Hamiltonian Equation (1) are integrated numerically with the help of the Stormer method of order six with the time step of 0.1 fs. The structure is considered to be stable if no structure transformations are observed within 100 ps. Structure transformations can be very well seen on the time dependencies of kinetic and potential energies during our simulations with NVE ensemble (constant number of particles, volume, and total energy). When structure changes, kinetic energy increases in expense of potential energy.

3. Results

In this section the equilibrium structures of CNT bundle under lateral compression are reported and their properties are analyzed. First, the potential energy and stress as functions of strain are given and then the change of CNT geometry with strain is presented. Finally, stability of equilibrium structures at $T = 300$ K is analyzed.

3.1. Energy and Stress in the System

We start with the analysis of potential energy per atom calculated for equilibrium structures at different values of compressive strain. In Figure 2a, total potential energy per atom is shown as a function of strain, while in Figure 2b–d this energy is decomposed into three parts: the energy of van der Waals interactions, the energy of valence bonds, and the energy of valence angles, respectively. Total potential energy in the range of strain below 3.75% increases with strain quadratically but for larger strain a linear increase of energy with strain can be observed. These two regimes are separated in Figure 2 by the vertical dashed line. This qualitative change in the behavior of potential energy is due to structural changes observed in the system for strain exceeding the critical value of 3.75%. Below this critical strain, all CNTs in the system have the same cross-section in the form of six-fold flattened cylinders (see Figure 3a), while above the critical value the CNTs become elliptic. Two different structures with elliptic CNTs can form, Structure I with the translational cell doubled in one direction (see Figure 3b) and Structure II with the translational cell doubled in two directions (see Figure 4). In Figure 2 results for Structure I are shown by red circles and for Structure II by black triangles.

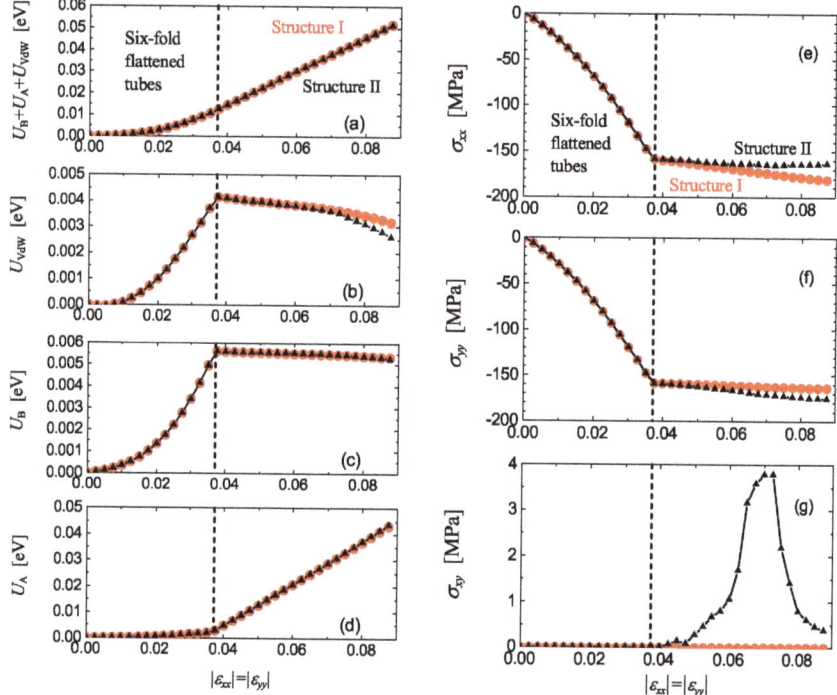

Figure 2. (a–d) Potential energy per atom and its parts as the functions of biaxial compressive strain. (e–g) Components of stress tensor as the functions of biaxial compressive strain. Results for Structure I [see Figure 3b] are shown by red circles and for Structure II (see Figure 4) by black triangles. These structures with elliptic nanotubes are stable for $0.0375 < |\varepsilon_{xx}| = |\varepsilon_{yy}| < 0.09$. For smaller values of compressive strain one has six-fold flattened nanotubes [see Figure 3a]. For compressive strain above 9% collapsed CNTs appear in the system (this regime is not studied here). The first critical value of strain is shown by the vertical dashed lines.

From Figure 2a–d it is clear that below the critical strain the potential energy of all three kinds increase with strain. At the critical value of strain just before the structure transformation one has $U_{vdW} = 0.0041$ eV, $U_B = 0.0056$ eV, and $U_A = 0.0029$ eV. The largest increase of energy is observed for the valence bonds because main mechanism of lattice deformation in this regime is contraction of valence bonds. The smallest contribution to the energy increase comes from the valence angles, since they do not change much during transformation of CNT cylindrical shape into six-fold flattened shape. This picture drastically changes for compressive strain larger than 3.75% when CNTs become elliptic. The deformation of structure in this regime is mainly due to change of valence angles and U_A increases rapidly with strain while other two components of energy decrease with strain. The decrease of U_{vdW} with strain in this regime is explained by formation of new van der Waals bonds with increasing ellipticity of CNTs. At strain of 8.75%, U_A is already one order of magnitude larger than two other components of energy. Note that both Structure I and Structure II have very close energies, and only for strain above 7% is U_{vdW} in Structure I slightly higher than in Structure II. Repetition of the simulations with different random atomic displacements has revealed that Structure I and Structure II are formed at the transition point with nearly equal probability, and this is because they have practically same energy.

Variation of stress components σ_{xx}, σ_{yy}, and σ_{xy} with strain is shown in Figure 2e–g, respectively. For strain below the critical level, compressive stress increases so that $\sigma_{xx} = \sigma_{yy}$ and $\sigma_{xy} = 0$. This is because the structure with six-fold symmetry is elastically isotropic (see Figure 3a). For strain above the critical level, deformation occurs at nearly constant pressure $p = -(\sigma_{xx} + \sigma_{yy})/2$. Both structures become anisotropic. In particular, Structure I is orthotropic with $|\sigma_{xx}| > |\sigma_{yy}|$ and $\sigma_{xy} = 0$. On the other hand, Structure II has general anisotropy with $|\sigma_{xx}| < |\sigma_{yy}|$ and $\sigma_{xy} \neq 0$.

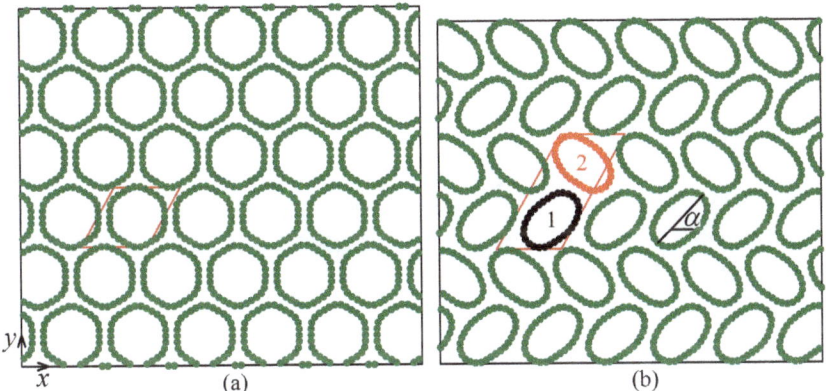

Figure 3. Equilibrium structures of CNT bundle observed at compressive strain of (**a**) 3.5% and (**b**) 5.5%. In (**a**) the displacements of atoms are multiplied by factor 4 to better reveal the six-fold flattened cylindrical shape of CNTs. In (**b**) the CNT cross-section is elliptic. Translation cells of the structures are shown by red lines. In (**a**) the cell includes single CNT, while in (**b**) period doubles in one direction. The latter structure is referred to as Structure I.

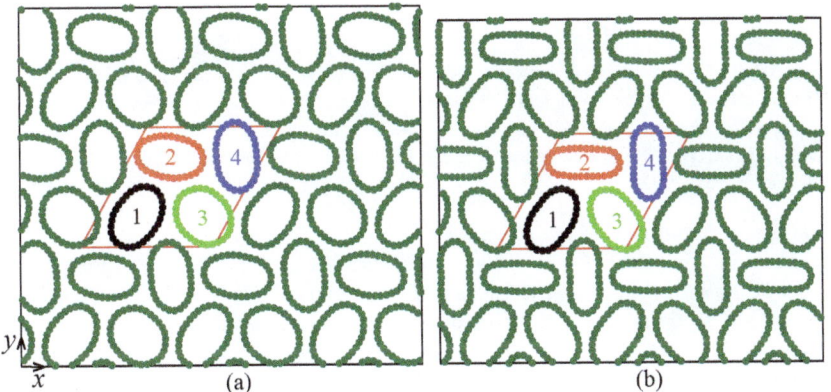

Figure 4. Equilibrium structures of CNT bundle observed at compressive strain of (**a**) 5.5% and (**b**) 8.5%. Translation cells of the structures are shown by red lines. Here, period doubles in two directions and this structure is called Structure II.

3.2. Geometry of CNTs

To better understand the relation between structure and macroscopic parameters of the system, let us quantify the geometry of CNTs in different structures. For each CNT we calculate its minimal and maximal diameters, D_{min} and D_{max}, and the angle α between the x-axis and the maximal diameter, as shown in Figure 3b.

In Figure 5a,b the ratio D_{min}/D_{max} is shown for Structures I and II, respectively. In Figure 5c,d the angle α is shown for Structures I and II, respectively.

Since the translation cell of Structure I includes two CNTs (numbered as 1 and 2 in Figure 3b), in Figure 5a,c, two different values of D_{min}/D_{max} and α can be seen for strain above the critical value.

77

The translation cell of Structure II includes four CNTs (numbered as 1 to 4 in Figure 4) and hence, in Figure 5b,d, four different values of D_{min}/D_{max} and α can be seen for strain above the critical value. For strain below the critical value, all curves merge into one since translation cell includes single CNT, see Figure 3a.

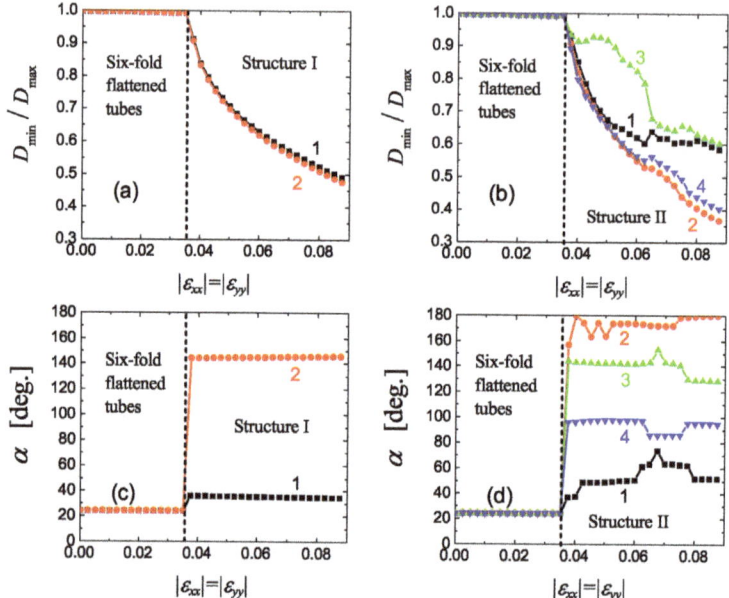

Figure 5. (**a**,**b**) The minimal to maximal diameter ratio for CNTs in Structures I and II, respectively, as functions of compressive strain. (**c**,**d**) Orientation angle of CNTs in Structures I and II, respectively, as functions of compressive strain. Numbers near the curves link them to the CNTs in translation cells of Structures I and II, as shown in Figures 3b and 4. The critical value of strain is shown by the vertical dashed lines.

From Figure 5a, one can see that in Structure I the ellipticity of SNTs gradually increases with increasing compressive strain. At the value of strain 8.75%, D_{min}/D_{max} is smaller than 0.5. Recall that further increase of strain above 9% results in the formation of collapsed CNTs with non-convex cross-section, but we do not analyze such structures here. As Figure 5c suggests, the orientation of elliptic CNTs in Structure I is practically strain-independent. Carbon nanotubes 1 and 2 have orientation angles $\alpha = 35$ and 145 deg., with the difference equal to 110 deg.

Structure II demonstrates more complicated evolution with strain.

Two different regimes can be distinguished looking at Figure 5b. For strain between 4% and 6.5%, three CNTs (1, 2, and 4) in the translation cell have nearly the same ellipticity, while CNT 3 is less elliptic (see Figure 4a). For larger strain, CNTs 2 and 4 become more elliptic than CNTs 1 and 3 (see Figure 4b). Orientation angles of CNTs also change with strain (see Figure 5d). At 8.75% strain, CNT 4 is nearly aligned with the y-axis (α is close to 90 deg.), while CNT 2 with the x-axis (α is close to 180 deg.). Carbon nanotubes 1 and 3 have angles of about 50 and 130 deg., with a difference of about 80 deg.

3.3. Temperature Effect

The stability of all three types of equilibrium structures reported in Section 3.1 with respect to thermal oscillations is investigated here at temperature $T = 300$ K.

It was found that the structure with all identical CNTs (see Figure 3a) is stable at room temperature up to compressive strain of about 4.0%, which slightly exceeds the stability range of this structure at 0 K (3.75% strain). At a strain of 4.0% and temperature 300 K, the cross-sectional shapes of CNTs fluctuate in time but on average all CNTs remain the same, preserving the high symmetry of the structure. The pressure-induced phase transition from high-symmetry structure to the structures with period doubling in one or two directions is the second-order phase transition. This is justified by the absence of the jumps in macroscopic properties at the transition point (3.75% compressive strain), see Figure 2. Low-symmetry Structures I and II under heating transform into high-symmetry structures, if the compressive strain is not too high.

Within the range of compressive strain from 4.1% to 6%, Structures I and II are stable at 300 K; they are preserved within the simulation time of 100 ps and no jumps of macroscopic parameters are observed.

Both Structures I and II become unstable at 300 K for compressive strain exceeding 6%. The instability of Structure I is illustrated in Figure 6 for compressive strain of 7%. In (a,b), one can see the time evolution of kinetic energy per atom and components of compressive stress, respectively. Until $t = 10$ ps, kinetic energy oscillates near the value of 0.0259 eV, which corresponds to 300 K, but then it starts to increase in expense of the potential energy (total energy is conserved in the system). Pressure drops at the transition point from 175 to 140 MPa. Jumps in the macroscopic parameters of the system indicate that this phase transition is of the first order. In (c), a snapshot of the structure is presented at $t = 20$ ps. One can see that the long-range crystal order is lost and an irregular structure that includes collapsed CNTs with non-convex cross-section is formed.

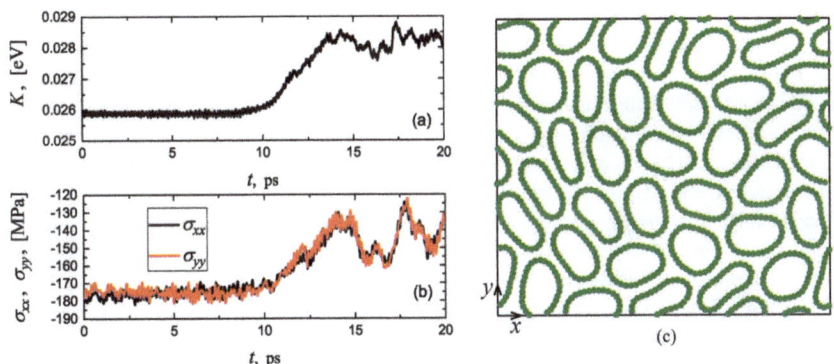

Figure 6. Instability of Structure I at compressive strain of 7% and temperature $T = 300$ K. (a,b) Time evolution of kinetic energy per atom and components of compressive stress, respectively. Structure transformation begins at $t \approx 10$ ps, which results in the change of macroscopic parameters. (c) Snapshot of structure at $t = 20$ ps. As a result of structure transformation, collapsed, non-convex CNTs appear in the system.

4. Discussion

The chain model introduced in the works [60,65] was developed here to enable simulation of the mechanical properties of CNT bundles under plane strain conditions. The model was applied to the analysis of structure transformations and mechanical properties of CNT crystal subjected to biaxial lateral compression. Carbon nanotube diameter is relatively small, so that the collapsed shape is unstable in the absence of external forces.

Three different crystalline structures stable at zero temperature have been found. For compressive strain $|\varepsilon_{xx}| = |\varepsilon_{yy}| < 3.75\%$, primitive translational cell of the crystal includes single CNT. In the range of compressive strain from 4% to 9%, two phases of nearly the same potential energy were found, one has two and the other one has four CNTs in a primitive translational cell. These structures are referred to as Structure I and II, respectively. Pressure-induced phase transition from the high-symmetry structure to the Structures I or II is of the second order because no jumps of macroscopic properties are seen in Figure 2 at the transition point at strain of 3.75%.

Thermal fluctuations increase the stability range of the high-symmetry structure with single CNT in the primitive translational cell. At 300 K, this structure is stable up to 4.0% compressive strain, while at 0 K the stability threshold is at 3.75% strain. The transformation of low-symmetry phase into high-symmetry phase under an increase in temperature is typical for the second-order phase transitions [66,67].

Thermal fluctuations reduce the stability range of Structures I and II. At zero temperature, they are stable from 3.75% to 8.75% of compressive strain, while at room temperature the stability range of compressive strain is from 4.1% to 6%. The transition above 6% strain is of the first order, it is accompanied by the jumps in macroscopic parameters (see Figure 6a,b), when crystalline structure with a long-range order transforms into an irregular structure (see Figure 6c).

As for the mechanical properties of CNT bundles under lateral compression, the transition to the structures with elliptic CNTs results in a considerable drop in rigidity of the bundle. Indeed, the deformation of the structure in the range from 3.75% to 9% compressive strain is at nearly constant pressure, see Figure 2e,f. Unloading of the system from any strain below 9% has shown that the structure is non-linear elastic with no hysteresis effect. Also note that the high-symmetry structure is elastically isotropic, while Structures I and II are anisotropic since for them $\sigma_{xx} \neq \sigma_{yy}$, see Figure 2e,f.

The results reported here are in agreement with the results of full atomic and continuum mechanics modelling [6,8,39,51,53,54,56–58], but they were obtained at a very low computational cost. For example, full atomic modelling with the same accuracy would require the use of the periodic boundary conditions in the z direction with at least one zigzag carbon chain within the translational cell. The calculation of interatomic forces between atoms within the considered cell and its translation images would require additional summation, which is absent in the chain model since it has been done in derivation of the effective potentials between the rigid atomic chains oriented along the z axis. For this reason, the chain model gives at least one order of magnitude acceleration of computations with the same accuracy, as compared to full atomic modelling.

Recall that the harmonic, unbreakable potential is used in this work to describe the valence interactions between carbon atoms, which is sufficient for modelling structure transformations at relatively small pressure considered here. In order to model irreversible deformation of CNT bundles under very high pressure, see, e.g., the work [39], the harmonic potential Equation (3) should be substituted with the breakable anharmonic potential, such as Morse potential [68].

In future works, it is important to study the effect of CNT diameter since new effects can be expected for larger diameter when collapsed isolated CNT is stable. Crystals composed of such CNTs can demonstrate irreversible plastic deformation with very peculiar mechanisms of plasticity. As for the applications, CNT bundles under lateral compression can show hysteresis effect, when it acts as an elastic damper [69], in a similar way to the compressed, vertically aligned CNT brushes [31].

The chain model proposed here can be readily adjusted to a number of newly found graphene-analogous 2D nanomaterials [70] by fitting model parameters to the results of first-principle or molecular dynamics simulations.

Consideration of bundles composed of different CNTs or multi-walled CNTs is straightforward. Substitution of the harmonic valence bond potential in Equation (3) with a suitable breakable anharmonic potential will enable the simulation of structure transformations in the CNT bundle under very high pressure.

5. Conclusions

We thus conclude that the chain model can be applied with a high numerical efficiency and sufficient accuracy to the analysis of structural and mechanical properties of CNT bundles under plane strain conditions.

The atomistic chain model proposed here, unlike continuum mechanics models, is able to describe high curvature of collapsed CNT wall and fracture of the walls under high pressure.

The chain model proposed here can be readily applied to the cases of CNTs of different diameter, collapsed CNTs, multi-walled CNTs, and even include graphene scrolls and cylindrically crumpled graphene.

Author Contributions: Design of the research, E.A.K. and S.V.D.; simulations, L.K.R. and E.G.S.; methodology, A.V.S.; writing—original draft preparation, E.A.K. and S.V.D.; discussion and analysis of results, E.G.S., E.G.E., M.A.I.

Funding: E.A.K. thanks financial support provided by the Russian Foundation for Basic Research, grant no. 18-32-20158.

Acknowledgments: For E.G.S. the research was carried out at Tomsk Polytechnic University within the framework of Tomsk Polytechnic University Competitiveness Enhancement Program grant. For E.G.E. the work was supported by Act 211 Government of the Russian Federation, contract No. 02.A03.21.0011. This work was partly supported by the State Assignment of IMSP RAS No. AAAA-A17-117041310220-8.

Conflicts of Interest: The authors declare no conflict of interest.

References

1. Pérez, E.M.; Martín, N. Curves ahead: Molecular receptors for fullerenes based on concave-convex complementarity. *Chem. Soc. Rev.* **2008**, *37*, 1512–1519. [CrossRef] [PubMed]
2. Lyapin, A.G.; Brazhkin, V.V.; Lyapin, S.G.; Popova, S.V.; Varfolomeeva, T.D.; Voloshin, R.A.; Pronin, A.A.; Sluchanko, N.E.; Gavrilyuk, A.G.; Trojan, I.A. Non-traditional carbon semiconductors prepared from fullerite C60 and carbyne under high pressure. *Phys. Status Solidi B* **1999**, *211*, 401–412. [CrossRef]
3. Popov, M.; Koga, Y.; Fujiwara, S.; Mavrin, B.N.; Blank, V.D. Carbon nanocluster-based superhard materials. *New Diam. Front. Carbon Technol.* **2002**, *12*, 229–260.
4. Ferrari, A.C. Raman spectroscopy of graphene and graphite: Disorder, electron-phonon coupling, doping and nonadiabatic effects. *Solid State Commun.* **2007**, *143*, 47–57. [CrossRef]
5. Ghosh, S.; Nika, D.L.; Pokatilov, E.P.; Balandin, A.A. Heat conduction in graphene: Experimental study and theoretical interpretation. *New J. Phys.* **2009**, *11*, 095012. [CrossRef]
6. Tersoff, J.; Ruoff, R.S. Structural properties of a carbon-nanotube crystal. *Phys. Rev. Lett.* **1994**, *73*, 676–679. [CrossRef]
7. Thess, A.; Lee, R.; Nikolaev, P.; Dai, H.; Petit, P.; Robert, J.; Xu, C.; Lee, Y.H.; Kim, S.G.; Rinzler, A.G.; et al. Crystalline ropes of metallic carbon nanotubes. *Science* **1996**, *273*, 483–487. [CrossRef]
8. Saether, E.; Frankland, S.J.V.; Pipes, R.B. Transverse mechanical properties of single-walled carbon nanotube crystals. Part I: Determination of elastic moduli. *Compos. Sci. Technol.* **2003**, *63*, 1543–1550. [CrossRef]
9. Rakov, E.G. Materials made of carbon nanotubes. The carbon nanotube forest. *Russ. Chem. Rev.* **2013**, *82*, 538–566. [CrossRef]
10. Chen, H.; Roy, A.; Baek, J.-B.; Zhu, L.; Qu, J.; Dai, L. Controlled growth and modification of vertically-aligned carbon nanotubes for multifunctional applications. *Mater. Sci. Eng. R Rep.* **2010**, *70*, 63–91. [CrossRef]
11. Bedewy, M.; Meshot, E.R.; Guo, H.; Verploegen, E.A.; Lu, W.; Hart, A.J. Collective mechanism for the evolution and self-termination of vertically aligned carbon nanotube growth. *J. Phys. Chem. C* **2009**, *113*, 20576–20582. [CrossRef]
12. Lan, Y.; Wang, Y.; Ren, Z.F. Physics and applications of aligned carbon nanotubes. *Adv. Phys.* **2011**, *60*, 553–678. [CrossRef]
13. Yu, M.-F.; Files, B.S.; Arepalli, S.; Ruoff, R.S. Tensile loading of ropes of single wall carbon nanotubes and their mechanical properties. *Phys. Rev. Lett.* **2000**, *84*, 5552–5555. [CrossRef] [PubMed]
14. Dhanabalan, S.C.; Dhanabalan, B.; Chen, X.; Ponraj, J.S.; Zhang, H. Hybrid carbon nanostructured fibers: Stepping stone for intelligent textile-based electronics. *Nanoscale* **2019**, *11*, 3046–3101. [CrossRef] [PubMed]

15. Bai, Y.; Zhang, R.; Ye, X.; Zhu, Z.; Xie, H.; Shen, B.; Cai, D.; Liu, B.; Zhang, C.; Jia, Z.; et al. Carbon nanotube bundles with tensile strength over 80 GPa. *Nat. Nanotechnol.* **2018**, *13*, 589–595. [CrossRef] [PubMed]
16. Qiu, L.; Wang, X.; Tang, D.; Zheng, X.; Norris, P.M.; Wen, D.; Zhao, J.; Zhang, X.; Li, Q. Functionalization and densification of inter-bundle interfaces for improvement in electrical and thermal transport of carbon nanotube fibers. *Carbon* **2016**, *105*, 248–259. [CrossRef]
17. Cho, H.; Lee, H.; Oh, E.; Lee, S.-H.; Park, J.; Park, H.J.; Yoon, S.-B.; Lee, C.-H.; Kwak, G.-H.; Lee, W.J.; et al. Hierarchical structure of carbon nanotube fibers, and the change of structure during densification by wet stretching. *Carbon* **2018**, *136*, 409–416. [CrossRef]
18. Fernández-Toribio, J.C.; Alemán, B.; Ridruejo, Á.; Vilatela, J.J. Tensile properties of carbon nanotube fibres described by the fibrillar crystallite model. *Carbon* **2018**, *133*, 44–52. [CrossRef]
19. Dang, Z.-M.; Yuan, J.-K.; Zha, J.-W.; Zhou, T.; Li, S.-T.; Hu, G.-H. Fundamentals, processes and applications of high-permittivity polymer-matrix composites. *Prog. Mater. Sci.* **2012**, *57*, 660–723. [CrossRef]
20. Bakshi, S.R.; Lahiri, D.; Agarwal, A. Carbon nanotube reinforced metal matrix composites—A review. *Int. Mater. Rev.* **2010**, *55*, 41–64. [CrossRef]
21. Dorri Moghadam, A.; Omrani, E.; Menezes, P.L.; Rohatgi, P.K. Mechanical and tribological properties of self-lubricating metal matrix nanocomposites reinforced by carbon nanotubes (CNTs) and graphene—A review. *Compos. Part B Eng.* **2015**, *77*, 448, 402–420. [CrossRef]
22. Reinert, L.; Lasserre, F.; Gachot, C.; Grützmacher, P.; Maclucas, T.; Souza, N.; Mücklich, F.; Suarez, S. Long-lasting solid lubrication by CNT-coated patterned surfaces. *Sci. Rep.* **2017**, *7*, 42873. [CrossRef] [PubMed]
23. Samsonidze, G.G.; Samsonidze, G.G.; Yakobson, B.I. Kinetic theory of symmetry-dependent strength in carbon nanotubes. *Phys. Rev. Lett.* **2002**, *88*, 065501:1–065501:4. [CrossRef] [PubMed]
24. Shenderova, O.A.; Zhirnov, V.V.; Brenner, D.W. Carbon nanostructures. *Crit. Rev. Solid State* **2002**, *27*, 227–356. [CrossRef]
25. Yu, M.-F. Fundamental mechanical properties of carbon nanotubes: Current understanding and the related experimental studies. *J. Eng. Mater. Trans. ASME* **2004**, *126*, 271–278. [CrossRef]
26. Yu, M.-F.; Lourie, O.; Dyer, M.J.; Moloni, K.; Kelly, T.F.; Ruoff, R.S. Strength and breaking mechanism of multiwalled carbon nanotubes under tensile load. *Science* **2000**, *287*, 637–640. [CrossRef]
27. Truong, T.K.; Lee, Y.; Suh, D. Multifunctional characterization of carbon nanotube sheets, yarns, and their composites. *Curr. Appl. Phys.* **2016**, *16*, 1250–1258. [CrossRef]
28. Yao, S.; Yuan, J.; Mehedi, H.-A.; Gheeraert, E.; Sylvestre, A. Carbon nanotube forest based electrostatic capacitor with excellent dielectric performances. *Carbon* **2017**, *116*, 648–654. [CrossRef]
29. Yao, X.; Hawkins, S.C.; Falzon, B.G. An advanced anti-icing/de-icing system utilizing highly aligned carbon nanotube webs. *Carbon* **2018**, *136*, 130–138. [CrossRef]
30. Yao, X.; Falzon, B.G.; Hawkins, S.C.; Tsantzalis, S. Aligned carbon nanotube webs embedded in a composite laminate: A route towards a highly tunable electro-thermal system. *Carbon* **2018**, *129*, 486–494. [CrossRef]
31. Cao, A.Y.; Dickrell, P.L.; Sawyer, W.G.; Ghasemi-Nejhad, M.N.; Ajayan, P.M. Super-compressible foamlike carbon nanotube films. *Science* **2005**, *310*, 1307–1310. [CrossRef] [PubMed]
32. Pathak, S.; Kalidindi, S.R. Spherical nanoindentation stress-strain curves. *Mater. Sci. Eng. R Rep.* **2015**, *91*, 1–36. [CrossRef]
33. Pathak, S.; Cambaz, Z.G.; Kalidindi, S.R.; Swadener, J.G.; Gogotsi, Y. Viscoelasticity and high buckling stress of dense carbon nanotube brushes. *Carbon* **2009**, *47*, 1969–1976. [CrossRef]
34. Maschmann, M.R.; Zhang, Q.; Du, F.; Dai, L.; Baur, J. Length dependent foam-like mechanical response of axially indented vertically oriented carbon nanotube arrays. *Carbon* **2011**, *49*, 386–397. [CrossRef]
35. Cao, C.; Reiner, A.; Chung, C.; Chang, S.-H.; Kao, I.; Kukta, R.V.; Korach, C.S. Buckling initiation and displacement dependence in compression of vertically aligned carbon nanotube arrays. *Carbon* **2011**, *49*, 3190–3199. [CrossRef]
36. Liang, X.; Shin, J.; Magagnosc, D.; Jiang, Y.; Jin Park, S.; John Hart, A.; Turner, K.; Gianola, D.S.; Purohit, P.K. Compression and recovery of carbon nanotube forests described as a phase transition. *Int. J. Solids Struct.* **2017**, *122–123*, 196–209. [CrossRef]
37. Koumoulos, E.P.; Charitidis, C.A. Surface analysis and mechanical behaviour mapping of vertically aligned CNT forest array through nanoindentation. *Appl. Surf. Sci.* **2017**, *396*, 681–687. [CrossRef]
38. Parisa Pour Shahid Saeed Abadi; Hutchens, S.B.; Greer, J.R.; Cola, B.A.; Graham, S. Buckling-driven delamination of carbon nanotube forests. *Appl. Phys. Lett.* **2013**, *102*, 223103.

39. Silva-Santos, S.D.; Alencar, R.S.; Aguiar, A.L.; Kim, Y.A.; Muramatsu, H.; Endo, M.; Blanchard, N.P.; San-Miguel, A.; Souza Filho, A.G. From high pressure radial collapse to graphene ribbon formation in triple-wall carbon nanotubes. *Carbon* **2019**, *141*, 568–579. [CrossRef]
40. Tangney, P.; Capaz, R.B.; Spataru, C.D.; Cohen, M.L.; Louie, S.G. Structural transformations of carbon nanotubes under hydrostatic pressure. *Nano Lett.* **2005** *5*, 2268–2273. [CrossRef]
41. Zhang, S.; Khare, R.; Belytschko, T.; Hsia, K.J.; Mielke, S.L.; Schatz, G.C. Transition states and minimum energy pathways for the collapse of carbon nanotubes. *Phys. Rev. B* **2006**, *73*, 075423. [CrossRef]
42. Shima, H.; Sato, M. Multiple radial corrugations in multiwalled carbon nanotubes under pressure. *Nanotechnology* **2008**, *19*, 495705. [CrossRef] [PubMed]
43. Zhao, Z.S.; Zhou, X.-F.; Hu, M.; Yu, D.L.; He, J.L.; Wang, H.-T.; Tian, Y.J.; Xu, B. High-pressure behaviors of carbon nanotubes. *J. Superhard Mater.* **2012**, *34*, 371–385. [CrossRef]
44. Islam, S.; Saleh, T.; Asyraf, M.R.M.; Mohamed Ali, M.S. An ex-situ method to convert vertically aligned carbon nanotubes array to horizontally aligned carbon nanotubes mat. *Mater. Res. Express* **2019**, *6*, 025019. [CrossRef]
45. Zhang, R.; Zhang, Y.; Wei, F. Horizontally aligned carbon nanotube arrays: Growth mechanism, controlled synthesis, characterization, properties and applications. *Chem. Soc. Rev.* **2017**, *46*, 3661–3715. [CrossRef] [PubMed]
46. Nam, T.H.; Goto, K.; Yamaguchi, Y.; Premalal, E.V.A.; Shimamura, Y.; Inoue, Y.; Naito, K.; Ogihara, S. Effects of CNT diameter on mechanical properties of aligned CNT sheets and composites. *Compos. Part A Appl. Sci. Manuf.* **2015**, *76*, 289–298. [CrossRef]
47. Qiu, L.; Wang, X.; Su, G.; Tang, D.; Zheng, X.; Zhu, J.; Wang, Z.; Norris, P.M.; Bradford, P.D.; Zhu, Y. Remarkably enhanced thermal transport based on a flexible horizontally-aligned carbon nanotube array film. *Sci. Rep.* **2016**, *6*, 21014. [CrossRef]
48. Qian, D.; Wagner, G.J.; Liu, W.K.; Yu, M.-F.; Ruoff, R.S. Mechanics of carbon nanotubes. *Appl. Mech. Rev.* **2002**, *55*, 495–532. [CrossRef]
49. Tang, J.; Sasaki, T.; Yudasaka, M.; Matsushita, A.; Iijima, S. Compressibility and polygonization of single-walled carbon nanotubes under hydrostatic pressure. *Phys. Rev. Lett.* **2000**, *85*, 1887–1889. [CrossRef]
50. Karmakar, S.; Sharma, S.M.; Teredesai, P.V.; Muthu, D.V.S.; Govindaraj, A.; Sikka, S.K.; Sood, A.K. Structural changes in single-walled carbon nanotubes under non-hydrostatic pressures: X-ray and Raman studies. *New J. Phys.* **2003**, *5*, 143.1–143.11. [CrossRef]
51. Wittmaack, B.K.; Volkov, A.N.; Zhigilei, L.V. Phase transformation as the mechanism of mechanical deformation of vertically aligned carbon nanotube arrays: Insights from mesoscopic modeling. *Carbon* **2019**, *143*, 587–597. [CrossRef]
52. Wittmaack, B.K.; Volkov, A.N.; Zhigilei, L.V. Mesoscopic modeling of the uniaxial compression and recovery of vertically aligned carbon. *Compos. Sci. Technol.* **2018**, *166*, 66–85. [CrossRef]
53. Yakobson, B.I.; Brabec, C.J.; Bernholc, J. Nanomechanics of carbon tubes: Instabilities beyond linear response. *Phys. Rev. Lett.* **1996**, *76*, 2511–2514. [CrossRef] [PubMed]
54. Rafii-Tabar, H.; Ghavanloo, E.; Fazelzadeh, S.A. Nonlocal continuum-based modeling of mechanical characteristics of nanoscopic structures. *Phys. Rep.* **2016**, *638*, 1–97. [CrossRef]
55. Harik, V.M. Ranges of applicability for the continuum beam model in the mechanics of carbon nanotubes and nanorods. *Solid State Commun.* **2001**, *120*, 331–335. [CrossRef]
56. Impellizzeri, A.; Briddon, P.; Ewels, C.P. Stacking- and chirality-dependent collapse of single-walled carbon nanotubes: A large-scale density-functional study. *Phys. Rev. B* **2019**, *100*, 115410. [CrossRef]
57. Chopra, N.G.; Benedict, L.X.; Crespi, V.H.; Cohen, M.L.; Louie, S.G.; Zettl, A. Fully collapsed carbon nanotubes. *Nature* **1995** *377*, 135–138. [CrossRef]
58. Chang, T. Dominoes in carbon nanotubes. *Phys. Rev. Lett.* **2008**, *101*, 175501. [CrossRef]
59. Ji, J.; Zhao, J.; Guo, W. Novel nonlinear coarse-grained potentials of carbon nanotubes. *J. Mech. Phys. Solids* **2019**, *128*, 79–104. [CrossRef]
60. Savin, A.V.; Korznikova, E.A.; Dmitriev, S.V. Scroll configurations of carbon nanoribbons. *Phys. Rev. B* **2015**, *92*, 035412. [CrossRef]
61. Savin, A.V.; Korznikova, E.A.; Dmitriev, S.V. Simulation of folded and scrolled packings of carbon nanoribbons. *Phys. Solid State* **2015**, *57*, 2348–2355. [CrossRef]

62. Savin, A.V.; Korznikova, E.A.; Lobzenko, I.P.; Baimova, Y.A.; Dmitriev, S.V. Symmetric scrolled packings of multilayered carbon nanoribbons. *Phys. Solid State* **2016**, *58*, 1278–1284. [CrossRef]
63. Savin, A.V.; Korznikova, E.A.; Dmitriev, S.V.; Soboleva, E.G. Graphene nanoribbon winding around carbon nanotube. *Comp. Mater. Sci.* **2017**, *135*, 99–108. [CrossRef]
64. Savin, A.V.; Mazo, M.A. 2D chain models of nanoribbon scrolls. *Adv. Struct. Mater.* **2019**, *94*, 241–262.
65. Savin, A.V.; Korznikova, E.A.; Dmitriev, S.V. Dynamics of surface graphene ripplocations on a flat graphite substrate. *Phys. Rev. B* **2019**, *99*, 235411. [CrossRef]
66. Dmitriev, S.V.; Abe, K.; Shigenari, T. One-dimensional crystal model for incommensurate phase. I. Small displacement limit. *J. Phys. Soc. Jpn.* **1996**, *65*, 3938–3944. [CrossRef]
67. Dmitriev, S.; Shigenari, T.; Abe, K. Mechanisms of transition between and incommensurate phases in a two-dimensional crystal model. *Phys. Rev. B* **1998**, *58*, 2513–2522. [CrossRef]
68. Brenner, W.; Shenderova, O.A.; Harrison, J.A.; Stuart, S.J.; Ni, B.; Sinnott, S.B. A second-generation reactive empirical bond order (rebo) potential energy expression for hydrocarbons. *J. Phys. Condens. Matter* **2002**, *14*, 783.
69. Evazzade, I.; Lobzenko, I.P.; Saadatmand, D.; Korznikova, E.A.; Zhou, K.; Liu, B.; Dmitriev, S.V. Graphene nanoribbon as an elastic damper. *Nanotechnology* **2018**, *29*, 215704. [CrossRef]
70. Liu, B.; Zhou, K. Recent progress on graphene-analogous 2D nanomaterials: Properties, modeling and applications. *Progr. Mater. Sci.* **2019**, *100*, 99–169. [CrossRef]

© 2019 by the authors. Licensee MDPI, Basel, Switzerland. This article is an open access article distributed under the terms and conditions of the Creative Commons Attribution (CC BY) license (http://creativecommons.org/licenses/by/4.0/).

Article

Homoepitaxy Growth of Single Crystal Diamond under 300 torr Pressure in the MPCVD System

Xiwei Wang [1], Peng Duan [1], Zhenzhong Cao [2], Changjiang Liu [2], Dufu Wang [1,2], Yan Peng [1,*] and Xiaobo Hu [1,*]

1. State Key Laboratory of Crystal Materials, Shandong University, Jinan 250100, China; xwang21@126.com (X.W.); 13256998266@163.com (P.D.); wangdufu@163.com (D.W.)
2. Jinan Diamond Technology Co. Ltd, Jinan 250101, China; 13608927651@163.com (Z.C.); owlchj@163.com (C.L.)
* Correspondence: pengyan@sdu.edu.cn (Y.P.); xbhu@sdu.edu.cn (X.H.)

Received: 30 October 2019; Accepted: 26 November 2019; Published: 28 November 2019

Abstract: The high-quality single crystal diamond (SCD) grown in the Microwave Plasma Chemical Vapor Deposition (MPCVD) system was studied. The CVD deposition reaction occurred in a 300 torr high pressure environment on a (100) plane High Pressure High Temperature (HPHT) diamond type II *a* substrate. The relationships among the chamber pressure, substrate surface temperature, and system microwave power were investigated. The surface morphology evolution with a series of different concentrations of the gas mixture was observed. It was found that a single lateral crystal growth occurred on the substrate edge and a systemic step flow rotation from the [100] to the [110] orientation was exhibited on the surface. The Raman spectroscopy and High Resolution X-Ray Diffractometry (HRXRD) prove that the homoepitaxy part from the original HPHT substrate shows a higher quality than the lateral growth region. A crystal lattice visual structural analysis was applied to describe the step flow rotation that originated from the temperature driven concentration difference of the C_2H_2 ion charged particles on the SCD center and edge.

Keywords: single crystal diamond; Homoepitaxy growth; 300 torr

1. Introduction

Diamond is gaining more attention for its outstanding optical, electrical, mechanical, and thermal properties. Since the first successful synthesis of diamond by General Electric more than half a century ago, diamond has attracted researchers to develop new growth technologies and extend multiple applications. Currently, High Pressure High Temperature (HPHT) and Microwave Plasma Chemical Vapor Deposition (MPCVD) technologies are widely used to manufacture large, single crystal diamonds with good quality and clarity. High-quality single crystal diamond (SCD) is available in superhard cutting tools, optical components, semiconductor and high-power electronics, and even in quantum applications [1–4].

Although the HPHT method still produces the overwhelming majority of a single crystal diamond, the metallic impurity incorporation and high dislocation density in crystal edges limits its further application in the semiconductor and quantum field. MPCVD technology has become a mature and reliable method for high quality SCD growth since it has the advantages of the control of the nitrogen concentration and dislocation density [5]. Many studies have indicated that the surface morphology, growth rate, and crystal quality depend on the internal reaction chamber structural parameters such as the substrate holder, input gas mixture, and chamber pressure. Accordingly, the growth mechanism of high quality SCD was proposed. Qi Lang et al. reported a growth rate of 165 µm/h by doped nitrogen into the reaction chamber [6]. Nad et al. compared different sizes of pocket holders to control the substrate surface morphology under a chamber pressure of 240 torr. Silva et al. showed that increasing the process pressure can increase the diamond growth rate and improve the crystal quality [7,8]. F. Lloret studied the stratigraphy of SCD lateral growth on a three-dimensional structure and designed a geometric model for the growth sector configuration [9–11]. Among these, the reaction pressure plays an important role in the overall carbon deposition and homoepitaxy growth because it determines the shape, distribution, and available area of the activated plasma. Due to the limitation of the CVD system parameter, achieving high chamber pressure during the growth meets a particular set of problems and uncertainties [12]. Most research studies have reported that pressure was controlled in a range of 100–250 torr [13–15] during diamond crystal growth. Growth pressures higher than 250 torr are still rarely seen.

In this study, we focused on SCD growth under a reaction pressure of 300 torr. The surface morphology and growth rate were studied in order to optimize the growth rate to ensure the crystal quality.

2. Materials and Methods

2.1. Preparation of the CVD Diamond Substrate

High quality HPHT diamond crystal, which is 7 × 7 mm in size, was grown in a cubic press system by Jinan Diamond Technology Co., Ltd. The origin diamond was type I b with a nitrogen concentration of about $10^{19}/cm^3$, as measured by Secondary Ion Mass Spectroscopy (SIMS) from EAG laboratories (Shanghai China) [16]. After the HPHT growth, the diamond crystal was boiled in the mixed solution of perchloric acid, sulfuric acid, and hydrochloric acid for 3 h to remove the dirt covered on the crystal surface (All the above acids are from Aladdin, Shanghai, China). The HPHT diamond substrate plate was cut parallel to the (100) surface by the laser sawing system. The substrate was 5.97 mm × 5.57 mm in size, and the thickness was 0.61 mm. Mechanical polishing was applied to remove the surface cracks generated during the sawing process. Ultrasonic cleaning of the methanol and acetone separately for half an hour was used to remove the organic substances on the surface of the substrate.

2.2. The Etching of the CVD Single Crystal Substrate

To achieve a nitrogen-free growth environment during the entire process, the chamber was evacuated to 0.1 torr by dry pump before the supply of the hydrogen. Etching was followed for half an hour under 300 torr pressure and 900 °C in a pure hydrogen environment to remove the impurities and defects caused by the mechanical polishing.

2.3. Growth Parameter Investigation

After the etching, the CVD diamonds were grown on the substrates with different parameters including the temperature, pressure, and methane concentration. The relationship between the temperature, pressure, and microwave power was studied systematically to understand the system's plasma energy distribution and absorbance. All diamond substrates were laser cut to 0.5 mm thickness and grown in the reaction chamber with a methane concentration of 3%.

The surface morphology and growth rate were investigated at different methane concentrations under a pressure of 300 torr and 1150 °C with hydrogen flow at 600 sccm for 4 h of growth. The methane concentration was varied from 2% to 5% in order to get an acceptable surface morphology and growth rate to optimize the SCD growth.

2.4. The Growth of the CVD Diamond Layer

After that, the SCD layer was grown for 4 h after hydrogen etching of the substrate at 900 °C for half an hour. The chamber pressure was 300 torr, while the substrate temperature was kept at 1150 °C with a gas mixture of $CH_4:H_2$ to 18:600 sccm. During the CVD reaction, the thickness of the sample increased and approached the lower edge of the plasma. The substrate surface temperature increased, which changed the thermodynamics balance for the carbon deposition and hydrogen etching. A thermo pyrometer was used to measure the in situ substrate temperature. Since the sample thickness increases during the CVD reaction, the self-control recipe mode was activated during the growth period in order to self-adjust the input microwave power to maintain the substrate surface temperature through the feedback of the thermo pyrometer. The temperature variation range was less than 10 °C throughout the whole period of growth.

2.5. The Surface Morphology and Crystal Quality Analysis

A confocal laser scanning microscope OLS-4000 from Olympus (Tokyo, Japan) was employed to examine the top surface morphology of the diamond. Single crystal quality was assessed via HRXRD by D8 Discover operating with CuKα_1 radiation with an anode on a fixed power supply at 40 kV/40 mA from Bruker. The room temperature Raman spectra were obtained by the LabRAM HR800 system of Horiba Jobin Yvon (Paris, France) with a 514 nm solid laser as the excitation source to determine the phase structure.

3. Results and Discussion

3.1. CVD System for SCD Growth

The System type Ardis 300 is equipped with a 2.45 GHz/6 kW microwave reactor by Optosystems Ltd. The microwaves enter the reaction chamber through the quartz window from the corn shape guide beneath the sample stage. Figure 1 shows a schematic diagram of the structure of the Ardis 300 reaction chamber. The sample stage is made of copper with water channels inside to allow cooling during the reaction. A molybdenum plate is placed on the top of the stage and a specially designed closed type sample holder is located in the center for the substrates [8,17]. Quartz windows are implanted into the stainless steel chamber shell on the top and side wall for inspection. A double interference infrared radiation thermo pyrometer is installed to measure the temperature of the substrate with an emissivity of 0.1 through a slit of 2 mm. In order to get a real-time substrate temperature measurement during the deposition reaction, the system was managed to carve a guide channel hole throughout the encircled ring, which is parallel to the ring surface. The pressure of the reactor was able to reach a maximum of 300.0 torr with an accuracy of 0.1 torr.

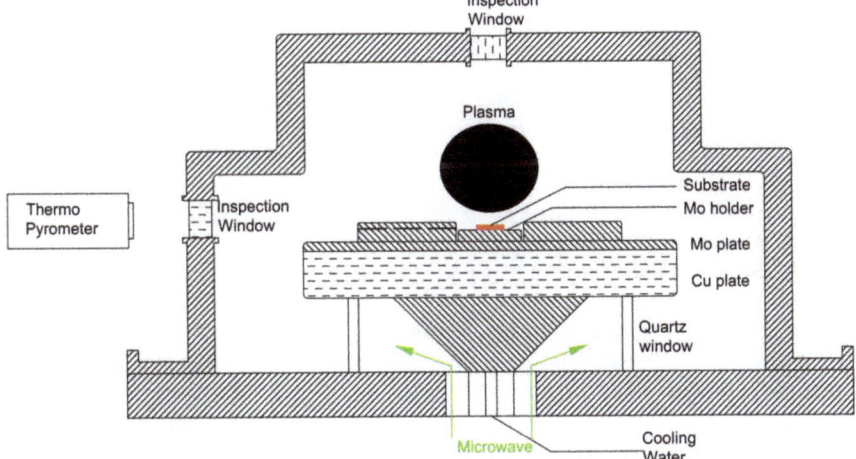

Figure 1. Schematic diagram of the Ardis 300 reaction chamber. The red rectangle on the Mo holder represents the High pressure High Temperature (HPHT) diamond substrate, and the green arrow stands for the direction and for the microwave transmission path.

3.2. The Influence of the Chamber Pressure on the Microwave Power and Substrate Temperature

The chamber pressure plays a significant role in determining the growth rate, morphology, and crystal quality prominently in terms of the homoepitaxy and lateral growth during the CVD reaction. In order to obtain optimized parameters for homoepitaxy growth, a stable thermal and electromagnetic field should be guaranteed by considering a series of deposition parameters such as microwave power, chamber pressure, and methane concentration. Figure 2a shows the relationship between microwave power and substrate temperature under different chamber pressures from 200 to 300 torr for Ardis 300. The results indicate that, under the same pressure of the reaction chamber, the substrate temperature increases linearly with the microwave power. The slopes under the different chamber pressures are calculated and fitted in Figure 2b. The slope of the curve increases from 0.134 (200 torr) to 0.234 (300 torr). The fitting curves follow the exponential function calculated as:

$$y = 0.13 + 6.2 \times 10^3 e^{\left[\frac{(x-197.87)}{36.64}\right]}. \tag{1}$$

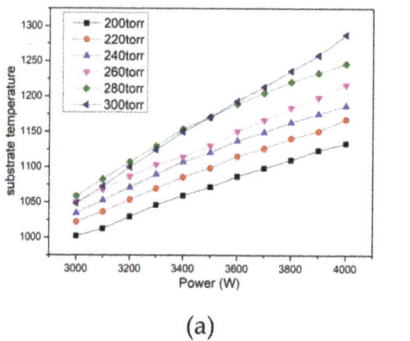
(a)

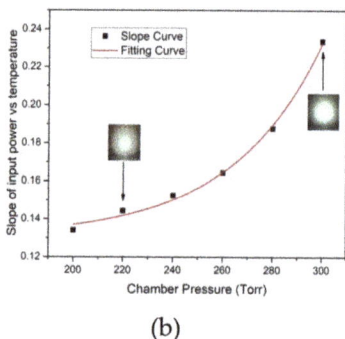
(b)

Figure 2. (a) Relationship of the microwave power to the substrate temperature during the CVD growth under different high chamber pressures. The chamber pressure varied from 200 to 300 torr. (b) Calculated slopes of the curve from Figure 2a at different chamber pressures with fitting. Two photos of the plasma were captured throughout the inspection window under 3800 kW, 1150 °C, 220 Torr, and 300 Torr, separately.

The entire CVD deposition reaction system is a strong self-adaption multi-field environment with different plasma particles generated by the microwave discharge in the chamber. The movement of all deposition-related ionized particles is violently affected by the constantly changing compound electromagnetic field provided by the reactor, which leads to different transmissions of the particles throughout the plasma generation to the deposition position on the diamond substrate. As the chamber pressure increases, the effective availability of the microwave increases to promote the methane and hydrogen to discharge, which causes a dynamic acceleration for both the carbon deposition and etching because of the different partial pressures [18]. The surface temperature difference of the substrate acts as the driving force for the material transformation in the thermodynamic routine. The present experimental result indicates a significant increase in the magnitude of the slope of microwave power over the substrate temperature since the reaction pressure increases. It reveals that there is enormous potential for plasma discharge absorption for the supplied power in an even higher pressure environment (over 200 torr).

The plasma photos in Figure 2b were captured under the same microwave power of 3800 W and temperature of 1150 °C. It shows that the plasma diminished in size and brightened in color from 220 to 300 torr separately. In this procedure, the plasma energy density increases. As a consequence, the charged concentration of the carbon-atom-related growth species such as CH_3 and C_2H_2 increases.

No clear point microwave discharge around the substrate and Molybdenum ring edge was found during the pressure change in the deposition reaction. The temperature gradient will be changed accordingly and will bring a different thermodynamic approach to the substrate homoepitaxy and the edge lateral overgrowth [19].

3.3. The Influence of the Methane Concentration on the Morphology and Growth Rate

As the only carbon atom provider, the methane concentration is particularly emphasized for the SCD growth with the CVD method. The optimal methane concentration leads to a polycrystalline-free layer surface with high quality and an acceptable growth rate. Besides the growth rate, another significant parameter, known as the edge boundary of the substrates, should be fully considered, which requires a controllable, sequential, and well organized epitaxial lateral overgrowth to exclude the growth of the polycrystalline. Several typical surface morphologies are exhibited in Figure 3 for methane concentrations from 2.0% to 5.0%. Harris and Goodwin [17,20] described the CVD diamond growth reaction using a simple model in which the growth rate depended on a metastable competition between the hydrogen etching and the methyl radical deposition. Figure 4 shows the relationship between the methane concentration and the growth rate, which were obtained at a pressure of 300 torr and 1150 °C and a hydrogen flow of 600 sccm throughout the entire experiment. An almost linear relationship for the methane concentration with the growth rate was present. When the methane concentration was 2%, which shows a recessive role in the competition, the substrate exhibited a multi-hole structure with plenty of etch pits on the diamond surface. The violent etching reaction restrained the appearance of the homoepitaxy step growth, which corresponded to a low growth rate of about 15.6 μm/h. When the methane concentration increased to 3%, the surface showed parallel and regular growth steps flow along the [100] direction, which indicates a well-controlled equilibrium of the carbon transportation to the substrate from the charge plasma and, meanwhile, indicates an organized arrangement of carbon atoms. As the methane concentration increased gradually, polycrystalline appeared and the different domains were present on the surface.

The surface was severely covered with the disordered polycrystalline scale clusters when the methane concentration increased to 5%, which proves that over-dosed methane led no approach to a compromised diamond quality, even though this sample achieved a high growth rate of 51 μm/h.

In order to obtain a higher quality homoepitaxy of a single crystal diamond layer with a reasonable growth rate, we decided to limit the methane concentration to 3% during the deposition growth.

Figure 3. Surface morphology evolution of the substrates grown under the environment of $CH_4:H_2$ at 2.0%, 3.0%, 4.0%, and 5.0%. The substrates were grown at 1150 °C and 300 torr. The images were captured by the confocal laser scanning mode of the confocal laser scanning microscope.

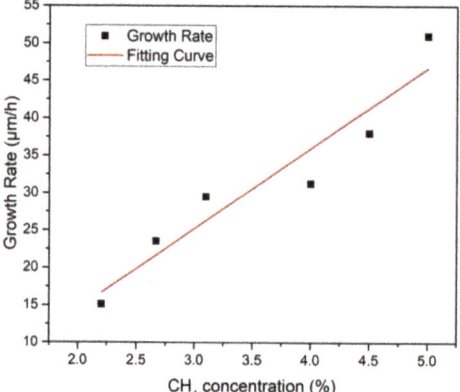

Figure 4. Relationship between the methane concentrations versus the deposition growth rate. The growth rate was calculated by the average measurement of the four corners and center after the growth and divided by the time. The red line is the fitting curve of the dot distribution.

3.4. The Homoepitaxy Growth of the Single Crystal Diamond

After determining the temperature, pressure, and the methane concentration, we tried to optimize the reaction parameter to achieve a high-quality CVD single crystal homoepitaxy diamond layer. The average growth rate was 27.12 μm/h, which almost matches the result we obtained from Figure 4.

Figure 5a is an image of the diamond surface taken by the confocal laser scanning microscope. Most of the surface area is covered with a regular step flow. Several large dark inclusions can be observed in the center of the sample, which represents the metal compound catalysts left in the crystal during the HPHT growth. The light yellow boundary shows the original HPHT diamond substrate edge beneath the as-grown sample. Lateral growth can be clearly observed by comparing the edge of the sample with the HPHT substrate boundary. Figure 5b–d show the photos captured in the laser scan mode. Two domains can be observed from Figure 5b,c in different positions of the sample. The center domain steps are along the [100] orientation, which can be easily observed on the surface of SCD homoepitaxy layers. The step width is about 10–16 μm. However, the steps on the edge domains go along the [110] orientation with a width of around 18–26 μm. Clear step merging along the [110] orientation can be observed on the edge lateral growth. After comparing the diameter of the as-grown sample with the original substrate before and after the growth, we confirmed that the [100] orientation domain on the edge is generated by the lateral growth. The right edges in Figure 5b show the generation of the polycrystalline. The polycrystalline can become an induction of the un-epitaxial features on the edges and corners and leads to the formation of asymmetric nuclei, which exhibit the pyramid structure revealed in the red block [21,22]. Shreya Nad et al. reported that the non-uniform surface temperature shall be generated at the sharp edges of the substrate during the CVD growth. This phenomenon may be caused by the difference in the plasma density in the corresponding ambient region between the substrate, holder surface, and the local heat transmission to the cooling water, which leads to the generation of higher index symmetric planes such as (110) and (111) in pyramid structures or polycrystalline [7].

Figure 5d shows the crack along the [100] direction, which may be caused by the metal inclusion. A. Ababou et al. reported that metal inclusions play a role in the CVD homoepitaxy growth by accelerating the formation of the sp^2 phases because of non-epitaxial growth. The dislocations brought by the metal inclusions will be propagated from the center of the substrate to the epitaxy layer throughout the entire growth process even if they are concealed inside. The stress will be relieved as the defects accumulate to generate enough lattice mismatch before breaking and cracking [23].

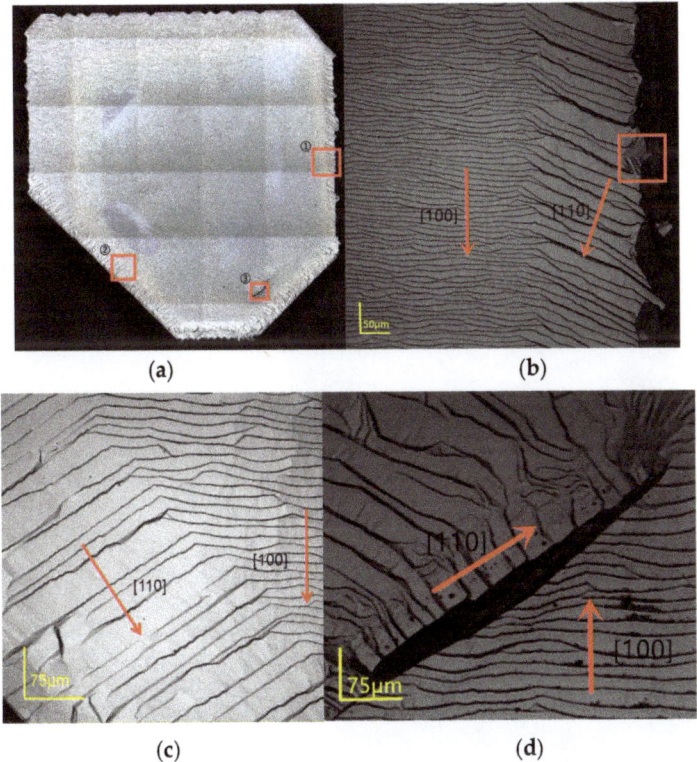

Figure 5. (a) Image of the entire diamond substrate surface after 4 h of growth at 300 torr/1150 °C, $CH_4:H_2$ = 16:600 sccm. Three detailed images are labeled as positions 1, 2, and 3. (b) The image of the sample of 50× amplification at position 1. The red box reveals the pyramid structure in the edge neighboring the polycrystalline. (c) The image of the 50× amplification at position 2. (d) The image of the 50× amplification at position 3 with a deep crack in the center. The red arrows show the direction of the step flow.

3.5. CVD Homoepitaxy Growth Diamond Quality

In order to understand the quality of the homoepitaxy layer and the lateral edge, we measured the rocking curves of the High Resolution X-Ray Diffractometry (HRXRD) and Raman spectra of the samples. Figure 6a shows the X-Ray Diffractometry (XRD) rocking curves of (400) reflection on different sections of the CVD substrate. The section positions are labeled from 1 to 7 corresponding to those points on the substrate. The distance between two measured points is 1 mm. Since the distance of the SCD layer in the labeled direction was about 6.13 mm, points 1 and 7 were on the lateral edges, and points 2 to 6 were on the homoepitaxy layer from the original substrate, as designed. The FWHM values of the rocking curve of points from 1 to 7 are shown in Figure 6b. The layer structural quality is similar to that of points 2 to 6. The FWHM values of the rocking curve for homoepitaxy layer were about 57–59″ while those of the edge section were 86.98″ and 70.98″ at points 1 and 7, respectively. The measurement was accomplished in a high-intensity configuration of the variable slit mode and indicates a high uniformity on the homoepitaxy growth and an imperfect crystalline quality on the lateral edges. Figure 7 shows the Raman spectra of the CVD layer at positions 1 and 4, which represents the structural quality of the homoepitaxy and lateral sections. Two spectra are dominated by a single sharp peak at 1332.43 cm^{-1}, which reveals that the internal stress did not increase by an order of magnitude. No peak related to the sp^2 carbon phase can be observed. The full width at the half maximum (FWHM) value of point 1 is somewhat larger than that of position 4, which indicates higher residual stress on the lateral edges than in the center. This matches the XRD rocking curve result. This means that the stress around position 4 is smaller than that of point 1. Both of the XRD rocking curves and Raman spectra prove that the sample layer shows a high crystal quality in both the center and edge parts of the sample, while the homoepitaxy region inherited from the HPHT substrate presents a high crystal quality with fewer defects and low internal stress compared with the edge overgrowth.

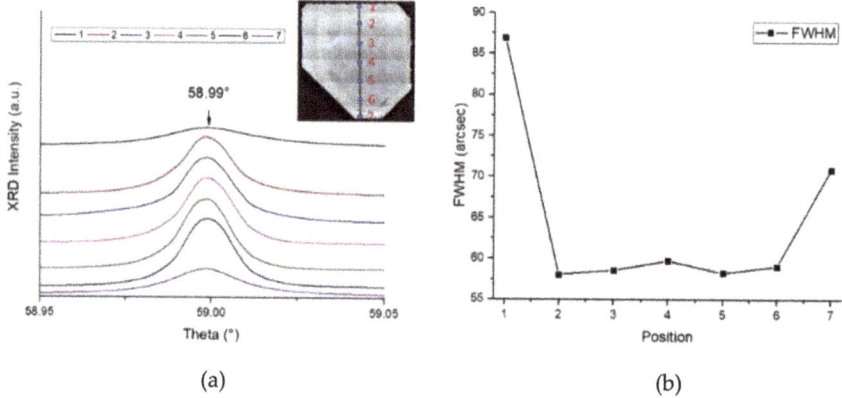

Figure 6. (a) HRXRD rocking curve of the (400) direction symmetric reflections of the location points from 1 to 7. The distance of each neighbor point was designed to be 1 mm. As a result, points 1 and 7 were located on the lateral growth area of the CVD layer. (b) The XRD FWHM of the location positions from points 1 to 7.

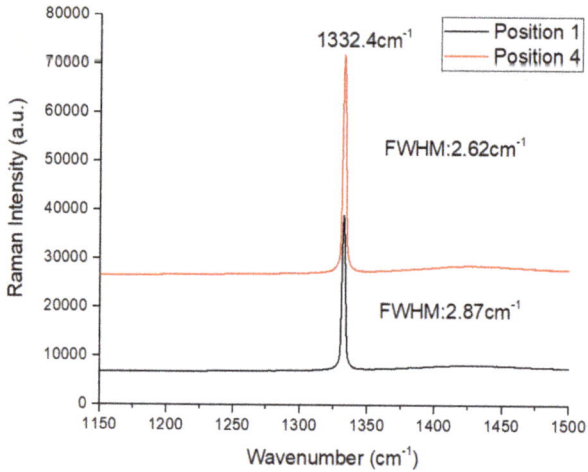

Figure 7. Raman spectrum of the CVD homoepitaxy layer at room temperature, the black and red lines correspond to location points 1 and 4, respectively.

3.6. Crystal Lattice Visual Structure Analysis of the Step Flow Rotation

In order to understand the growth mechanism of the homoepitaxy and lateral growth as well as the step flow rotation on the edges, diamond lattice structure analysis was used to describe what happened on the edges. Figure 8a shows the basic double unit cells of the carbon atom's cubic region at the center of the face during diamond phase formation. The blue atoms form the (100) unit face while the red ones are in (110) faces. After the deposition of the (100) surface, following the original step flow, the step flow rotates 45° in the [110] direction on the edge of the sample. Figure 8b simulates the step flow of the CVD sample edge to describe the appearance of the step rotation. The black atoms are on the sample surface linked by the blue lines, forming the [100] orientation step surface. The green surface represents the (110) surface. Alexandre Tallaire et al. reported this phenomenon generated from the (100) face to the (110) face on the lateral side, which is almost the same as our result of the transformation of the step flow. Their team explained that the face rotation should evolve due to the temperature difference between the center and the edge [13,24]. E.V. Bushuev et al. also considered the dependence of the diamond epitaxial growth rate on the substrate temperature. They predicted that the microwave power would play a minor role in the growth kinetics via the plasma distribution [25]. C.C. Battaile et al. reported the CH_3 and C_2H_2 groups are the main source for the growth of the (100) and (110) surfaces, respectively, during the CVD diamond deposition. Furthermore, acetylene is particularly important for aiding in the formation of multi-carbon clusters on the nuclei on the different orientation surfaces [26,27]. In sum, we predict that the step flow rotation of our sample was caused by the anomalous distribution of the C_2H_2 on the substrate surface. Different from the CH_3 guided nuclei, which dominated the primary surface for the crystal growth, for the C_2H_2 guiding mode, two carbon atoms were first combined to form a double bond bridge onto the ravine of the (110) surface before further deposition, which resulted in a 45° guide to the lattice formation. As a result, the generation of the C_2H_2 cluster needed extra time and energy to generate equivalent growth on the (110) surface. The thermodynamic of the growth of the center and edge of the sample were driven by the temperature gradient provided by the optimized microwave plasma. The higher temperature of the edge provided the required surface energy for the growth of the (110) surface. According to the Gibbs–Wulff principle, the (110) plane has a lower surface free energy than that of the (100) surface. Meanwhile, it has a lower growth rate as well [24]. As the time elapsed during the reaction, the (110) surface was found to have a lower surface energy level than that of the (100) surface on the edge and lateral growth regions. Additionally, the lower growth rate of the (110) surface led to the step merging at the rotation site from the [100] to [111] steps. Similar morphology can be observed in Figure 5d because of the poor thermo conductivity of the surface crack. The temperature on the crack edge was higher than that of the other regions. It may have also been caused by the plasma point charge.

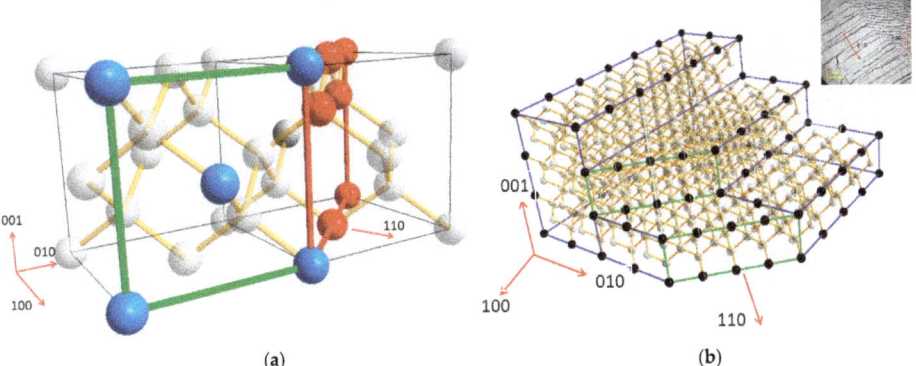

Figure 8. (a) Atom lattice visual structure of the diamond double unit molecule cell. The green lines are links of the carbon molecules on the (100) surface. The red lines are the (110) surface. (b) The lattice visual structure indicates the edge rotation transformation from the direction [100] to [110]. The black atoms are the carbon atoms on the surface of the substrate. The blue lines contact the atoms on the (100), (010), and (001) surfaces. The green lines link the carbon atoms and form the faces vertical to the [110] direction.

4. Conclusions

In this paper, we introduced an HPHT diamond substrate as the seed of the homoepitaxy growth in a CVD system under a high chamber pressure of 300 torr. The relationships among the input power, substrate temperature, and chamber pressure were first studied in order to understand the plasma behavior during the deposition reaction. A series of methane concentrations, growth rates, and surface morphologies were investigated to optimize the growth parameter. After that, a high quality SCD layer was achieved with a 27.12 μm/h growth rate under a 300 torr high pressure environment. The sample surface presented a regular uniform [100] orientation surface step flow in the original substrate area and a [110] orientation step in the lateral growth edges. Raman spectroscopy and the HRXRD rocking curve were used to investigate the internal stresses and crystal quality of both the homoepitaxy and lateral layers. The lattice visual structure model was employed to explain the step flow rotation in the lateral growth edge. Different from the CH_3 guided [100] orientation deposition model, C_2H_2 acted as the preponderant growth source of the [100] orientation step flow, which relies on higher energy provided by the higher temperature on the substrate edges.

Author Contributions: Conceptualization, X.W. and Y.P.; methodology, X.W. and P.D.; software, X.W. and Y.P.; validation, D.W., Y.P., X.H.; formal analysis, X.W. and P.D.; investigation, X.W. and Z.C.; resources, X.W., Y.P. and C.L.; data curation, X.W. and Y.P.; writing—original draft preparation, X.W.; writing—review and editing, Y.P. and X.H.; visualization, X.W.; supervision, X.H.; project administration, D.W. and C.L.; funding acquisition, D.W., C.L. and X.H.

Funding: This research was funded by The National Key R&D Program of China, grant number 2018YFB0406501, and the Key Basic Research Projects of Shandong Province Natural Science Foundation, grant number ZR2017ZA010.

Acknowledgments: We would like to thank Jinan Diamond Technology Co. Ltd for providing the HPHT diamond substrate that was used in the experiments.

Conflicts of Interest: The authors declare no conflict of interest.

References

1. Li, F.; Zhang, J.; Wang, X.; Lin, Z.; Wang, W.; Li, S.; Wang, H.X. X-ray photoelectron Spectrosc. study of Schottky junctions based on oxygen-/fluorine-terminated (100) diamond. *Diam. Relat. Mater.* **2016**, *63*, 180–185. [CrossRef]
2. Lu, H.-C.; Peng, Y.-C.; Lin, M.-Y.; Chou, S.-L.; Lo, J.-I.; Cheng, B.-M. Analysis of boron in diamond with UV photoluminescence. *Carbon* **2017**, *111*, 835–838. [CrossRef]
3. Zhang, X.M.; Wang, S.Y.; Shi, Y.B.; Yuan, H.; Ma, Z.M.; Wang, F.; Lin, Z.D.; Qin, L.; Liu, J. Quantitative analysis of spectral characteristics and concentration of ensembles of NV− centers in diamond. *Diam. Relat. Mater.* **2017**, *76*, 21–26. [CrossRef]
4. Shikata, S. Single crystal diamond wafers for high power electronics. *Diam. Relat. Mater.* **2016**, *65*, 168–175. [CrossRef]
5. Angus, J.C. Diamond synthesis by chemical vapor deposition: The early years. *Diam. Relat. Mater.* **2014**, *49*, 77–86. [CrossRef]
6. Liang, Q.; Chin, C.Y.; Lai, J.; Yan, C.S.; Meng, Y.; Mao, H.K.; Hemley, R.J. Enhanced growth of high quality single crystal diamond by microwave plasma assisted chemical vapor deposition at high gas pressures. *Appl. Phys. Lett.* **2009**, *94*, 024103. [CrossRef]
7. Silva, F.; Achard, J.; Brinza, O.; Bonnin, X.; Hassouni, K.; Anthonis, A.; De Corte, K.; Barjon, J. High quality, large surface area, homoepitaxial MPACVD diamond growth. *Diam. Relat. Mater.* **2009**, *18*, 683–697. [CrossRef]
8. Nad, S.; Gu, Y.; Asmussen, J. Growth strategies for large and high quality single crystal diamond substrates. *Diam. Relat. Mater.* **2015**, *60*, 26–34. [CrossRef]
9. Lloret, F.; Araújo, D.; Eon, D.; Bustarret, E. Three-Dimensional Diamond MPCVD Growth over MESA Structures: A Geometric Model for Growth Sector Configuration. *Cryst. Growth Des.* **2018**, *18*, 7628–7632. [CrossRef]

10. Lloret, F.; Araujo, D.; Eon, D.; Villar, M.d.; Gonzalez-Leal, J.-M.; Bustarret, E. Influence of methane concentration on MPCVD overgrowth of 100-oriented etched diamond substrates. *Phys. Status Solidi A* **2016**, *213*, 2570–2574. [CrossRef]
11. Lloret, F.; Fiori, A.; Araujo, D.; Eon, D.; Villar, M.P.; Bustarret, E. Stratigraphy of a diamond epitaxial three-dimensional overgrowth using doping superlattices. *Appl. Phys. Lett.* **2016**, *108*, 181901. [CrossRef]
12. Muehle, M.; Asmussen, J.; Becker, M.F.; Schuelke, T. Extending microwave plasma assisted CVD SCD growth to pressures of 400 Torr. *Diam. Relat. Mater.* **2017**, *79*, 150–163. [CrossRef]
13. Tallaire, A.; Achard, J.; Silva, F.; Sussmann, R.S.; Gicquel, A. Homoepitaxial deposition of high-quality thick diamond films: Effect of growth parameters. *Diam. Relat. Mater.* **2005**, *14*, 249–254. [CrossRef]
14. Mokuno, Y.; Chayahara, A.; Yamada, H.; Tsubouchi, N. Improving purity and size of single-crystal diamond plates produced by high-rate CVD growth and lift-off process using ion implantation. *Diam. Relat. Mater.* **2009**, *18*, 1258–1261. [CrossRef]
15. Chen, J.; Wang, G.; Qi, C.; Zhang, Y.; Zhang, S.; Xu, Y.; Hao, J.; Lai, Z.; Zheng, L. Morphological and structural evolution on the lateral face of the diamond seed by MPCVD homoepitaxial deposition. *J. Cryst. Growth* **2018**, *484*, 1–6. [CrossRef]
16. Xie, X.; Wang, X.; Peng, Y.; Cui, Y.; Chen, X.; Hu, X.; Xu, X.; Yu, P.; Wang, R. Synthesis and characterization of high quality {100} diamond single crystal. *J. Mater. Sci. Mater. Electron.* **2017**, *28*, 9813–9819. [CrossRef]
17. Charris, A.; Nad, S.; Asmussen, J. Exploring constant substrate temperature and constant high pressure SCD growth using variable pocket holder depths. *Diam. Relat. Mater.* **2017**, *76*, 58–67. [CrossRef]
18. Mokuno, Y.; Chayahara, A.; Soda, Y.; Yamada, H.; Horino, Y.; Fujimori, N. High rate homoepitaxial growth of diamond by microwave plasma CVD with nitrogen addition. *Diam. Relat. Mater.* **2006**, *15*, 455–459. [CrossRef]
19. Gu, Y.; Lu, J.; Grotjohn, T.; Schuelke, T.; Asmussen, J. Microwave plasma reactor design for high pressure and high power density diamond synthesis. *Diam. Relat. Mater.* **2012**, *24*, 210–214. [CrossRef]
20. Silva, F.; Hassouni, K.; Bonnin, X.; Gicquel, A. Microwave engineering of plasma-assisted CVD reactors for diamond deposition. *J Phys Condens Matter* **2009**, *21*, 364202. [CrossRef]
21. Achard, J.; Silva, F.; Brinza, O.; Tallaire, A.; Gicquel, A. Coupled effect of nitrogen addition and surface temperature on the morphology and the kinetics of thick CVD diamond single crystals. *Diam. Relat. Mater.* **2007**, *16*, 685–689. [CrossRef]
22. Bogdan, G.; Nesládek, M.; D'Haen, J.; Maes, J.; Moshchalkov, V.V.; Haenen, K.; D'Olieslaeger, M. Growth and characterization of near-atomically flat, thick homoepitaxial CVD diamond films. *Phys. Status Solidi A* **2005**, *202*, 2066–2072. [CrossRef]
23. Ababou, B.C.A.; Gortz, G. Surface characterization of microwave-assisted chemically vapour deposited carbon deposits on silicon and transition metal substrates. *Diam. Relat. Mater.* **1992**, *1*, 875–881. [CrossRef]
24. Tallaire, A.; Mille, V.; Brinza, O.; Thi, T.N.T.; Brom, J.M.; Loguinov, Y.; Katrusha, A.; Koliadin, A.; Achard, J. Thick CVD diamond films grown on high-quality type IIa HPHT diamond substrates from New Diamond Technology. *Diam. Relat. Mater.* **2017**, *77*, 146–152. [CrossRef]
25. Bushuev, E.V.; Yurov, V.Y.; Bolshakov, A.P.; Ralchenko, V.G.; Khomich, A.A.; Antonova, I.A.; Ashkinazi, E.E.; Shershulin, V.A.; Pashinin, V.P.; Konov, V.I. Express in situ measurement of epitaxial CVD diamond film growth kinetics. *Diam. Relat. Mater.* **2017**, *72*, 61–70. [CrossRef]
26. Battaile, C.C. Atomic scale simulations of chemical vapor deposition on flat and vicinal diamond substrates. *J. Cryst. Growth* **1998**, *194*, 353–368. [CrossRef]
27. Battaile, C.C. Morphologies of diamond films from atomic-scale simulations of Chemical Vapor Deposition. *J. Cryst. Growth* **1998**, *6*, 198–1206. [CrossRef]

© 2019 by the authors. Licensee MDPI, Basel, Switzerland. This article is an open access article distributed under the terms and conditions of the Creative Commons Attribution (CC BY) license (http://creativecommons.org/licenses/by/4.0/).

Article
Modeling of One-Side Surface Modifications of Graphene

Alexander V. Savin * and Yuriy A. Kosevich

N.N. Semenov Federal Research Center for Chemical Physics of Russian Academy of Sciences, 4 Kosygin str., 119991 Moscow, Russia; yukosevich@gmail.com
* Correspondence: asavin@center.chph.ras.ru

Received: 21 October 2019; Accepted: 9 December 2019; Published: 12 December 2019

Abstract: We model, with the use of the force field method, the dependence of mechanical conformations of graphene sheets, located on flat substrates, on the density of unilateral (one-side) attachment of hydrogen, fluorine or chlorine atoms to them. It is shown that a chemically-modified graphene sheet can take four main forms on a flat substrate: the form of a flat sheet located parallel to the surface of the substrate, the form of convex sheet partially detached from the substrate with bent edges adjacent to the substrate, and the form of a single and double roll on the substrate. On the surface of crystalline graphite, the flat form of the sheet is lowest in energy for hydrogenation density $p < 0.21$, fluorination density $p < 0.20$, and chlorination density $p < 0.16$. For higher attachment densities, the flat form of the graphene sheet becomes unstable. The surface of crystalline nickel has higher adsorption energy for graphene monolayer and the flat form of a chemically modified sheet on such a substrate is lowest in energy for hydrogenation density $p < 0.47$, fluorination density $p < 0.30$ and chlorination density $p < 0.21$.

Keywords: graphene; one-side modification; hydrogenation; fluorination; chlorination

1. Introduction

Recently, intensive studies have been performed of various derivatives of graphene (hexagonal monolayer of carbon atoms) [1–4], such as graphane CH and fluorographene CF (a monolayer of graphene, completely saturated on both sides with hydrogen or fluorine) [5–8], grafone C_2H (graphene monolayer saturated with hydrogen on one side) [9–12], one-side fluorinated graphene C_4F [13,14], chlorinated graphene C_4Cl [15]. Valence attachment of an external atom to a graphene sheet leads to the local convexity of the sheet as result of the appearance of the sp^3 hybridization at the joining point [16,17]. Therefore, if hydrogen atoms are attached on one side in the finite domain of the sheet, creating a local peace of graphone on the sheet, a characteristic convex deformation of the sheet occurs in this region [18]. If the hydrogen, fluorine or chlorine atoms are attached uniformly to one whole side of the sheet, the whole small sheet will take a convex shape while a large sheet will fold into a roll [19–22].

From the macroscopic point of view, the bending of a graphene sheet under the effect of one-side chemical modification is similar to the bending of a thin solid film caused by the difference of surface stresses on its sides. Such difference is created during the thin film growth on a substrate with lattice mismatch [23] or during the one-side epitaxial growth of a surfactant on the film [24,25]. This effect is used, for example, for the measurements with optical technique of the change in surface stress caused by the monolayer and sub-monolayer adsorption of a surfactant [26,27]. On the other hand, the equal surface stresses on both sides of elastic thin film cause the change of the film thickness inversely proportional to its thickness [28–31], which can also be used for the characterization of the modification of film surfaces state induced by surface treatment. Corrugated conformation of a graphene sheet can also be controlled by the patterned vacancy 'defects' in its hexagonal lattice [32].

The most convenient way to obtain graphene sheet modified on one side is to attach the sheet to a flat substrate and further chemically modify its outer surface. For getting a graphone sheet, the latter needs to be hydrogenated being attached to a substrate. Modeling of the hydrogenation of a graphene sheet [33,34] has shown that the substrate has a significant effect on this process, and it is very difficult to obtain a perfect graphone-like structure with the formula C_2H (one hydrogen atom per two carbon atoms). Hydrogenation leads to the formation of randomly distributed uncorrelated domains with average hydrogenation density of the sheet $p < 0.5$.

Important experimentally observable consequence of the coupling of two-dimensional atomic layer with elastic substrate is the appearance of the gapped resonance modes in the vibrational spectrum of the two-dimensional system of distributed oscillators on elastic substrate [35–40]. In Section 2 we use the value of the spectral gap in transverse oscillations of the monolayer on elastic substrate for the evaluation of the coupling strength of carbon, fluorine and chlorine atoms with nickel substrate.

In this paper, we model with the use of the force field method, the dependence of the mechanical conformations (formation of secondary structures) of graphene sheets, placed on flat substrates (flat surfaces of molecular crystals), on the density of one-side attachment of hydrogen, fluorine or chlorine atoms.

2. Model of Modified Graphene Sheet

In our modeling, we use the force field AMBER (Assisted Model Building with Energy Refinement) [41]. To model the chemically modified graphene sheet, we use the force field in which distinct potentials describe the deformation of valence bonds and of the valence, torsion and dihedral angles, and of non-valent atomic interactions [42]. In this model, the strain energy of the valence sp^2 and sp^3 C–C and C–CR bonds, and of O–H, C–R bonds (here an atom or group of atoms R = H, F) is described by the Morse potential:

$$V(\rho) = \epsilon_b \left[e^{-\alpha(\rho-\rho_0)} - 1 \right]^2, \tag{1}$$

where ρ and ρ_0 are the current and equilibrium bond lengths, ϵ_b is the binding energy, and the parameter α sets the bond stiffness $K = 2\epsilon_b \alpha^2$. The values of potential parameters for various valence bonds are presented in Table 1.

Table 1. Values of the Morse potential parameters (1) for different valence bonds X—Y (C and C' are carbon atoms involved in the formation of the sp^2 and sp^3 bonds).

X—Y	ϵ_b (eV)	ρ_0 (Å)	α (Å^{-1})
C—C	4.9632	1.418	1.7889
C—C'	4.0	1.522	1.65
C—H	4.28	1.08	1.8
C'—F	5.38	1.36	2.0
C'—Cl	3.40	1.761	2.0

Energy of the deformation of the valence angles X–Y–Z is described by the potential

$$U(\mathbf{u}_1, \mathbf{u}_2, \mathbf{u}_3) = U(\varphi) = \epsilon_a(\cos \varphi - \cos \varphi_0), \tag{2}$$

where the cosine of the valence angle is defined as $\cos \varphi = -(\mathbf{v}_1, \mathbf{v}_2)/|\mathbf{v}_1||\mathbf{v}_2|$, with vectors $\mathbf{v}_1 = \mathbf{u}_2 - \mathbf{u}_1$, $\mathbf{v}_2 = \mathbf{u}_3 - \mathbf{u}_2$, the vectors $\mathbf{u}_1, \mathbf{u}_2, \mathbf{u}_3$ specify the coordinates of the atoms forming the valence angle φ, φ_0 is the value of equilibrium valence angle. Values of potential parameters used for various equilibrium valence angles are presented in Table 2.

Table 2. Values of the parameters of the potential of the valence angle X–Y–Z (2) for different atoms (atom W = H, F, Cl).

X—Y—Z	ϵ_a (eV)	φ_0 (°)
C—C—C	1.3143	120.0
C—C'—C	1.3	109.5
C'—C'—C'	1.3	109.5
C—C—H	0.8	120.0
C—C'—W	1.0	109.5
C'—C'—H	1.0	109.5
H—C'—H	0.7	109.5

Deformations of the torsion and dihedral angles, in the formation of which edges carbon atoms with attached external atoms do not participate (torsion angles around sp^2 C–C bonds), are described by the potential:

$$W_1(\mathbf{u}_1, \mathbf{u}_2, \mathbf{u}_3, \mathbf{u}_4) = \epsilon_{t,1}(1 - z \cos \phi), \qquad (3)$$

where $\cos \phi = (\mathbf{v}_1, \mathbf{v}_2)/|\mathbf{v}_1||\mathbf{v}_2|$, with vectors $\mathbf{v}_1 = (\mathbf{u}_2 - \mathbf{u}_1) \times (\mathbf{u}_3 - \mathbf{u}_2)$, $\mathbf{v}_2 = (\mathbf{u}_3 - \mathbf{u}_2) \times (\mathbf{u}_3 - \mathbf{u}_4)$, the factor $z = 1$ for the dihedral angle (the equilibrium angle $\phi_0 = 0$) and $z = -1$ for the torsion angle (the equilibrium angle $\phi_0 = \pi$), the energy $\epsilon_{t,1} = 0.499$ eV (the vectors $\mathbf{u}_1,...,\mathbf{u}_4$ determine equilibrium positions of the atoms, which form the angle). More detailed description of the deformation of the torsion and dihedral angles is given in [43].

Deformations of the angles around sp^3 bonds C–C' are described by the potential:

$$W_2(\mathbf{u}_1, \mathbf{u}_2, \mathbf{u}_3, \mathbf{u}_4) = \epsilon_{t,2}(1 + \cos 3\phi), \qquad (4)$$

with energy $\epsilon_{t,2} = 0.03$ eV.

It is worth mentioning that the attachment of two hydrogen atoms on one side of the sheet to the carbon atoms bonded by valence bond is not energetically favorable [44]. Therefore, we will consider such attachment configurations on one side of the sheet of X atoms (X = H, F, Cl), in which if an X atom is attached to one carbon atom, the X atoms are not attached to the three neighboring carbon atoms. The valence bonds CX–CX, and the valence and torsion angles formed by these bonds, are absent in such structures. Therefore the corresponding potentials can be omitted.

The nonvalent van der Waals interactions of atoms are described by the Lennard-Jones potential

$$W_0(r) = 4\epsilon_0[(\sigma/r)^{12} - (\sigma/r)^6], \qquad (5)$$

where r is the distance between interacting atoms, ϵ_0 is the interaction energy (equilibrium bond length $r_0 = 2^{1/6}\sigma$). The used values of the potential parameters for different pairs of atoms are presented in Table 3. The values of the potential parameters for carbon atoms of the graphene layer are taken from [45], for the remaining atoms are taken from [46].

Table 3. Values of the parameters of the Lennard-Jones potential (5) for different pairs of interacting atoms X, Y.

X, Y	ϵ_0 (eV)	σ (Å^{-1})
C, C	0.002757	3.393
H, H	0.000681	2.471
C, H	0.001369	2.932
F, F	0.002645	3.118
C, F	0.002700	3.256
Cl, Cl	0.009843	3.516
C, Cl	0.005209	3.455

In the simulation, the polarization of the C–F and C–Cl valence bonds was taken into account. At the atoms forming these bonds, the charges $-q$, $+q$ were used from the PCFF force field (for the first bond, the charge $q = 0.25e$, for the second bond $q = 0.184e$). The interaction of two hydroxyl groups (hydrogen bond OH··· OH) was described with the use of the potentials from the PCFF force field.

We define the interaction of the sheet with a flat substrate using the potential $W_s(h)$, which describes the dependence of the atomic energy on its distance to the substrate plane h. For a flat surface of a molecular crystal, the energy of the interaction of an atom with a surface can be described with a good accuracy by the (k, l) Lennard-Jones potential [47]:

$$W_s(h) = \epsilon_s [k(h_0/h)^l - l(h_0/h)^k]/(l-k), \tag{6}$$

where $l > k$ is assumed for the exponents. The potential (6) has a minimum value $W_s(h_0) = -\epsilon_s$ (ϵ_s is the binding energy of an atom with the substrate). For a flat surface of crystalline graphite, the exponents in the potential are $l = 10$, $k = 3.75$. The binding energies are $\epsilon_s = 0.052, 0.0187, 0.0465$ and 0.1026 eV for the C, H, F, and Cl atoms, respectively, the corresponding equilibrium distances are $h_0 = 3.37, 2.92, 3.24$, and 3.435 Å.

When graphene is located on the surface of crystalline nickel, a stronger chemical interaction of carbon atoms with the atoms of the substrate occurs. Therefore, the interaction of the carbon atom in the graphene sheet with the (111) surface of the Ni crystal is more convenient to describe by the Morse potential:

$$W_s(h) = \epsilon_s \{\exp[-\beta(h-h_0)] - 1\}^2 - \epsilon_s. \tag{7}$$

For a carbon atom, the interaction energy with the nickel surface is $\epsilon_s = 0.133$ eV [48] and the equilibrium distance to the substrate plane is $h_0 = 2.135$ Å [49].

In result of the interaction of a graphene sheet with a crystal surface, a gap of the magnitude $\omega_0 = 240$ cm^{-1} appears at the bottom of the frequency spectrum of transverse oscillations of the sheet [40]. From this we can estimate the harmonic coupling parameter of the interaction of the sheet atom with the substrate $K_0 = \omega_0^2 M = 41$ N/m (M is a mass of carbon atom), see also [35], as well as the value of the parameter $\beta = \sqrt{K_0/2\epsilon_s} = 3.1$ Å^{-1}. For the fluorine atom we obtain $\epsilon_s = 0.13$ eV, $h_0 = 1.655$ Å, $\beta = 3.75$ Å^{-1}, for the chlorine atom we obtain $\epsilon_s = 0.299$ eV, $h_0 = 2.115$ Å, $\beta = 3.17$ Å^{-1}.

In the following we will consider only these two substrate potentials. The first potential describes the weak interaction of the sheet with the substrate while the second potential describes the strong interaction. Other commonly-used substrates (surfaces of crystalline silicon Si, silicon dioxide SiO$_2$, silicon carbide 6H-SiC, hexagonal boron nitride h-BN and gold Au) are characterized by the intermediate values of the coupling parameters—see Table 4.

Table 4. Values of the parameters of the substrate potential $W_s(h)$ for different flat substrates (harmonic coupling parameter $K_0 = W_s''(h_0)$, frequency gap $\omega_0 = \sqrt{K_0/2M}$).

	ϵ_s (eV)	h_0 (Å^{-1})	l	k	K_0 (N/m)	ω_0 (cm^{-1})
SiO$_2$	0.037	4.13	16	3.75	2.1	54.5
graphite	0.052	3.37	10	3.75	2.8	62.4
Si (100)	0.061	3.85	14	3.75	3.4	69.8
6H-SiC	0.073	4.19	17	3.75	4.2	77.5
Au (111)	0.073	2.96	10	3.5	4.7	81.4
h-BN	0.090	3.46	10	3.75	4.5	80.1
Ni (111)	0.133	2.14	-	-	41.0	240

3. Stationary Structures of Square Sheet

To find the stationary state of the modified graphene sheet, it is necessary to find the minimum of the potential energy

$$E \to \min : \{\mathbf{u}_n\}_{n=1}^N, \tag{8}$$

where N is the total number of atoms on the sheet, $\mathbf{u}_n$ is a three-dimensional vector defining position of the nth atom, E is a total potential energy of the molecular system (given by the sum of all interaction potentials of atoms in the system). The minimization problem (8) will be solved numerically by the conjugate gradient method. Choosing the starting point of the minimization procedure, one can obtain all the main stationary states of the modified sheet bonded with a flat substrate.

Consider a square graphene sheet of size 8.47×8.37 nm^2, consisting of $N_c = 2798$ carbon atoms. The sheet has $N_b = 148$ edge atoms. To simplify the model, we assume that only one hydrogen atom is always attached to each edge carbon atom, see Figure 1a. With maximum one-side hydrogenation of the sheet, $N_m = 1324$ hydrogen atoms can be attached (for every two internal carbon atoms there is one hydrogen atom). Thus, the graphene sheet under consideration can be described by the formula $C_{N_c}H_{N_b} = C_{2798}H_{148}$, and the corresponding graphone sheet by the formula $C_{N_c}H_{N_b+N_m} = C_{2798}H_{1472}$. If $0 \leq N_h \leq N_m$ of hydrogen atoms are one-side attached to the graphene carbon atoms, the dimensionless concentration of attached hydrogen atoms (hydrogenation density) is $p = N_h/(N_c - N_b) \in [0, 0.5]$.

The size of the graphene sheet was chosen on the basis of the possibilities of computer modeling. For the larger sheet, we can reach higher precision in determining the maximal possible density of the one-side hydrogenation (fluorination or chlorination). On the other hand, the increase in the sheet size results in considerable increase in computation time. The used sheet size, 8.47×8.37 nm^2, provides a compromise between the computation time and the resulting precision. The use of a sheet with half the area, with the size 6.02×5.81 nm^2 which includes $N_c = 1398$ carbon atoms, gives practically the same results.

For the modeling of random hydrogenation (fluorination, chlorination) of a graphene sheet, first we consider the perfect graphone sheet, namely the graphene sheet with N_m hydrogen (fluorine, chlorine) atoms attached to its outer surface. Then we randomly remove N_0 atoms and get a sheet with dimensionless hydrogenation density $p = (N_m - N_0)/(N_c - N_b)$. After that, having solved the problem for the minimum of the energy (8), we find possible stationary structures of the modified sheet. Each structure (atomic packaging) will be characterized by the specific energy $E_c = E/N_c$. To evaluate this energy, the removal of N_0 atoms will be carried out by 128 independent random ways. This allows you to find the average value and standard deviation of the specific energy over 128 independent random implementations of the hydrogenation of the sheet with a fixed attachment density p.

A free graphone sheet can have two main structures: the single roll and the double roll [11,22]. Two more stable structures are possible on a flat substrate: a flat form of a sheet placed parallel to the surface of the substrate and a convex form of the sheet partially torn off the substrate with folded edges attached to the substrate. The characteristic appearance of these four stable structures of a sheet on flat substrate is shown in Figure 1. For the flat form, the sheet is located parallel to the substrate plane and hydrogen atoms are randomly attached to its outer side. The structure of a single roll has the form of densely-packed roll (scroll) of a sheet with an external hydrogenated side, lying on a flat substrate. Double roll structure is realized by folding of the sheet into two scrolls simultaneously from two opposite edges.

Dependencies of the normalized energy of the sheet E_c on the density of its hydrogenation p for four main structures of a sheet located on the flat surface of a graphite crystal and on (111) surface of nickel crystal are shown in Figure 2. On the surface of graphite, pure graphene sheet (hydrogenation density $p = 0$) has only one stable flat structure. Sustainable roll structure can exist only for the hydrogenation density $p > 0.018$, a stable double roll structure can exist only for $p > 0.075$, and partially convex structure can exist only for $p > 0.26$. For lower hydrogenation density (for $p < 0.018, 0.075, 0.26$), these structures become unstable on the surface of graphite and the graphene takes the form of a flat sheet.

The flat sheet form remains stable for $p \in [0, 0.49]$. With full hydrogenation (for $p = 0.5$), the flat form becomes unstable. The flat structure is most energetically favorable only for hydrogenation densities $p \in [0, 0.21]$, single roll structure—for $p \in [0.21, 0.41]$, and the double roll structure—for

$p \in [0.41, 0.5]$. Therefore, when a graphene sheet is located on a flat surface of crystalline graphite, it is hardly possible to achieve its one-sided hydrogenation with density $p > 0.21$ because of the folding of the sheet into a roll.

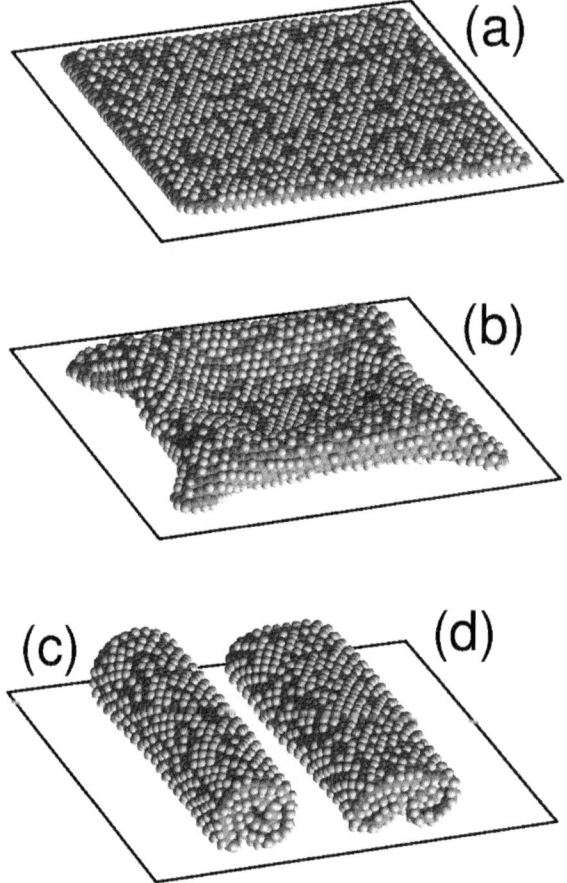

Figure 1. View of a square graphene sheet of size 8.47×8.37 nm^2 (the number of carbon atoms N_c =2798), placed on a flat surface of a graphite crystal with attachment density of hydrogen atoms to graphene outer surface $p = 0.3019$ (the number of hydrogen atoms attached to the surface is $N_h = 800$) in (**a**) planar structure parallel to the substrate, (**b**) convex structure with edges attached to the substrate, and in structures of (**c**) single roll and (**d**) double roll. Dark (light) beads show carbon (hydrogen) atoms.

The surface of crystalline nickel has a higher energy of the interaction with graphene sheet. Therefore, the flat form of the graphene sheet remains stable in this case for any density of hydrogenation of its outer side (for $p \in [0, 0.5]$). The flat structure is energetically favorable for $p < 0.47$. Roll folding of the sheet will be stable only for hydrogenation density $p > 0.038$, and for $p \in [0.47, 0.5]$ it becomes the most energetically favorable, see Figure 2b. Double roll form can exist for any hydrogenation density $p \in [0, 0.5]$, but it will always be energetically disadvantageous. The partially convex sheet structure is always the most energetically disadvantageous, and it is stable for hydrogenation density $p > 0.32$. Therefore, the location of the graphene sheet on a flat surface of nickel crystal substrate allows to achieve its hydrogenation with the density close to maximal possible.

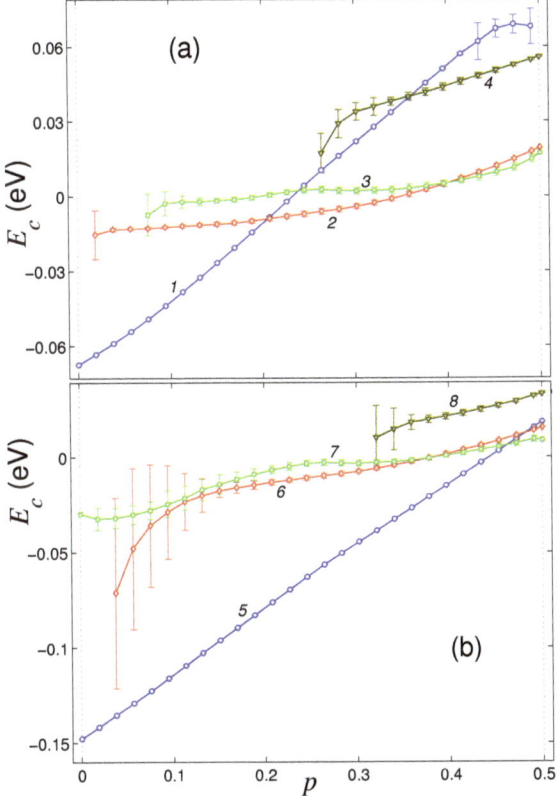

Figure 2. The dependence of the normalized energy $E_c = E/N_c$ of a square graphene sheet of size 8.47×8.37 nm^2 on the dimensionless hydrogenation density p for a sheet located on a flat surface of (a) graphite crystal and (b) (111) surface of nickel crystal. Curves 1, 5 give the dependencies for the flat structure, curves 2, 6 and 3, 7—for the single and double roll structures, curves 4, 8—for partially convex structure with the edges attached to the substrate.

Four similar stable structures of the graphene sheet are obtained by its unilateral fluorination (attaching fluorine atoms to the outer surface of the sheet with density $p \in [0, 0.5]$). The characteristic appearance of these four stable structures of fluorinated graphene sheet on flat substrate is shown in Figure 3. Dependencies of the normalized energy of the sheet E_c on the density of its fluorination p for four main structures of the sheet located on the flat surface of graphite crystal and on (111) surface of the nickel crystal are shown in Figure 4. Stable roll structures can exist on these substrates only for fluorination density $p > 0.018$.

On the surface of crystalline graphite, the planar structure retains its stability for fluorination density $p \in [0, 0.42]$. With higher fluorination density, the flat form becomes unstable. The partially convex shape of the sheet is stable only for $p \in [0.3, 0.5]$. The flat structure is the most energetically favorable only for fluorination density $p \in [0, 0.20]$, the single roll structure – for density $0.20 < p < 0.35$, and the double roll structure – for density $0.35 < p \leq 0.5$, see Figure 4a. Therefore, when a graphene sheet is located on a flat surface of crystalline graphite, it is impossible to achieve its one-side fluorination with density $p > 0.20$ (this will be prevented by the folding of the sheet into a roll).

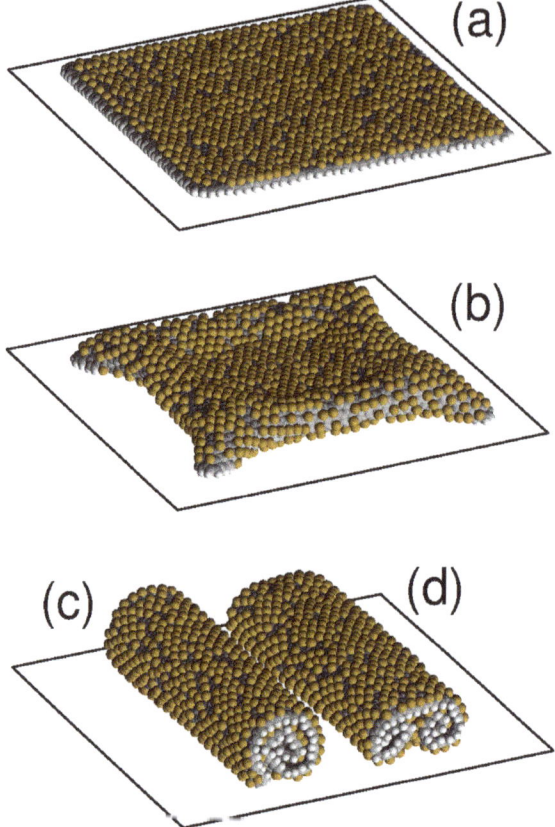

Figure 3. View of a square graphene sheet located on flat surface of nickel crystal with density of fluorine atoms attached to its outer side $p = 0.3396$ (the number of fluorine atoms is $N_h = 900$) in (**a**) planar structure parallel to the substrate, (**b**) convex structure with the edges attached to the substrate, and in structures of (**c**) single roll and (**d**) double roll. Gray/white/yellow beads show carbon/hydrogen/fluorine atoms.

On the surface of crystalline nickel, the flat structure of the sheet remains stable for fluorination density $p \in [0, 0.5]$, and partially convex form remains stable only for $0.30 < p \leq 0.5$. Here, the flat shape is the most energetically favorable for fluorination density $p \in [0, 0.30]$, and the double roll structure—for $0.30 < p \leq 0.5$, see Figure 4b. Therefore by placing graphene sheet on a flat surface of a nickel crystal, the higher density of one-side fluorination, $p < 0.30$, can be achieved. The decrease of the maximal fluorination density with respect to the maximally achievable hydrogenation density is related with the larger size of the fluoride atom.

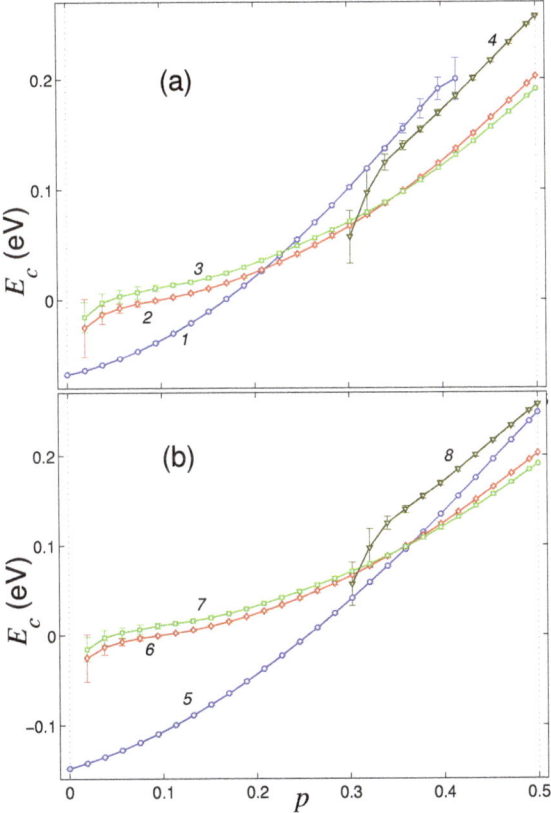

Figure 4. Dependence of the normalized energy E_c of a square graphene sheet on the dimensionless fluorination density p for a sheet located on (**a**) flat surface of graphite crystal and (**b**) (111) surface of nickel crystal. Curves 1, 5 give the dependencies for the flat structure, curves 2, 6 and 3, 7—for the single and double roll structures, curves 4, 8—for partially convex structure with the edges attached to the substrate.

Chlorine atoms are larger than the fluorine and hydrogen atoms, which should hinder the chlorination of graphene. Characteristic view of four stable structures of the chlorinated graphene sheet on a flat substrate shown in Figure 5. Dependencies of the normalized sheet energy E_c on the chlorination density p for four main structures of a sheet located on the flat surface of a graphite crystal and on (111) the surface of the crystal nickel are shown in Figure 6. On these substrates, a stable single-roll structures can exist only for chlorination density $p \geq 0.019$.

On the surface of crystalline graphite, the planar structure retains its stability for $p \in [0, 0.25]$. With higher chlorination densities, the flat form becomes unstable. The double-roll structure is stable only for $p \geq 0.076$, and the partially convex shape of the sheet is stable for $p > 0.13$. The flat structure is the most energy-efficient only for chlorination densities $p \in [0, 0.16]$, a single-roll structure—for $0.16 < p < 0.285$, and the double-roll structure—for $p > 0.285$, see Figure 6a. Therefore, when a graphene sheet is located on a flat surface of crystalline graphite, only the one-side chlorination with the density $p < 0.16$ can be achieved.

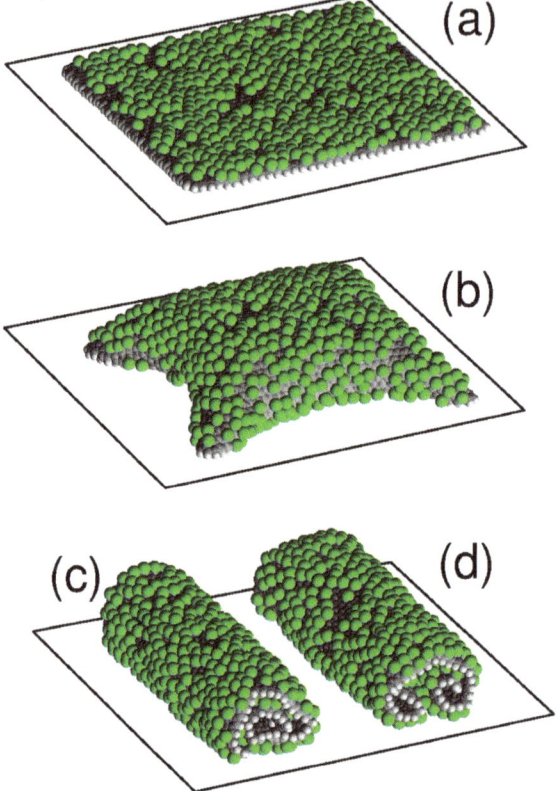

Figure 5. View of square graphene sheet located on a flat surface of nickel crystal at attachment density of chlorine atoms to its outer side $p = 0.2075$ (the number of chlorine atoms is $N_h = 550$) in (**a**) planar structure parallel to the substrate, (**b**) convex structure with the edges adjacent to the substrate, and in structures of (**c**) single roll and (**d**) double roll. Gray/white/green beads show carbon/hydrogen/chlorine atoms.

On the surface of crystalline nickel, the flat structure of the sheet remains stable for $p \in [0, 0.264]$. For the higher chlorination, the flat form becomes unstable. The double-roll structure is stable only for $p \geq 0.094$, and the partially convex shape of the sheet is stable for $p > 0.15$. The flat structure is the most energy-efficient for chlorination densities $p \in [0, 0.21]$, the single-roll structure—for $0.21 < p < 0.264$, and the double-roll structure—for $p > 0.264$, see Figure 6b. Therefore for the graphene sheet placed on a flat surface of nickel crystal, the one-side chlorination can only be achieved for densities $p < 0.21$.

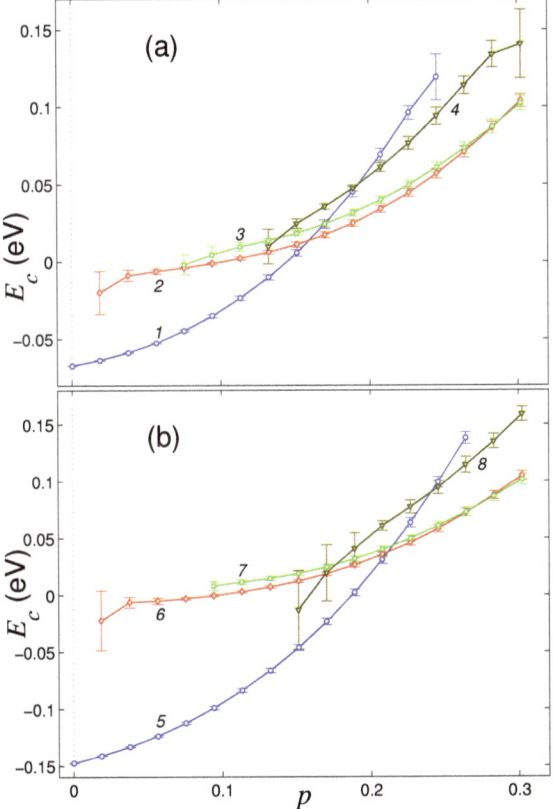

Figure 6. Dependence of the normalized energy E_c of a square graphene sheet on the dimensionless chlorination density p for a sheet located on (**a**) flat surface of graphite crystal and (**b**) (111) surface of nickel crystal. Curves 1, 5 give the dependencies for the flat structure, curves 2, 6 and 3, 7—for the single and double roll structures, curves 4, 8—for partially convex structure with the edges attached to the substrate.

4. Conclusions

We have performed numerical modeling, based on the force field method, of mechanical conformations of graphene sheets placed on flat substrates, caused by the change in the density of one-side attachment of hydrogen, fluorine or chlorine atoms to the sheet. It is shown that the chemically modified graphene sheet can take four main forms on the flat substrate: the form of a flat sheet parallel to the surface of the substrate, the form of a sheet with convex shape partially detached from the substrate with bent edges attached to the substrate, and the form of single or double rolls on the substrate. On the surface of crystalline graphite, the flat sheet form is the most favorable in energy for hydrogenation density $p < 0.21$, fluorination density $p < 0.20$ and chlorination density $p < 0.16$. The surface of crystalline nickel has higher energy of graphene adsorption, here the flat form of chemically modified sheet is the lowest in energy for hydrogenation density $p < 0.47$, fluorination density $p < 0.30$ and chlorination density $p < 0.21$. Our modeling and quantitative estimates of the attachment densities can serve for the determination of the maximal possible one-side chemical modifications of the flat graphene sheets. At higher attachment densities, the rolling of the sheet will prevent it from further chemical modification.

Author Contributions: All authors have read and agree to the published version of the manuscript. Design of the research, A.V.S. and Y.A.K.; simulations, A.V.S.; writing—original draft preparation, A.V.S. and Y.A.K.; writing—review and editing, A.V.S. and Y.A.K.

Funding: This research was funded by the Russian Science Foundation (award No. 16-13-10302).

Acknowledgments: This research was carried out using supercomputers at the Joint Supercomputer Center of the Russian Academy of Sciences (JSCC RAS).

Conflicts of Interest: The authors declare no conflict of interest.

References

1. Novoselov, K.S.; Geim, A.K.; Morozov, S.V.; Jiang, D.; Zhang, Y.; Dubonos, S.V.; Grigorieva, I.V.; Firsov, A.A. Electric Field Effect in Atomically Thin Carbon Films. *Science* **2004**, *306*, 666. [CrossRef] [PubMed]
2. Geim, A.K.; Novoselov, K.S. The rise of graphene. *Nat. Mater.* **2007**, *6*, 183. [CrossRef] [PubMed]
3. Soldano, C.; Mahmood, A.; Dujardin, E. Production, properties and potential of graphene. *Carbon* **2010**, *48*, 2127. [CrossRef]
4. Peng, Q.; Dearden, A.K.; Crean, J.; Han, L.; Liu, S.; Wen, X.; De, S. New materials graphyne, graphdiyne, graphone, and graphane: Review of properties, synthesis, and application in nanotechnology. *Nanotechnol. Sci. Appl.* **2014**, *7*, 1. [CrossRef] [PubMed]
5. Sofo, J.O.; Chaudhari, A.S.; Barber, G.D. Graphane: A two-dimensional hydrocarbon. *Phys. Rev. B* **2007**, *75*, 153401. [CrossRef]
6. Elias, D.C.; Nair, R.R.; Mohiuddin, T.M.G.; Morozov, S.V.; Blake, P.; Halsall, M.P.; Ferrari, A.C.; Boukhvalov, D.W.; Katsnelson, M.I.; Geim, A.K.; et al. Control of Graphene's Properties by Reversible Hydrogenation: Evidence for Graphane. *Science* **2009**, *323*, 610. [CrossRef]
7. Nair, R.R.; Ren, W.; Jalil, R.; Riaz, I.; Kravets, V.G.; Britnell, L.; Blake, P.; Schedin, F.; Mayorov, A.S.; Yuan, S.; et al. Fluorographene: A two-dimensional counterpart of Teflon. *Small* **2010**, *6*, 2877–2884. [CrossRef]
8. Leenaerts, O.; Peelaers, H.; Hernández-Nieves, A.D.; Partoens, B.; Peeters, F.M. First-principles investigation of graphene fluoride and graphane. *Phys. Rev. B* **2010**, *82*, 195436. [CrossRef]
9. Zhou, J.; Wang, Q.; Sun, Q.; Chen, X.C.; Kawazoe, Y.; Jena, P. Ferromagnetism in Semihydrogenated Graphene Sheet. *Nano Lett.* **2009**, *9*, 3867. [CrossRef]
10. Balog, R.; Jorgensen, B.; Nilsson, L.; Andersen, M.; Rienks, E.; Bianchi, M.; Fanetti, M.; Lagsgaard, E.; Baraldi, A.; Lizzit, S.; et al. Bandgap opening in graphene induced by patterned hydrogen adsorption. *Nat. Mater.* **2010**, *9*, 315. [CrossRef]
11. Neek-Amal, M.; Beheshtian, J.; Shayeganfar, F.; Singh, S.K.; Los, J.H.; Peeters, F.M. Spiral graphone and one-sided fluorographene nanoribbons. *Phys. Rev. B* **2013**, *87*, 075448. [CrossRef]
12. Zhao, W.; Gebhardt, J.; Späth, F.; Gotterbarm, K.; Gleichweit, C.; Steinrück, H.-P.; Görling, A.; Papp, C. Reversible Hydrogenation of Graphene on Ni(111)—Synthesis of "Graphone". *Chem. Eur. J.* **2015**, *21*, 3347. [CrossRef] [PubMed]
13. Robinson, J.T.; Burgess, J.S.; Junkermier, C.E.; Badescu, S.C.; Reinecke, T.L.; Perkins, F.K.; Zalatudinov, M.K.; Baldwin, J.W.; Culbertson, J.C.; Sheehan, P.E.; et al. Properties of Fluorinated Graphene Films. *Nano Lett.* **2010**, *10*, 3001–3005. [CrossRef] [PubMed]
14. Enyashin, A.N.; Ivanovskii, A.L. Layers and tubes of fluorographene C_4F: Stability, structural and electronic properties from DFTB calculations. *Chem. Phys. Lett.* **2013**, *576*, 44–48. [CrossRef]
15. Sahin, H.; Ciraci, S. Chlorine Adsorption on Graphene: Chlorographene. *J. Phys. Chem. C* **2012**, *116*, 24075–24083. [CrossRef]
16. Ruffieux, P.; Gröning, O.; Bielmann, M.; Mauron, P.; Schlapbach, L.; Gröning, P. Hydrogen adsorption on sp^2-bonded carbon: Influence of the local curvature. *Phys. Rev. B* **2002**, *66*, 245416. [CrossRef]
17. Pei, Q.X.; Zhang, Y.W.; Shenoy, V.B. A molecular dynamics study of the mechanical properties of hydrogen functionalized graphene. *Carbon* **2010**, *48*, 898. [CrossRef]
18. Reddy, C.D.; Zhang, Y.W.; Shenoy, V.B. Patterned graphene—A novel template for molecular packing. *Nanotechnology* **2012**, *23*, 165303. [CrossRef]
19. Liu, Z.; Xue, Q.; Xing, W.; Dub, Y.; Han, Z. Self-assembly of C_4H-type hydrogenated graphene. *Nanoscale* **2013**, *5*, 11132. [CrossRef]

20. Zhu, S.; Li, T. Hydrogenation enabled scrolling of graphene. *J. Phys. D Appl. Phys.* **2013**, *46*, 075301. [CrossRef]
21. Liu, Z.; Xue, Q.; Tao, Y.; Li, X.; Wu, T.; Jin, Y.; Zhang, Z. Carbon nanoscroll from C4H/C4F-type graphene superlattice: MD and MM simulation insights. *Phys. Chem. Chem. Phys.* **2015**, *17*, 3441. [CrossRef] [PubMed]
22. Savin, A.V.; Sakovich, R.A.; Mazo, M.A. Using spiral chain models for study of nanoscroll structures. *Phys. Rev. B* **2018**, *97*, 165436. [CrossRef]
23. Stoney, G.G. The tension of metallic films deposited by electrolysis. *Proc. R. Soc. Lond. Ser. A* **1909**, *82*, 172. [CrossRef]
24. Martinez, R.E.; Augustyniak, W.M.; Golovchenko, J.A. Direct Measurement of Crystal Surface Stress. *Phys. Rev. Lett.* **1990**, *64*, 1035. [CrossRef]
25. Schell-Sorokin, A.J.; Tromp, R.M. Mechanical Stresses in (Sub) monolayer Epitaxial Films. *Phys. Rev. Lett.* **1990**, *64*, 1039. [CrossRef]
26. Schell-Sorokin, A.J.; Tromp, R.M. Measurement of surface stress during epitaxial growth of Ge on arsenic terminated Si(001). *Surf. Sci.* **1994**, *319*, 110. [CrossRef]
27. Zahl, P.; Kury, P.; von Hoegen, M.H. Interplay of surface morphology, strain relief, and surface stress during surfactant mediated epitaxy of Ge on Si. *Appl. Phys. A* **1999**, *69*, 481. [CrossRef]
28. Kosevich, Y.A.; Kosevich, A.M. On the possibility of measuring the tensor of surface stress in thin crystalline plates. *Solid State Commun.* **1989**, *70*, 541. [CrossRef]
29. Cammarata, R.C.; Sieradzki, K. Effects of Surface Stress on the Elastic Moduli of Thin Films and Superlattices. *Phys. Rev. Lett.* **1989**, *62*, 2005. [CrossRef]
30. Cammarata, R.C. Surface and interface stress in thin films. *Progr. Surf. Sci.* **1994**, *46*, 1. [CrossRef]
31. Kosevich, Y.A. Capillary phenomena and macroscopic dynamics of complex two-dimensional defects in crystals. *Progr. Surf. Sci.* **1997**, *55*, 1. [CrossRef]
32. Grima, J.N.; Grech, M.C.; Grima-Cornish, J.N.; Gatt, R.; Attard, D. Giant Auxetic Behaviour in Engineered Graphene. *Ann. Phys.* **2018**, *530*, 1700330. [CrossRef]
33. Woellner, C.F.; Autreto, P.A.S.; Galvao, D.S. Graphone (one-side hydrogenated graphene) formation on different substrates. *arXiv* **2016**, arXiv:1606.09235.
34. Woellner, C.F.; Autreto, P.A.S.; Galvao, D.S. One Side-Graphene Hydrogenation (Graphone): Substrate Effects. *MRS Adv.* **2016**, *1*, 1429. [CrossRef]
35. Kosevich, Y.A.; Syrkin, E.S. Long wavelength surface oscillations of a crystal with an adsorbed monolayer. *Phys. Lett. A* **1989**, *135*, 298. [CrossRef]
36. Garova, E.A.; Maradudin, A.A.; Mayer, A.P. Interaction of Rayleigh waves with randomly distributed oscillators on the surface. *Phys. Rev. B* **1999**, *59*, 13291. [CrossRef]
37. Maznev, A.A.; Gusev, V.E. Waveguiding by a locally resonant metasurface. *Phys. Rev. B* **2015**, *92*, 115422. [CrossRef]
38. Boechler, N.; Eliason, J.K.; Kumar, A.; Maznev, A.A.; Nelson, K.A.; Fang, N. Interaction of a contact resonance of microspheres with surface acoustic waves. *Phys. Rev. Lett.* **2013**, *111*, 036103. [CrossRef]
39. Beltramo, P.J.; Schneider, D.; Fytas, G.; Furst, E.M. Anisotropic hypersonic phonon propagation in films of aligned ellipsoids. *Phys. Rev. Lett.* **2014**, *113*, 205503. [CrossRef]
40. Dahal, A.; Batzill, M. Graphene–nickel interfaces: A review. *Nanoscale* **2014**, *6*, 2548. [CrossRef]
41. Cornell, W.D.; Cieplak, P.; Bayly, C.I.; Gould, I.R.; Merz, K.M.; Ferguson, D.M.; Spellmeyer, D.C.; Fox, T.; Caldwell, J.W.; Kollman, P.A. A second generation force field for the simulation of proteins, nucleic acids, and organic molecules. *J. Am. Chem. Soc.* **1995**, *117*, 5179. [CrossRef]
42. Savin, A.V.; Mazo, M.A. Simulation of Scrolled Packings of Graphone Nanoribbons. *Phys. Solid State* **2017**, *59*, 1260. [CrossRef]
43. Savin, A.V.; Kivshar, Y.S.; Hu, B. Suppression of thermal conductivity in graphene nanoribbons with rough edges. *Phys. Rev. B* **2010**, *82*, 195422. [CrossRef]
44. Casolo, S.; Lovvik, O.M.; Martinazzo, R.; Tantardini, G.F. Understanding adsorption of hydrogen atoms on graphene. *J. Chem. Phys.* **2009**, *130*, 054704. [CrossRef] [PubMed]
45. Setton, R. Carbon nanotubes—II. Cohesion and formation energy of cylindrical nanotubes. *Carbon* **1996**, *34*, 69–75. [CrossRef]
46. Rappé, A.K.; Casewit, C.J.; Colwell, K.S.; Goddard, W.A., III; Skiff, W.M. UFF, a full periodic table force field for molecular mechanics and molecular dynamics simulations. *J. Am. Chem. Soc.* **1992**, *114*, 10024–10035.

47. Savin, A.V.; Savina, O.I. Bistability of Multiwalled Carbon Nanotubes Arranged on Plane Substrates. *Phys. Solid State* **2019**, *61*, 2241–2248. [CrossRef]
48. Lahiri, J.; Miller, T.S.; Ross, A.J.; Adamska, L.; Oleynik, I.I.; Batzill, M. Graphene growth and stability at nickel surfaces. *New J. Phys.* **2011**, *13*, 025001. [CrossRef]
49. Gamo, Y.; Nagashima, A.; Wakabayashi, M.; Terai, M.; Oshima, C. Atomic structure of monolayer graphite formed on Ni(111). *Surf. Sci.* **1997**, *374*, 61–64. [CrossRef]

© 2019 by the authors. Licensee MDPI, Basel, Switzerland. This article is an open access article distributed under the terms and conditions of the Creative Commons Attribution (CC BY) license (http://creativecommons.org/licenses/by/4.0/).

Article

Surface Morphology of the Interface Junction of CVD Mosaic Single-Crystal Diamond

Xiwei Wang [1], Peng Duan [1], Zhenzhong Cao [2], Changjiang Liu [2], Dufu Wang [1,2], Yan Peng [1,*], Xiangang Xu [1,*] and Xiaobo Hu [1,*]

- [1] State Key Laboratory of Crystal Materials, Shandong University, Jinan 250100, China; xwang21@126.com (X.W.); 13256998266@163.com (P.D.); wangdufu@163.com (D.W.)
- [2] Jinan Diamond Technology Co. Ltd., Jinan 250101, China; m13608927651@163.com (Z.C.); owlchj@163.com (C.L.)
- * Correspondence: pengyan@sdu.edu.cn (Y.P.); xxu@sdu.edu.cn (X.X.); xbhu@sdu.edu.cn (X.H.)

Received: 20 November 2019; Accepted: 19 December 2019; Published: 23 December 2019

Abstract: The diamond mosaic grown on the single-crystal diamond substrates by the microwave plasma chemical vapor deposition (MPCVD) method has been studied. The average growth rate was about 16–17 µm/h during 48 hours' growth. The surface morphologies of the as-grown diamond layer were observed. It was found that the step flow was able to move across the substrates and cover the junction interface. Raman spectroscopic mapping in the central area of the junction revealed the high stress region movement across the junction interface from one substrate to the other for about 200–400 µm. High-resolution X-ray diffractometry (HRXRD) results proved that the surface step flow movement direction had nothing to do with the off-axis directions of the original substrates. It was found that the surface height difference of substrate was the main driving force for the step flow movement, junction combination and surface morphology changing. The mechanism of the mosaic interface junction combination and step flow transformation on the mosaic surface was proposed.

Keywords: CVD mosaic; single crystal diamond; surface morphology

1. Background

The development of diamond growth technology makes diamond a promising material not only in superhard industry, optical components, semiconductor and high power electronics but even also a promising application in the quantum information and computing field [1–6]. To realize the actual application of diamond, many researchers focus on the growth of the single-crystal diamond with large size and high quality by the chemical vapor deposition (CVD) method. Due to the low lateral growth rate during the CVD reaction, one can rarely achieve a large single-crystal diamond with small substrates. Although 10 mm × 10 mm single-crystal diamond substrates are commercially available, the size of the CVD single-crystal diamond is still not competitive with other wide gap semiconductor materials, such as SiC and GaN. Furthermore, the quality problem and expensive price of the large diamond substrate also makes it difficult for the further application [7,8].

One of the breakthroughs regarding the size limitation is called "mosaic" technology. It was first proposed by Geis et al. in 1990s and the diamond was grown by a hot-filament CVD system. However, due to the limitation of the system ability, the early mosaic wafers had poor crystal quality with pin holes at the corner and junction boundaries. Yamada et al. created the "Clone" technology to prepare free-standing diamond wafers in the MPCVD equipment and achieved a 2-inch mosaic plate later, but the mosaic boundaries were easily identified by naked eyes and interface crystal quality remained low [9]. Muchnikov et al. studied mosaic CVD diamond crystal growth and paid attention to the crystallographic orientation inherits [10]. No specific correlation between the misorientation of the diamond seed and the substrates junction boundary morphology was found. Shu et al. mapped the stress and defect distributions in the cross-section of the mosaic junctions with Raman spectroscopy, measured and calculated the maximum stress value and corresponding position distance away from the interface [11]. However, few studies focus on the relationship between the surface morphology variation and the internal stress of the junction. Meanwhile, a lack of an evaluation standard for the crystal quality of the junction also increase the difficulty in studying the crystal growth behavior among the mosaic plates. As we know, a uniform step flow generation and movement on the interface junction are quite important to achieve a high quality mosaic connection with low internal stress and little dislocations.

In this paper, we fabricated a CVD single crystal mosaic layer on four seed pieces of high temperature high pressure (HTHP) single crystal substrates. The substrate surface was roughly parallel to (100) plane with a 2–4 °C misorientation. After the CVD growth for 24 and 48 hours, the mosaic top surface morphology was observed by optical microscopy. Internal stress and crystal quality were assessed by Raman spectroscopy and high resolution X-ray diffractometry.

2. Experiment

2.1. The Preparation of the Diamond Substrates

High-quality IIa type HTHP single crystal diamonds were grown in the cubic press system with a solvent of metal Fe-Ni-Al system in 0.9 GPa by Jinan Diamond Technology Co., Ltd. (Jinan, China) After HTHP growth, the diamond crystals were sawed parallel to the (100) surface with around 3° miscut angle (Please check the text below for the detail angles) by a laser-cutting system [12]. The side faces are polished parallel to (010) or (001) plane which is perpendicular to the growth surface. Mechanical polishing was employed to remove the amorphous and graphite carbon, and the remaining thickness was about 0.3 mm (Table 1). Eventually, the surface roughness average (Ra) reaches to 2 nm confirmed by Atomic Force Microscope (Veeco, Plainview, NY, USA). Before the CVD growth, the substrates were cleaned in a mixture of the sulfuric acid, nitric acid and perchloric acid for three hours. Then they were placed into the acetone and methanol for ultra-sonic cleaning to remove the organic material from the surfaces.

2.2. Mosaic Growth of the Chemical Vapor Deposition (CVD) Diamond

The CVD mosaic growth was operated in Arids-300 type MPCVD system (Optosystems Co. Ltd., Moscow, Russia) equipped with a 6 kW/2.45 GHz microwave reactor. In order to increase charged plasma density and the crystal growth rate, closed type substrate pocket holder was used throughout all the reaction. Four pieces of HTHP diamond substrates of 5 mm × 5 mm were placed in the center of the substrate pocket holder next to each other in a 2 × 2 matrix in order to achieve a 10 mm × 10 mm size mosaic plate. Hydrogen etching was operated before the growth to remove the surface impurity and mechanical polishing scratching in 900 °C/300 torr for 30 min. The CVD reaction was performed at 1100 °C/300 torr with a gas mixture of CH_4/H_2 at 16/600 sccm flow after the hydrogen etching.

2.3. Mosaic Growth Analysis

In the experiment, sample thicknesses were measured before and after 24 and 48 hours' growth respectively to calculate the growth rate. The top surface morphologies were captured by the confocal laser scanning microscope with the type of Olympus OLS4000 (Tokyo, Japan). Raman mapping for the sample surface was measured by the LabRAM HR800 Horiba Jobin Yvon (Kyoto, Japan) with 532 nm solid laser system at room temperature. The mapping region was located in the center of the sample covering the interface junction and vicinity homoepitaxy area from all the mosaic substrates. Substrate off-axis angle directions and crystal quality were investigated by a high-resolution X-ray diffractometer (HRXRD) by D8 Discover from Bruker (Karlsruhe, Germany). The system was operated at 40 KV and 40 mA with $CuK_{\alpha 1}$ radiation. The X-ray was generated as a scattered beam and was aligned parallel

2.4. Experiment Result

Homoepitaxy Growth and Morphology of the CVD Mosaic Layer

After the CVD mosaic growth, the sample did not show any crack on the junction edge and the entire wafer appeared like a solid plate. The size of the mosaic diamond increased to 10.95 mm × 10.94 mm and 11.75 mm × 11.75 mm after 24 and 48 hours' growth due to the lateral growth. The Table 1 is the substrates thickness and calculated growth rates of the mosaic after 24 and 48 hours' growth. The substrate identification number is labeled in the clockwise direction as shown in Figure 1a. The lateral overgrowth rates were calculated to be 16.98 µm/h and 16.25 µm/h while the vertical growth rates of all substrates are about 16–17 µm/h.

Table 1. Thickness and calculated growth rate of the mosaic sample before and after chemical vapor deposition (CVD) growth.

Thickness (mm)	Substrate 1	Substrate 2	Substrate 3	Substrate 4
Before Growth	0.27	0.28	0.32	0.27
Growth 24 h	0.65	0.7	0.72	0.69
Growth 48 h	1.07	1.09	1.13	1.09
Growth Rate (µm/h)	**Substrate 1**	**Substrate 2**	**Substrate 3**	**Substrate 4**
Growth 24 h	16.18	17.71	17.61	16.77
Growth 48 h	17.50	16.25	17.08	16.67

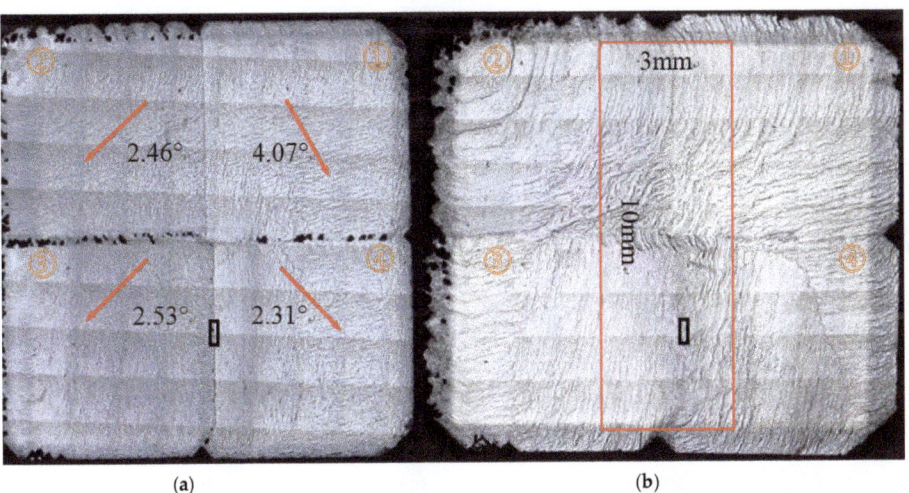

Figure 1. (a,b) The confocal laser-scanning images of the entire as grown mosaic sample after growth of 24 and 48 hours respectively, the substrate identification numbers are labeled as 1 to 4 in the clockwise direction. The red arrows in Figure 1a corresponded to the off-axis directions of the (100) planes of the original substrate 1 to 4 with the misorientation angles. The red block in Figure 1b shows the region with a diameter of 3 mm × 10 mm area for the Raman mapping measurement, the red lines correspond to the height scanning routine below.

Figure 1 shows the confocal laser scanning images of the entire as-grown mosaic sample after 24 and 48 hours' growth. A clear brighter rim area can be observed near the edge of the mosaic sample in Figure 1b, which refers to the crystal lateral growth region over the original diamond substrates. After the CVD growth, most of the area was covered by regular step flow on the top surface. Small polycrystalline clusters with diameter ranges from 30 μm to 150 μm appeared after 24 hours' growth on the center junction edges among the substrates, however, they disappeared after growth of 48 hours as shown in Figure 1b. Same phenomena can be observed for the polycrystalline on the lateral growth rim region.

The off-axis direction angles of the (100) faces were also measured by the HRXRD and shown in Figure 1a. The arrows on the four substrates corresponded to the (100) crystallographic axis off directions. Thus, the substrate surfaces were not strictly parallel to the (100) planes instead of having different off-axis angles and directions. However, the entire mosaic surface morphology showed anisotropy step flow domains along the orientations to the substrates off-axis direction after the CVD growth.

Figure 2a,b are the images of mosaic interface area between substrate 3 and 4 after the growth of 24 and 48 h with the height scanning images in the lower parts. The junction assumed a "Notch" shape and appeared on the substrate 3 edge. Its length is about 43 μm along the substrate edge. An obvious gap could be found between the two substrates and the height difference was about 9.045 μm. There was a narrow area with the width of 30 μm located in the substrate 4 near the junction area showing a step kink with the rotation angle for about 35° to 50°, which probably corresponded to the lateral growth region of the substrate 1 in the center. After another 24 hours' growth, the entire junction region was covered by the (100) step with the same step flow direction as shown in Figure 2a,b provides a clear image proving that the gap in the center disappeared. In this figure, the rotation step flow can also be seen clearly. The height scanning image attached also revealed a uniform continuous curve across the junction. The junction "Notch" on the former image transformed to a disturbance of the step flow in the red block. The step flow exhibited a "zigzag" pattern. In contrast, the step flow in other area showed a fluent and uniform pattern.

Figure 3 is the comparison of the surface morphologies from substrate 4 after the growth for 24 and 48 h. One can easily see the transformation of surface step flow directions and domains. The blue lines separate two surface domains marked with red arrows presenting the step flow moving directions. While, the yellow arrows correspond to the off-axis directions of the diamond seeds. Both step flow directions in the two domains are different from the off-axis directions of the original seeds, i.e., the new step flow moving direction has nothing to do with the off-axis direction of the original seed in the procedure of growth. However, the primary step flow almost paralleled to the interface between substrates 3 and 4 "invaded" into the secondary step flow region generated from the original seed from the left to the right.

Figure 4a,b are the images of the height distribution of the cross junction covering four substrate regions after the 24 and 48 hours' growth respectively. The images are obtained by the laser height scanning microscope in 1279 μm × 1279 μm area. From Figure 4a, an obvious height gap appeared between the edges of substrate 3 and 4 which matched the result of Figure 2a. However, the height difference of the substrate 1 and 2 junction was difficult to determine. The height order of the four substrates in the interface junction ranks as 3 > 4 > 2 > 1. The entire junction area was covered with the step flow along the same direction from the left to the right. The step terrace width ranges from 9 to 12 μm after 24 hours' growth. However, the junction was covered with the step rotation in the interface of the edge, the step bunching happened while the terrace width broadened to 36–55 μm after growth of 48 hours.

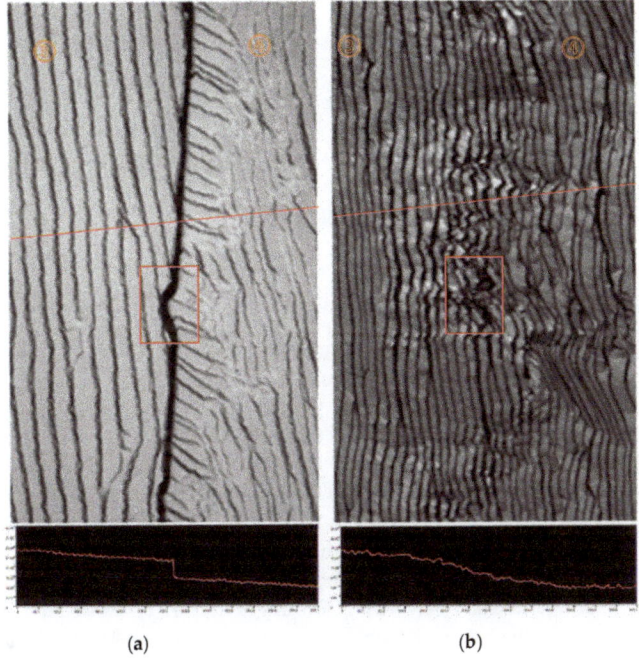

Figure 2. (a,b) are the images of the junction between the substrates 3 and 4 after the growth of 24 and 48 h from the black block of the Figure 1a,b respectively. The red lines across are the surface height scanning routine and height distributions are attached below.

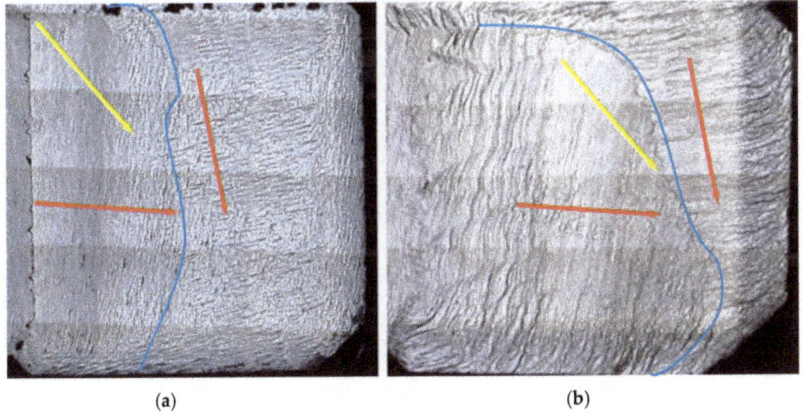

Figure 3. (a,b) are the surface step flow transformation of substrate 4 after the growth of 24 and 48 hours' growth. The blue lines mark the boundary of the two surface domains with different step flow directions labeled by red arrows, while the yellow arrows are the (100) off direction of substrate 4.

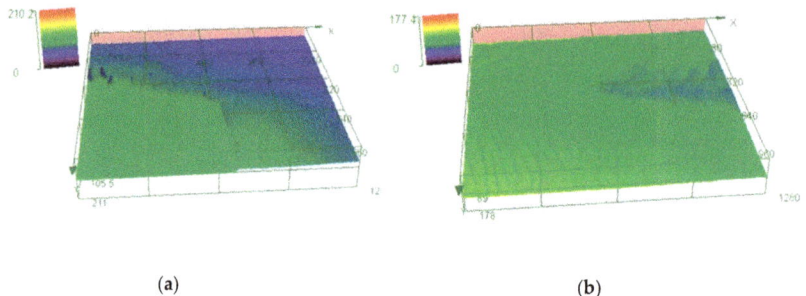

Figure 4. (a,b) are the height distribution images of the junction center of the substrates after the growth of 24 and 48 hours captured by confocal laser scanning.

In order to investigate the internal stress and interior defect distribution of the mosaic sample, Raman mapping was performed on the top surface of the center area for the sample after 48 hours' growth. The mapping region is 10 mm × 3 mm and covers the cross junction of the homoepitaxy layer from the four substrates, as shown in the Figure 1b. The mapping measurement was performed with the scanning step of 50 μm and 200 μm respectively in the X and Y axes, while the integral time lasted 1.2 second for each point. Figure 5a is the full width at half maximum (FWHM) distribution image of the Raman sp^3 diamond carbon phase whose unstressed Raman peak is about 1332.5 cm^{-1}. The result was fitted by Lorentz simulation and exhibited color ranges from 1.760–7.960 cm^{-1}. A clear narrow "cross" shape region to a width of about 400 μm was revealed in the center with a larger FWHM than that in the vicinity area, corresponding to the crystal quality distribution on interface junction and the layer grown from the original substrates respectively. The FWHM of Raman peak at the central junction illustrated indirectly that the crystal layer created by the lateral growth in the center was single crystal with acceptable quality. The joint boundary between substrates 3 and 4 moved about 200–400 μm towards substrate 4 because of the step flow movement, which was confirmed by the surface morphology in Figure 2. The FWHM of the Raman peak of the homoepitaxy layer grown from original substrates was smaller and approximate to 3 cm^{-1}, which represented the good quality of the crystal layer grown on substrates. The FWHM of the junction area ranges from 3.000 to 8.000 cm^{-1} in the center, only several small areas in the interface showed FWHM larger than 4 cm^{-1}. Figure 5b shows the Raman peak shift of diamond sp^3 peak of the entire mapping area. The peak shift from 1332.37 to 1331.49 cm^{-1}, i.e., the peak in all area shifts to the lower wave number in junction area compared with the unstressed diamond peak of 1332.5 cm^{-1}. The peak shift mapping distribution also present a "cross" pattern similar with the FWHM distribution which shows a higher wave number shift from the center to the periphery area. Steep discoloration can be observed in several areas on the central junction and the peak shift changed for about 1–2 cm^{-1}. Same spots appeared on the largest peak shift area between the substrate 3 and 4 moves to the direction of substrate 4 for about 200–400 μm which match the FWHM distribution image in the Figure 5a.

Figure 5c shows the internal stress distribution of the scanning area. The internal stress was calculated by the following formula according to the result of Figure 5b:

$$P = 0.34 \frac{GPa}{cm^{-1}} \times (v - v_0)$$

where v_0 refers to the theoretical value of the unstressed diamond layer Raman peak, normally is 1332.5 cm^{-1}. v is the measured position of the Raman peak [13].

According to the calculated result of Figure 5c, the internal stress of the mapping region ranged from −0.356 to −0.056 GPa. Therefore, the whole region was filled with a tensile stress. The central area between substrate 1 and 2 shows the strong stress fluctuation, meaning the release of the accumulated tensile stress. The largest difference value of the stress is about 0.295 GPa from the red block in Figure 5c.

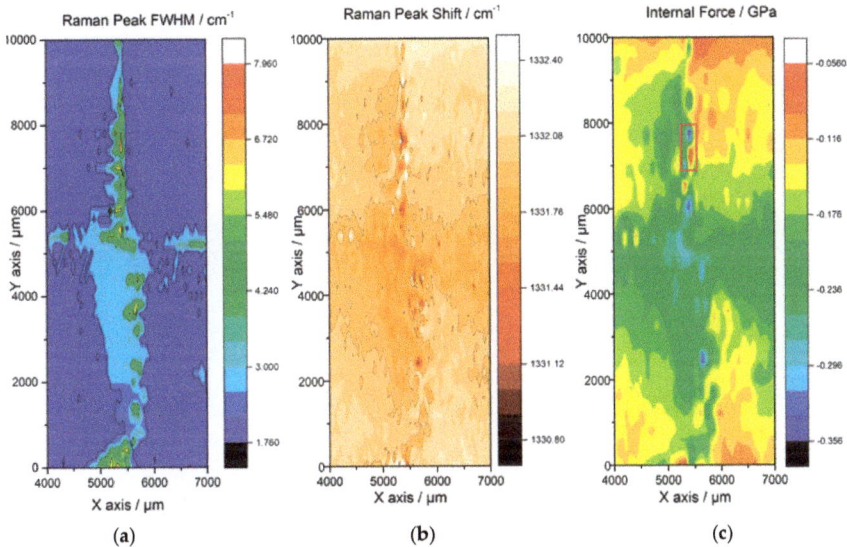

Figure 5. The Raman mapping area of the mosaic sample, the scanning region covered the junction interface edge in the center for the 3 mm × 10 mm region. (**a**) The Raman full width at half maximum (FWHM) of the mapping region, the FWHM ranges from 3.000–8.000 cm^{-1} and verified from the color bright red to deep blue. (**b**) The Raman sp^3 diamond peak shift of the mapping region, ranging from 1332.37 to 1331.45 cm^{-1}. (**c**) The calculation result of the internal stress distribution of the mapping region based on the result of (**b**).

In order to understand the evolution of the surface morphology and assess the layer crystal quality, HRXRD was used to measure the rocking curves of (100) reflection in the different positions of the mosaic sample. The FWHM of rocking curve reflects homoepitaxy layer quality. Figure 6 shows the rocking curves of the (400) reflection measured from the mosaic layer grown from four substrates. All the measuring results show a single narrow rocking curve peak and the position of the angle ω varies from 55.43° to 57.19° which corresponded to a 2.31° to 4.07° misorientation generated from the cutting and polishing. The FWHM are in the range of 60–90 arcsec, which reveals that the homoepitaxy layer has good quality. However, substrate 2 and 3 possessed the smaller misorientation angles than substrate 1 and 4, the step flows on substrate 2 and 3 showed the most aggression for the movement of the interface in the center area after the CVD growth.

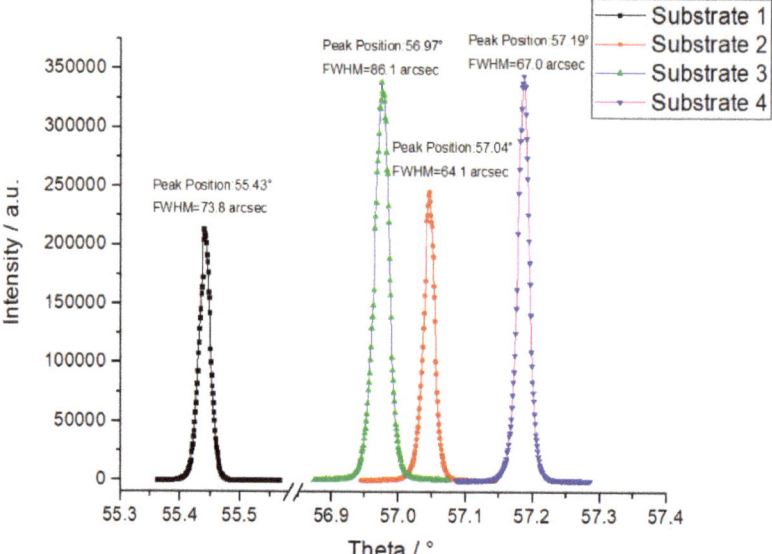

Figure 6. The high-resolution X-ray diffractometry (HRXRD) rocking curves of the (400) reflections from four positions on the mosaic homoepitaxy layer grown on the different substrates, the peak positions reflected the misorientation of the original substrates.

3. Result Discussion

The above diamond homoepitaxy layers are different from those grown by the "clone" method. The key of the "clone" technology is to apply the exact one single diamond plate to create numbers of repetitive substrates with same crystallographic morphology on the top surface for the mosaic growth to avoid crack of the junction boundary. However, unlike the method for "clone" technology, the off-axis directions of substrates in our experiment were not controlled to the same crystallographic direction before the growth since they may be not from one HTHP diamond. By contrast with the report from Hideaki Yamada et al., we did not see the similar morphology and crack on the mosaic junction area, since the angle between two off-axis directions for the substrate 3 and 4 was about to be 90°. The off-axis direction from the original substrate did not affect the surface step flow direction for the CVD mosaic growth [14,15]. The observation on the surface morphology in our experiment indicated that the step flow across the central junction showed more correlation with the thickness of all substrates used in mosaic growth.

As we know, unlike other crystal growth, such as liquid phase or melting method, MPCVD for diamond growth is a method with growth rate violently influenced by the plasma generated by the microwave and gas mixture instead of temperature gradient and material concentration. The single diamond CVD deposition is a continuous reaction in a stable hydrogen/methane environment with a self-coupled electromagnetic field. The chamber shape and gas pressure play an important role in the transmission of the microwave to create plasma with acceptable shape and density for the carbon deposition. The surface morphology could present a uniform step flow along the same direction covering the entire substrate with an excellent control in the system parameters [16–18].

Many studies have focused on the environment design for the plasma generation and deposition condition, including the shape of the chamber, gas pressure and substrate pocket holder [1,19–22]. Several reports simulated the substrate surface condition and found that the plasma-discharged electric field intensity and surface temperature were much higher on the surface for the thicker substrates than the thinner one [20,23,24]. This implies that more carbon ions with higher free energy may overcome the potential barrier to deposit on the diamond substrate when a thicker substrate is used. It causes an excessive high growth rate compared to the lower area and generates the new step flow with the direction perpendicular to the contour line of height distribution.

However, for the MPCVD diamond mosaic growth, more parameters should be considered, such as the gap width, substrate shape and thickness difference. In this way, the substrate surface morphology could receive secondary effects introduced by the other substrates during the mosaic growth. A new step flow may generate along the different direction over the original one when the substrate edges meet each other or polycrystalline was grown in the junctions [25,26]. Shu et al. studied the stress and defects distributions in the cross-section of the mosaic junctions with Raman spectroscopy, measured and calculated the maximum stress value and corresponding position distance away from the interface [11,27]. They reported that the location of the major part of the defected zone might move from one crystal to the neighbor crystal across the junction, which matches our result in Raman surface mapping and the surface step flow movement.

In previous research, it was found that the internal stress of the diamond homoepitaxy growth layer could be inherited from the defects of substrate [28,29]. The substrate defects could be introduced from the internal inclusions of the HTHP single-crystal diamond or the cutting and polishing process after the growth. Anatoly B. Muchnikov et al. pointed out that the surface stresses generated by the thermochemical polishing should be also taken into consideration for the formation of the surface step direction when preparing the diamond mosaic [10]. We believe that the scratches might introduce an anomalous charge ion concentration and temperature gradient on the substrate surface. As a result, step flow direction may be changed. In our experiment, a high-quality HTHP diamond was used for the substrates preparation. The fine polishing was performed after the laser cutting and a scratching-free upper and side surface with Ra smaller than 2 nm was achieved. In this way, the internal stress in both the homoepitaxy and lateral growth area would be reduced during the CVD growth since anomalous charged electric field on the deposition surface and edges was avoided in a smooth surface. As a result, the central area eventually shows the uniform step flow without any cracking even though the clear height gap appeared after the mosaic growth for 24 hours.

Mechanism of the Mosaic Junction Creation and Interface Step Flow Movement

Based on the surface morphology, Raman scanning and HRXRD result of the substrate interface, we schematically proposed the mechanism to describe the combination of high quality mosaic interface junction and the step flow movement and transformation across the junction on the deposition layer as shown in Figure 7. Before the growth, the substrates were placed into the reaction chamber, with narrow gap space among the diamond seeds. Even with carefully cutting and polishing, thickness differences still existed among them. When the CVD reaction started, a homoepitaxy growth occurred on the substrate surface and the original step flow was formed, which is normally caused by thickness difference of the substrate surfaces. Simultaneously, lateral overgrowth happened on the substrates edge and the gap distance between the substrates reduced persistently when the deposition proceeded. With continuous lateral growth, the two edges contact and form interface junctions between the substrates. Meanwhile the homoepitaxy growth causes the increase of the sample thickness, the mosaic shall penetrate into the plasma in the chamber. As a consequence, the charged electric field density will increase on the edges and corners of the sample, which results in a higher temperature and the concentration of charged carbon on the interface. This leads to the slightly increase of the crystal growth rate in center junction and the formation of the new step flow parallel to the interface and perpendicular to the original surface step. The new step flow moves across the gap and consolidates the two individual substrates into an entire block plate. After the combination of the interface junction, the new step continues to move on the samples and cover the original step flow, and the whole mosaic plate surface was eventually covered by a uniform surface step.

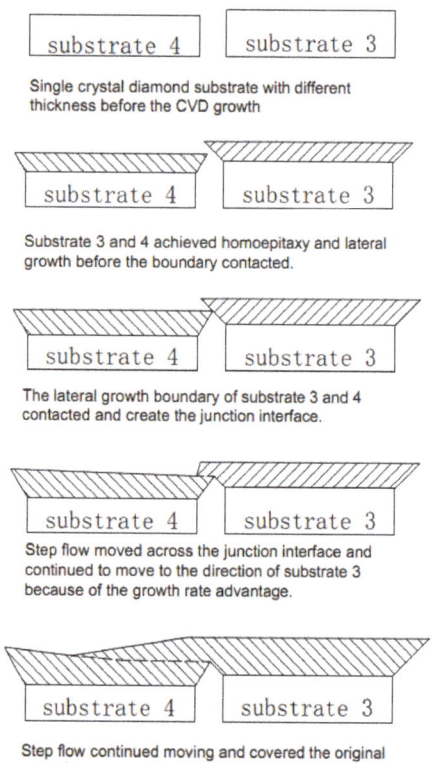

Figure 7. The mechanism of the CVD diamond mosaic junction interface formation and the movement of the surface step flow.

4. Conclusions

In this paper, a mosaic diamond was grown by MPCVD on HTHP single-crystal diamond plates. After diamond growth, we observed the morphologies of the mosaic edges and center junction and analyzed the step flow movement across the junction interface. Raman mapping images proved that the mosaic layer in the center junction possessed high crystal quality and the junction stress field moved for about 200–400 µm from one substrate to another. The HRXRD result indicated that the mosaic surface step flow direction had nothing to do with the off-axis direction of the original substrates. Furthermore, the surface height distribution proved that the surface step flow generation and movement was driven by the height difference of the entire mosaic surface. The mechanism of the junction interface formation and the mosaic surface step morphology transformation was proposed.

Author Contributions: Conceptualization, X.W. and Y.P.; methodology, X.W. and P.D.; validation, D.W. and X.X.; formal analysis, X.W.; investigation, X.W. and Z.C.; resources, X.W., Y.P. and C.L.; data curation, X.W. and P.D.; writing—original draft preparation, X.W.; writing—review and editing, X.H.; supervision, X.H. and X.X.; project administration, D.W. and C.L.; funding acquisition, X.H. and X.X. All authors have read and agreed to the published version of the manuscript.

Funding: This research was funded by National Key R & D Program of China (Grant No. 2018YFB0406501).

Acknowledgments: The authors were thankful to Jinan Diamond Technology Co., Ltd for the growth of HTHP single crystal diamonds and substrates preparation.

Conflicts of Interest: The authors declare no conflict of interest.

References

1. Tallaire, A.; Brinza, O.; Mille, V.; William, L.; Achard, J. Reduction of Dislocations in Single Crystal Diamond by Lateral Growth over a Macroscopic Hole. *Adv. Mater.* **2017**, *29*, 1604823. [CrossRef]
2. Fairchild, B.A.; Olivero, P.; Rubanov, S.; Greentree, A.D.; Waldermann, F.; Taylor, R.A.; Walmsley, I.; Smith, J.M.; Huntington, S.; Gibson, B.C. Fabrication of Ultrathin Single-Crystal Diamond Membranes. *Adv. Mater.* **2008**, *20*, 4793–4798. [CrossRef]
3. Volpe, P.-N.; Pernot, J.; Muret, P.; Omnès, F. High hole mobility in boron doped diamond for power device applications. *Appl. Phys. Lett.* **2009**, *94*, 092102. [CrossRef]
4. Zong, W.; Zhang, J.; Liu, Y.; Sun, T. Achieving ultra-hard surface of mechanically polished diamond crystal by thermo-chemical refinement. *Appl. Surf. Sci.* **2014**, *316*, 617–624. [CrossRef]
5. Lu, H.-C.; Peng, Y.-C.; Lin, M.-Y.; Chou, S.-L.; Lo, J.-I.; Cheng, B.-M. Analysis of boron in diamond with UV photoluminescence. *Carbon* **2017**, *111*, 835–838. [CrossRef]
6. Nesladek, M.; Bogdan, A.; Deferme, W.; Tranchant, N.; Bergonzo, P. Charge transport in high mobility single crystal diamond. *Diam. Relat. Mater.* **2008**, *17*, 1235–1240. [CrossRef]
7. Mokuno, Y.; Chayahara, A.; Yamada, H. Synthesis of large single crystal diamond plates by high rate homoepitaxial growth using microwave plasma CVD and lift-off process. *Diam. Relat. Mater.* **2008**, *17*, 415–418. [CrossRef]
8. Mokuno, Y.; Chayahara, A.; Yamada, H.; Tsubouchi, N. Improving purity and size of single-crystal diamond plates produced by high-rate CVD growth and lift-off process using ion implantation. *Diam. Relat. Mater.* **2009**, *18*, 1258–1261. [CrossRef]
9. Yamada, H.; Chayahara, A.; Mokuno, Y.; Kato, Y.; Shikata, S. A 2-in. mosaic wafer made of a single-crystal diamond. *Appl. Phys. Lett.* **2014**, *104*, 102110. [CrossRef]
10. Muchnikov, A.B.; Radishev, D.B.; Vikharev, A.L.; Gorbachev, A.M.; Mitenkin, A.V.; Drozdov, M.N.; Drozdov, Y.N.; Yunin, P.A. Characterization of interfaces in mosaic CVD diamond crystal. *J. Cryst. Growth* **2016**, *442*, 62–67. [CrossRef]
11. Shu, G.; Bing, D.; Ralchenko, V.G.; Khomich, A.A.; Ashkinazi, E.E.; Bolshakov, A.P.; Bokova-Sirosh, S.N.; Liu, K.; Zhao, J.; Han, J. Epitaxial growth of mosaic diamond: Mapping of stress and defects in crystal junction with a confocal Raman spectroscopy. *J. Cryst. Growth* **2017**, *463*, 19–26. [CrossRef]
12. Xie, X.; Wang, X.; Peng, Y.; Cui, Y. Synthesis and characterization of high quality {100} diamond single crystal. *J. Mater. Sci. Mater. Electron.* **2017**, *28*, 9813–9819. [CrossRef]
13. Widmann, C.J.; Müller-Sebert, W.; Lang, N.; Nebel, C.E. Homoepitaxial growth of single crystalline CVD-diamond. *Diam. Relat. Mater.* **2016**, *64*, 1–7. [CrossRef]
14. Yamada, H.; Chayahara, Y.M.; Shikata, S. Effects of crystallographic orientation on the homoepitaxial overgrowth on tiled single crystal diamond clones. *Diam. Relat. Mater.* **2015**, *57*, 17–21. [CrossRef]
15. Yamada, H.; Chayahara, A.; Mokuno, Y.; Umezawa, H.; Shikata, S.-I.; Fujimori, N. Fabrication of 1 Inch Mosaic Crystal Diamond Wafers. *Appl. Phys. Express* **2010**, *3*, 051301. [CrossRef]
16. Bushuev, E.V.; Yurov, V.Y.; Bolshakov, A.P.; Ralchenko, V.G.; Khomich, A.A.; Antonova, I.A.; Ashkinazi, E.E.; Shershulin, V.A.; Pashinin, V.P.; Konov, V.I. Express in situ measurement of epitaxial CVD diamond film growth kinetics. *Diam. Relat. Mater.* **2017**, *72*, 61–70. [CrossRef]
17. Charris, A.; Nad, S.; Asmussen, J. Exploring constant substrate temperature and constant high pressure SCD growth using variable pocket holder depths. *Diam. Relat. Mater.* **2017**, *76*, 58–67. [CrossRef]
18. Chen, J.; Wang, G.; Qi, C.; Zhang, Y.; Zhang, S.; Xu, Y.; Hao, J.; Lai, Z.; Zheng, L. Morphological and structural evolution on the lateral face of the diamond seed by MPCVD homoepitaxial deposition. *J. Cryst. Growth* **2018**, *484*, 1–6. [CrossRef]
19. Nad, S.; Gu, Y.; Asmussen, J. Growth strategies for large and high quality single crystal diamond substrates. *Diam. Relat. Mater.* **2015**, *60*, 26–34. [CrossRef]
20. Gu, Y.; Lu, J.; Grotjohn, T.; Schuelke, T.; Asmussen, J. Microwave plasma reactor design for high pressure and high power density diamond synthesis. *Diam. Relat. Mater.* **2012**, *24*, 210–214. [CrossRef]
21. Wu, G.; Chen, M.H.; Liao, J. The influence of recess depth and crystallographic orientation of seed sides on homoepitaxial growth of CVD single crystal diamonds. *Diam. Relat. Mater.* **2016**, *65*, 144–151. [CrossRef]

22. Mokuno, Y.; Chayahara, A.; Soda, Y.; Horino, Y.; Fujimori, N. Synthesizing single-crystal diamond by repetition of high rate homoepitaxial growth by microwave plasma CVD. *Diam. Relat. Mater.* **2005**, *14*, 1743–1746. [CrossRef]
23. Su, J.J.; Li, Y.F.; Ding, M.H.; Li, X.L.; Liu, Y.Q.; Wang, G.; Tang, W.Z. A dome-shaped cavity type microwave plasma chemical vapor deposition reactor for diamond films deposition. *Vacuum* **2014**, *107*, 51–55. [CrossRef]
24. Silva, F.; Hassouni, K.; Bonnin, X.; Gicquel, A. Microwave engineering of plasma-assisted CVD reactors for diamond deposition. *J. Phys. Condens. Matter* **2009**, *21*, 364202. [CrossRef] [PubMed]
25. Muehle, M.; Asmussen, J.; Becker, M.F.; Schuelke, T. Extending microwave plasma assisted CVD SCD growth to pressures of 400 Torr. *Diam. Relat. Mater.* **2017**, *79*, 150–163. [CrossRef]
26. Tallaire, A.; Achard, J.; Silva, F.; Sussmann, R.S.; Gicquel, A. Homoepitaxial deposition of high-quality thick diamond films: Effect of growth parameters. *Diam. Relat. Mater.* **2005**, *14*, 249–254. [CrossRef]
27. Shu, G. Vertical-substrate epitaxial growth of single-crystal diamond by microwave plasma-assisted chemical vapor deposition. *J. Cryst. Growth* **2018**, *486*, 104–110. [CrossRef]
28. Naamoun, M.; Tallaire, A.; Doppelt, P.; Gicquel, A.; Legros, M.; Barjon, J.; Achard, J. Reduction of dislocation densities in single crystal CVD diamond by using self-assembled metallic masks. *Diam. Relat. Mater.* **2015**, *58*, 62–68. [CrossRef]
29. Silva, F.; Achard, J.; Brinza, O.; Bonnin, X.; Hassouni, K.; Anthonis, A.; de Corte, K.; Barjon, J. High quality, large surface area, homoepitaxial MPACVD diamond growth. *Diam. Relat. Mater.* **2009**, *18*, 683–697. [CrossRef]

© 2019 by the authors. Licensee MDPI, Basel, Switzerland. This article is an open access article distributed under the terms and conditions of the Creative Commons Attribution (CC BY) license (http://creativecommons.org/licenses/by/4.0/).

Article

On the Impact of Substrate Uniform Mechanical Tension on the Graphene Electronic Structure

Konstantin P. Katin [1,2], Mikhail M. Maslov [1,2,*], Konstantin S. Krylov [1] and Vadim D. Mur [1,*]

1 National Research Nuclear University "MEPhI", Kashirskoe Shosse 31, 115409 Moscow, Russia; KPKatin@mephi.ru (K.P.K.); krylov.const@gmail.com (K.S.K.)
2 Laboratory of Computational Design of Nanostructures, Nanodevices, and Nanotechnologies, Research Institute for the Development of Scientific and Educational Potential of Youth, Aviatorov str. 14/55, 119620 Moscow, Russia
* Correspondence: Mike.Maslov@gmail.com (M.M.M.); VDMur@mephi.ru (V.D.M.)

Received: 30 September 2020; Accepted: 19 October 2020; Published: 21 October 2020

Abstract: Employing density functional theory calculations, we obtain the possibility of fine-tuning the bandgap in graphene deposited on the hexagonal boron nitride and graphitic carbon nitride substrates. We found that the graphene sheet located on these substrates possesses the semiconducting gap, and uniform biaxial mechanical deformation could provide its smooth fitting. Moreover, mechanical tension offers the ability to control the Dirac velocity in deposited graphene. We analyze the resonant scattering of charge carriers in states with zero total angular momentum using the effective two-dimensional radial Dirac equation. In particular, the dependence of the critical impurity charge on the uniform deformation of graphene on the boron nitride substrate is shown. It turned out that, under uniform stretching/compression, the critical charge decreases/increases monotonically. The elastic scattering phases of a hole by a supercritical impurity are calculated. It is found that the model of a uniform charge distribution over the small radius sphere gives sharper resonance when compared to the case of the ball of the same radius. Overall, resonant scattering by the impurity with the nearly critical charge is similar to the scattering by the potential with a low-permeable barrier in nonrelativistic quantum theory.

Keywords: graphene; substrates; energy gap; Dirac velocity; mechanical deformation; critical charge; supercharged impurity; resonant scattering

---

1. Introduction

Since its experimental synthesis in 2004 [1], graphene has attracted considerable attention from researchers due to its promising electronic properties. In terms of electronic characteristics, graphene is a gapless semiconductor [2]. While a traditional semiconductor such as silicon or germanium has an energy (semiconductor) gap, then graphene has a zero gap. In other words, the bottom of the conduction band and the ceiling of the valence band converge at one point in graphene, which is called the Dirac point. Moreover, ideal free-standing graphene possesses a linear dispersion relation in the vicinity of the Dirac point [3]. Charge carriers in graphene behave like massless relativistic particles that are called Dirac fermions, which determines their extremely high mobility. For example, the mobility of charge carriers in graphene reaches extremely high values, up to ~200,000 $cm^2 V^{-1} s^{-1}$ [4]. In other semiconductors (such as silicon or germanium), it is fifty or more times lower. At the same time, graphene has a number of other unique qualities. For example, Young's modulus of graphene, which characterizes the material's resistance to mechanical deformation, is higher than the corresponding values of steel and tungsten [5,6], and its thermal conductivity is significantly higher than the conductivity of traditional conductors such as copper or silver [7]. Unfortunately, the absence of a finite bandgap makes it impossible to get rid of leakage currents. This means that the current through graphene can never be turned off completely. The latter is a severe barrier to the graphene use in logic devices, e.g., transistors. Therefore, for the practical applications of graphene in nanoelectronics, a considerable energy gap should be opened. Various methods are proposed for introducing the semiconductor gap in graphene. Among them are chemical functionalization [8–10], the formation of graphene nanoribbons [11,12], mechanical strain [13,14], and the use of suitable substrates. It should be noted that the most proposed substrate for creating a semiconducting gap in graphene is a boron nitride substrate [15–19], but it is not the only possible one. Thus, properly chosen substrates can significantly change the electronic band structure of graphene [20], and, along with the mechanical stretching, it can become an alternative approach for straightforward tuning the electronic properties of graphene and obtaining a required bandgap (see Review [21] for details). Moreover, there are approaches that allow one to generate asymmetric deformation or doping between layers and methods for its quantitative determination using a supported isotopically labeled bilayer graphene studied by in situ Raman spectroscopy [22]. The mechanically controlled bandgap was previously observed in other 2D materials [23–26]. However, the exceptional mechanical properties of graphene [27,28] provide outstanding efficiency of mechanical strain engineering in this material.

In addition, graphene with a gap in the electronic spectrum in the presence of a multiple-charge impurity is of particular interest. Such a system is similar to the relativistic Coulomb problem [29]. In the two-dimensional case, the total angular momentum $\hbar J = \hbar(M + 1/2)$, where $\hbar M$ is the orbital angular momentum is a good quantum number by virtue of the axial symmetry [30]. In the three-dimensional problem, the Dirac quantum number is conserved due to spherical symmetry [31].

Since graphene is an almost ideal planar system, the orbital angular momentum may be quantized fractionally [32,33]. In addition, if one takes into account the motion reversal invariance, then M is limited to half-integer values only [34]. Therefore, the possible values of the total angular momentum $J = M + 1/2$ in graphene are either half-integers or integers with the zero included [35]. Therefore, the radial two-dimensional Dirac equation, in the case of the Coulomb impurity, is identical to the three-dimensional one only for the integer nonzero values $J = -= \pm 1, \pm 2, \ldots$.

It has long been known that pure Coulomb potential is an excessive idealization in the three-dimensional problem, which (for the ground state $-$ -1) loses its meaning for a nucleus charge $Z > \alpha^{-1} \approx 137$ [36]. Here, $\alpha = e^2/\hbar c$ is the Sommerfeld fine structure constant, which, in graphene, is replaced by the effective fine structure constant $\alpha_D = e^2/\hbar v_D$, where v_D is the Dirac velocity. Since $\alpha_D \sim 1$ in graphene, regularization of the Coulomb potential is required already for values of the impurity charge $Z \equiv Z_0/\epsilon \gtrsim 1$, where we take dielectric properties of the environment into account by means of the effective dielectric constant ϵ. Here, Z_0 is the "bare" impurity charge. According to Reference [37], the finite impurity sizes r_0 should be taken into account, which can be achieved by modifying the Coulomb potential at small distances.

The particular value $Z_s \approx 137$ in the three-dimensional problem corresponds to the singular effective charge value $Z_s \alpha_D = |J|$ in graphene. In two dimensions, the lowest energy level of the states with the total angular momentum J, described by the equivalent of the Sommerfeld formula [38,39] $E = m^* v_D^2 \sqrt{J^2 - Z^2 \alpha_D^2}$ (m^* is the effective mass of the charge carriers), vanishes at $Z = Z_s$ and is imaginary for $Z > Z_s$. Thus, to eliminate this difficulty and make the effective radial Dirac Hamiltonian self-adjoint, it is necessary to regularize the Coulomb potential [37].

After the regularization, a further increase in the charge results in the energy level below zero and, at the critical charge value [37] $Z = Z_{cr}$, crossing the boundary of the lower continuum of the Dirac equation solutions, which, for the gapped graphene, is the ceiling of the valence band. The electron level disappears from the spectrum at $Z > Z_{cr}$ [37], and the quasi-stationary state of the hole appears in the lower continuum instead. This picture is described in detail in Reference [39], and, in the semiclassical approximation, it is described in Reference [40]. In the latter case, the Dirac equation is reduced to the Schrödinger equation [41,42] in which the effective potential has a low-permeable barrier at $E < 0$ [43].

Since the radial Dirac Hamiltonian is self-adjointed, one can interpret the solution of the Dirac equation with the energy $E < -m^*v_D^2$ as the state of a hole with positive energy $\overline{E} = -E > 0$ scattering by a supercritical impurity. In this case, the complex energies $E_p = E_0 - i\Gamma/2$ of quasidiscrete levels, where E_0 and Γ are their positions and widths, respectively, are given by the poles of the unitary partial elastic scattering matrix $S_J(\overline{E}; r_0) = \exp\{2i\delta_J(\overline{E}; r_0)\}$.

The Coulomb barrier is broad enough so that the quasistationary states have the small widths $\Gamma \ll E_0 \sim m^*v_D^2$, and Breit–Wigner resonances can arise in the scattering of holes, as in the nonrelativistic theory of scattering by a potential barrier with a low permeability [44]. More specifically, the elastic scattering phase $\delta_J(\overline{E}; r_0)$ sharply changes when the hole energy $\overline{E}$ is within the width Γ near the position E_0 of the quasidiscrete level [39], and the holes scattering partial cross-section comes close to the unitary limit. This can be detected in the current-voltage characteristics obtained experimentally by means of scanning tunneling microscopy [45].

In the gapless case, one can legitimately call Z_s the critical charge Z_c, as is often found in the literature [46,47], since the lower continuum states at $Z > Z_s$ are the scattering states of holes. The validity of the approach to describe the dynamics of charge carriers in the gapless graphene with Coulomb impurities using the radial Dirac equation was established in References [46,47], where theoretical calculations based on this assumption are in good agreement with the spectra of current-voltage characteristics dI/dV measured experimentally near the Dirac point. This is especially clearly manifested by the presence of a peak in the current-voltage characteristics measured near the center of a cluster containing five calcium dimers, which correspond to the scattering of a hole with $J = 1/2$ by a supercritical impurity (see Figure 1E in Reference [46]). The emerging second peak is possibly related to the resonance in the state with $J = 1$, i.e., with the half-integer value $M = 1/2$ of the orbital angular momentum. There is not even a hint of its existence in theoretical calculations. Apparently, this is due to the fact that the scattering states considered in these calculations do not include ones with half-integer values of the orbital angular momentum, which are possible in two-dimensional problems (half-integer quantization of the orbital angular momentum is realized in circular quantum dots with an odd number of electrons [48,49]). In the gapless graphene, as opposed to the gapped one, the holes scattering by a charged impurity in the state with $J = 0$ cannot lead to Breit–Wigner resonances and, therefore, to the peaks in the dI/dV spectra since the scattering phase in these states $\delta_0(E; r_0)$ is smoothly dependent on energy [39]. Therefore, in the current work, we mainly focus on the state with $J = 0$.

In the presented study, we analyze the electronic behavior of monolayer graphene on the hexagonal boron nitride (hBN) and graphitic carbon nitride (gC$_3$N$_4$) substrates under the mechanical biaxial strain in the frame of density functional theory, supplemented by the van der Waals dispersion corrections. It was obtained that the weak van der Waals interlayer interactions, alongside the uniform strain of the graphene sheet on the substrate, leads to the considerable bandgap. Moreover, the value of bandgap increases monotonically with an increasing stretch and vice versa. Note that, due to the remarkable mechanical properties, some substrates allow reversible stretching of graphene up to 30% [50,51]. Therefore, the gap can vary in a broad region. It should be noted that the hBN or gC$_3$N$_4$ substrate has a crucial role since free-standing graphene does not have an energy gap under uniform mechanical stretching. We discuss a method to smoothly change the electronic properties of graphene, which has a gap in the electronic spectrum by mechanical stretching or compression, which includes tuning the critical value $q_{cr} = Z_{cr}e^2/\hbar v_D$ of the effective charge $q = Z\alpha_D$ due to the change in the Dirac velocity v_D as well as the effective mass m^* of charge carriers. We suppose that the data obtained in the presented study can allow one to create an effective way of tuning the electronic characteristics of graphene for its further use in micro-electronic and nano-electronic as well as straintronics devices.

The rest of the presented article is organized as follows. In Section 2, we describe the ab initio technique that is used to analyze the electronic characteristics of graphene deposited on substrates and also explain the atomic structure of the samples. Section 3 gives a brief description of the Dirac equation properties and analytical results in the case of a supercritically charged Coulomb impurity.

In Section 4, we discuss the theoretical results obtained. Finally, Section 5 provides concluding remarks on the presented study.

2. Materials and Methods

To study the geometry and electronic structure properties of graphene on the different substrates, we used the ab initio approach, namely density functional theory (DFT) and its implementation in the program Quantum ESPRESSO ver. 6.5 [52,53]. We consider Perdew-Burke-Ernzerhof (GGA-PBE) functional for the description of exchange-correlation energy [54], and, for the electron-ion interaction, we use the projector-augmented-wave method (PAW) [55,56]. The kinetic energy cutoff of 120 Ry (1632 eV) was chosen. The weak van der Waals interactions between the non-covalently bound graphene sheet and substrate are taken into the D3 Grimme (DFT-D3) dispersion corrections [57]. DFT-D3 approach possess improved accuracy due to the use of environmentally dependent dispersion coefficients and the inclusion of a three-body component to the dispersion correction energy term. The interlayer distance between the "graphene-on-the-substrate" layers is equal to 20 Å, which provides sufficient space separation to avoid unphysical interactions. Thus, the lattice parameter optimization along the Z-axis becomes unnecessary. The geometry optimization of the unstressed graphene deposited on the substrate was carried out without symmetry constrains until the Hellman-Feynman forces acting on the atoms became smaller than 10^{-6} hartree/bohr. The parameters of the supercells were also optimized. However, we perform the structural relaxation of the samples strained uniformly along the lattice vectors in the XY-plane with fixed parameters of the supercell. For the k-point sampling of the Brillouin zone integrations, the $12 \times 12 \times 1$ Monkhorst-Pack mesh grid [58] is used. For the non-self-consistent field calculations, the k-point grid size has been increased to $24 \times 24 \times 1$. For the structural relaxation, the Methfessel-Paxton smearing [59] technique with the width of 0.02 eV was used, and, for the density of electronic states calculation, the Böchl tetrahedron method [60] was applied. The properties of the electronic structure were elucidated by analyzing the band structure of the samples and their density of electronic states.

First, we attempted to select such substrates that would facilitate the opening of an energy gap in graphene without imposing additional conditions, such as an external electric field or mechanical stresses. We examined about fifty different substrates, including the common SiO_2 and SiC, as well as more complex systems such as Te_2Mo, hBN/Ni(111), or C_3B/C_3N. At the level of theory used, we obtained that, only quasi-2D hexagonal boron nitride (hBN) and graphitic carbon nitride (gC_3N_4), opened the band gap in graphene. In all other cases, the "Dirac cone" on the electronic band structure conserved its original shape, and graphene remained a gapless semiconductor, or the Fermi level shifted to the conduction band. Thus, graphene began to exhibit a metallic nature.

Nevertheless, some previous studies revealed that the external electric field could effectively tune the energy gap in graphene on the SiO_2 and SiC substrates [61–63] by the Fermi level shifting to the forbidden energy band. However, there are no additional conditions for bandgap opening in the case of hBN and gC_3N_4 substrates. Peculiarities of the interlayer interactions between the graphene and these substrates lead to the "pure gap" that is characteristic of the narrow-gap semiconductor.

We represent the graphene on the hBN and gC_3N_4 substrates by the hexagonal supercells. In the case of the hBN substrate, the supercell contains eight carbon atoms, and, in the case of the gC_3N_4 substrate, the graphene sheet inside the supercell is represented by 18 carbon atoms (see Figure 1). In both cases, periodical boundary conditions are applied. Such a size and shape of the supercells are suitable for modeling the uniform graphene sheet stretching on the substrate. We carried out the uniform stretching and compression of the graphene sheet by simultaneously increasing the lattice parameters of the supercell in the XY-plane and further optimizing the atomic positions in the supercell.

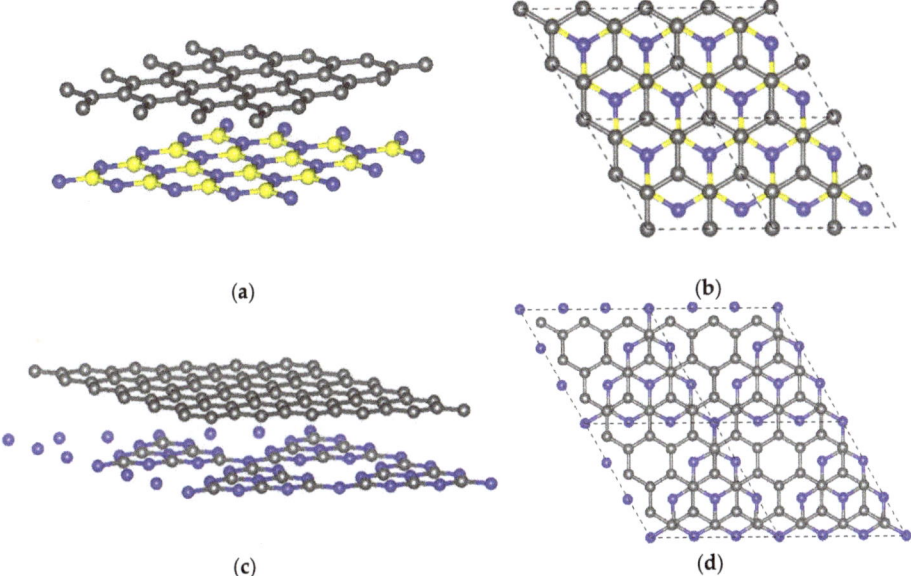

Figure 1. The atomic structures of hexagonal graphene supercells: graphene deposited on the hexagonal boron nitride (perspective view (**a**), top view (**b**)) and graphitic carbon nitride (perspective view (**c**), top view (**d**)). Gray, blue, and yellow balls represent C, N, and B atoms, respectively.

3. Theoretical Background

The dynamics of massive charge carriers in the state with energy E and total angular momentum J in graphene in the presence of a multiply-charged impurity near the Dirac point is described in the continuous limit by the effective two-dimensional radial Dirac equation.

$$H_D \Psi_{\varepsilon,J}(\rho) = \varepsilon\, \Psi_{\varepsilon,J}(\rho), \quad \Psi_{\varepsilon,J} = \begin{pmatrix} F(\rho) \\ G(\rho) \end{pmatrix},$$
$$H_D = \begin{pmatrix} 1 + V_R(\rho) & \frac{J}{\rho} + \frac{d}{d\rho} \\ \frac{J}{\rho} - \frac{d}{d\rho} & -1 + V_R(\rho) \end{pmatrix}, \qquad (1)$$

See References [35,39,64]. Here:

$$E = m^* v_D^2 \varepsilon, \quad J = M + \frac{1}{2} = 0, \pm\frac{1}{2}, \pm 1, \ldots, \quad |\mathbf{x}| = l_D \rho, \qquad (2)$$

where $\mathbf{x} = (x,y)$ is a 2D position vector, $l_D = \hbar/m^* v_D$ is the "Compton length" in graphene, and

$$V_R(\rho) = -\frac{q}{R}\begin{pmatrix} R/\rho, & \rho \geqslant R, \\ f(\rho/R), & \rho \leqslant R \end{pmatrix} \qquad (3)$$

is modified at small distances ($\rho \leqslant R \ll 1$) of the Coulomb attraction potential [37].

$$V_C(\rho) = -\frac{q}{\rho}, \quad q = Z\alpha_D, \quad \alpha_D = \frac{e^2}{\hbar v_D}, \quad q > 0, \qquad (4)$$

where $Z = Z_0/\epsilon$ is the effective charge, Z_0 is the "bare" impurity charge, $\epsilon = (1+\epsilon_s)/2$ is the effective dielectric constant, and ϵ_s is the dielectric constant of the substrate. Passing to the continuous limit is justified if

$$a_{CC} \ll r_0 \ll l_D, \quad r_0 = Rl_D, \tag{5}$$

where a_{CC} is the distance between carbon atoms and r_0 is the cutoff radius.

The rectangular cutoff $f(\rho/R) \equiv 1$ allows one to solve the system (1) analytically for any $J \neq 0$. For the case $J = 0$, the analytical solution is possible with an arbitrary cutoff function [39]. Regularization of the Coulomb potential at small distances ensures the self-adjointness of the Dirac Hamiltonian H_D and allows one to trace the motion of the level with the given E and J as the effective charge Z increases. In particular, at some critical value Z_{cr} [37], the electron level reaches the boundary of the lower continuum of the Dirac equation solutions, i.e., the upper boundary of the valence band.

The work [64] considers the effect of a supercharged Coulomb impurity $Z > Z_{cr}$ on the system of Dirac charge carriers in graphene with a gap in the electronic spectrum. It particularly discusses the screening of the impurity charge by electrons produced together with holes from the Dirac sea, which is in complete analogy with the spontaneous production of the electron-positron pairs at $Z > Z_{cr}$ in the relativistic Coulomb problem [43,65–67]. According to the review [67], the electron level collides with the positron level at $Z = Z_{cr}$, and with a further increase in the charge $Z > Z_{cr}$, it disappears from the spectrum [37], plunging into the lower continuum. The creation of a virtual electron-positron pair does not require the energy. The electron "lands" on the supercharged nucleus, becoming "superbound" [68], and the positron goes to infinity. Therefore, the issue "cannot be solved in the framework of the quantum mechanics of a single particle" [29].

However, due to the self-adjointness of the radial Dirac Hamiltonian, see Reference [39] and the mathematically rigorous paper [69], the one-particle approximation for the Dirac equation is valid not only for $Z \leqslant Z_{cr}$, but also for $Z > Z_{cr}$ (see References [35,70] for the case of short-range potential [71]). Qualitatively, this can be understood using the semiclassical approximation, when the system (1) near the boundary of the lower continuum of solutions to the Dirac equation is equivalent to the Schrödinger equation [41,42] with effective energy $E_{\text{eff}} = (\bar{\varepsilon}^2 - 1)/2$ and potential

$$U_{\text{eff}}(\rho; \bar{\varepsilon}; J) = -\frac{1}{2} V_R^2(\rho) - \bar{\varepsilon} V_R(\rho) + \frac{J^2}{2\rho^2}, \quad \bar{\varepsilon} = -\varepsilon > 1, \tag{6}$$

(see Figure 1 in Reference [40]). This means that, at small distances, both the particle with energy ε and the anti-particle with energy $\bar{\varepsilon} = -\varepsilon$ is attracted to the center, in contrast to the Schwinger mechanism of the pair production [72], when a constant electric field breaks the virtual e^+e^--loop. Therefore, the impurity charge screening mechanism specified in Reference [64], according to the scenario described in the review [67], cannot be realized [35]. Therefore, the one-particle approach for the effective radial Dirac equation (1) remains valid in the supercritical region $Z > Z_{cr}$. This is necessary so that the local density of states in graphene can be directly extracted from the experiment [45].

All these conclusions remain valid for the gapless radial Dirac equation in the presence of a supercharged impurity $Z > Z_c = |J|\alpha_D$. Noteworthy is the work [46], which shows that theoretical calculations describing the electronic properties of gapless graphene agree with the experimental data on the spectra of current-voltage characteristics obtained by the scanning tunneling microscopy. Thus, for example, one can see a peak in the dI/dV spectra measured near the center of a cluster of five calcium dimers corresponding to the resonant scattering of a hole in the state with $J = 1/2$, i.e., $M = 0$ by a supercharged impurity (see Figure 1E in Reference [46]). However, there is no peak corresponding to the value $J = 0$, i.e., $M = -1/2$ since the scattering phase, in this case, is a smooth function of the hole energy $\bar{\varepsilon} = -\varepsilon > 0$ [39].

$$\delta_0(\bar{\varepsilon}; R) = Z\alpha_D[\ln(2\bar{\varepsilon}R) - c]. \tag{7}$$

Here, the values $c = 1$ and $c = 4/3$ correspond to a uniform distribution of the impurity charge over the sphere and the ball of radius r_0, respectively.

The scattering phase for states with the total angular momentum $J = 0$ is different for the graphene with a gap in the electronic spectrum. In this case, $\delta_0(\bar{\varepsilon}; q; R)$ as a function of the hole energy $\bar{\varepsilon}$ abruptly changes when q is close to its critical value, which can correspond to Breit–Wigner resonances in holes scattering by the supercritical impurity (see Figure 7 in Reference [39]). Therefore, in the rest of this paper, we will focus on discussing states with $J = 0$, i.e., half-integer orbital angular momentum $M = -1/2$. The electron level with such a value of J first descends to the boundary of the lower continuum, i.e., for the top of the valence band, see Figure 1 in Reference [39].

The critical charge $q_{cr}(n) = Z_{cr}(n)\alpha_D$ at which the n-th level with the total angular momentum $J = 0$ reaches the boundary $\varepsilon = -1$, is found from the equation below.

$$\arg\Gamma[2iq_{cr}(n)] + q_{cr}(n)\{f_0 - \ln[2q_{cr}(n)R]\} = \pi n, \quad n = 0, 1, 2, \ldots \tag{8}$$

Here, $\Gamma(z)$ is the Euler gamma function, and the values $f_0 = 1$ and $f_0 = 2$ correspond to a uniform charge distribution over the sphere and the ball of radius $r_0 = R l_D$, respectively.

At $Z > Z_{cr}$ the electron level disappears from the spectrum [37], and the system (1) at $\bar{\varepsilon} = -\varepsilon > 1$ describes the scattering of a hole by a supercharged impurity. For the partial elastic scattering matrix, we have the following [35,39].

$$S_0(k) = e^{2i\delta_0(k)} = \frac{\mathcal{F}_0^*(k)}{\mathcal{F}_0(k)}, \tag{9}$$

where

$$\mathcal{F}_0(k) = -i[e^{\frac{\pi q}{2}-i\eta} a - e^{-\frac{\pi q}{2}+i\eta} b^*] \tag{10}$$

is the analog of the Jost function in the nonrelativistic scattering theory [44], and the following notation is used.

$$a = \frac{1+\bar{\varepsilon}-k}{\Gamma[1-iq(1-\bar{\varepsilon}/k)]}, \quad b = \frac{1+\bar{\varepsilon}+k}{\Gamma[1-iq(1+\bar{\varepsilon}/k)]}, \tag{11}$$
$$e^{2i\eta} = e^{2iq[f_0-\ln(2kR)]} \frac{\Gamma(2iq)}{\Gamma(-2iq)}, \quad k = \sqrt{\bar{\varepsilon}^2 - 1}.$$

The poles of the scattering matrix $\mathcal{F}_0(k) = 0$ lead to the equation for the spectrum of the complex energies

$$\frac{1+\bar{\varepsilon}-k}{1+\bar{\varepsilon}+k} \cdot \frac{\Gamma[1+iq(1+\bar{\varepsilon}/k)]}{\Gamma[1-iq(1-\bar{\varepsilon}/k)]} = e^{-\pi q + 2i\eta}. \tag{12}$$

The solutions give both the positions ε_0 and the widths γ of the quasi-discrete levels.

$$\bar{\varepsilon} = \varepsilon_0 - i\gamma/2, \quad \varepsilon_0 > 1, \quad \gamma > 0. \tag{13}$$

Near the top of the valence band $Z - Z_{cr} \ll 1$, due to the low permeability of the Coulomb barrier at $\rho \gg R$ in the effective potential (6), the width of the quasistationary state is small $\gamma \ll \varepsilon_0 \sim 1$.

As in the nonrelativistic theory of scattering, see Chapter 13 in Reference [41], quasistationary states can manifest themselves as resonances in the hole scattering if its energy $\bar{\varepsilon}$ is within the region of a sharp change in the scattering phase. In this case, if the background phase is absent $\delta_{bg} = 0$, the partial cross-section is described by the Breit–Wigner formula.

$$\sigma_0(\bar{\varepsilon}) = \sin^2 \delta_0 = \frac{(\gamma^*/2)^2}{(\bar{\varepsilon} - \varepsilon_0^*)^2 + (\gamma^*/2)^2}, \tag{14}$$

The position of the resonance ε_0^* follows from the equality $\delta_0(\varepsilon_0^*; q; R) = \pi/2$ when the scattering phase changes abruptly from 0 to π over the width γ^*. At the same time, the width γ^* is determined from Equation (12) for such an effective charge q close to q_{cr}, which corresponds to the value of ε_0^*. This leads to the appearance of peaks in the current-voltage characteristics measured near the center of the supercharged impurity in graphene. However, when the background phase $\delta_{bg} \neq 0$, and then near the resonance $\bar{\varepsilon} \approx \varepsilon_0^*$, the total phase $\delta_0(\bar{\varepsilon}; q; R)$ increases sharply from δ_{bg} to $\delta_{bg} + \pi$, which leads to a change in the shape of the resonance, and the Breit–Wigner formula is replaced by a more general one (see Reference [44]).

In the review [67], resonant scattering by supercritical nuclei was associated with the production of e^+e^--pairs, and the width γ was considered as the probability of such a process. This point of view is shared by the authors of Reference [73]. However, due to the self-adjointness of the radial Dirac Hamiltonian, the scattering phase $\delta_0(k; q; R)$ is real, and the partial elastic scattering matrix $S_0(k) = \exp[2i\delta_0(k)]$ is unitary. Therefore, there are no inelastic processes, including spontaneous pair production. This statement applies to any value of the total angular momentum J [39]. In the gapless case, this fact was confirmed experimentally [46] for the states with $J = 1/2$.

4. Results and Discussion

First of all, we determined the unstrained reference configuration of the graphene sheet on the hBN and gC$_3$N$_4$ substrates by optimizing the supercell parameters as well as atomic positions inside the supercell. For considered samples (see Figure 1), we chose the types of packing, that is, the mutual arrangement of graphene atoms and the substrate, which correspond to the lowest total energy and, therefore, are more thermodynamically stable [74,75]. The structure of graphene on the hBN is characterized as follows: one carbon atom is located directly above the boron atom, and the other carbon atom is located above the center of the hBN hexagon (see Figure 1). In the most energetically favorable graphene/gC$_3$N$_4$ system, graphene carbon atoms are located directly above the carbon atoms of the substrate and above the center of the C-N hexagon (see Figure 1). It should be noted that graphene conserves its plane structure, and there is no covalent bonding between graphene and substrate. In the case of the hBN substrate, the intercarbon bond length in the unstrained graphene is equal to 1.435 Å, and the distance between the graphene sheet and substrate is equal to 3.429 Å (experimental value is equal to 3.3 Å [76]). In the case of the gC$_3$N$_4$ substrate, the intercarbon bond length in the unstrained graphene is equal to 1.407 Å, and the distance between the graphene sheet and substrate is equal to 3.386 Å (experimental value is equal to 3.325 Å [77]). It should be noted that the lattice parameters of gC$_3$N$_4$ are about three times larger than the corresponding values for graphene. This means that for 1 × 1 supercell of the substrate corresponds to 3 × 3 supercells of graphene. Taking this into account, the mismatch of the lattice parameters of the supercells for graphene and gC$_3$N$_4$ substrate is about 3.6%. The lattice constants for graphene and hexagonal boron nitride correspond much better to each other. Their discrepancy is only 1.8%. In addition, the gC$_3$N$_4$ substrate contains large holes. It is "atomically inhomogeneous" in contrast to the hBN substrate, which leads to slight distortions of the graphene sheet. However, the intercarbon bond length in the unstrained graphene is in good agreement with the previously obtained experimental data [78,79]. All subsequent deformations and strains are defined with respect to this equilibrium honeycomb structure.

At the next step, we analyze the electronic characteristics, namely band structure and density of electronic states, of the equilibrium honeycomb structure of graphene sheet on the hBN and gC$_3$N$_4$ substrates (see Figure 1). In both cases, we observe the energy gap. The presence of the hBN substrate leads to the semiconducting gap in graphene that is equal to 22.9 meV, and the presence of the gC$_3$N$_4$ substrate opens the semiconducting gap that is equal to 22.2 meV. Therefore, hBN and gC$_3$N$_4$ substrates contribute to a zero-gap semiconductor to a narrow-gap semiconductor transition. In the case of the hexagonal boron nitride substrate, it was previously shown that the mechanism of the energy gap appearing is concerned with the charge redistribution in the graphene layer and charge transfer between graphene and boron nitride layers by modifying the on-site energy difference of carbon p orbitals at two different sublattices [80]. It should be noted that, contrary to the free-standing graphene or the graphene on the hBN substrate, the band structure of the graphene on the gC$_3$N$_4$ substrate possesses a characteristic feature: the Dirac cone at the K point in the unit cell of graphene moves to the Γ point. The corresponding data are shown in Figure 2.

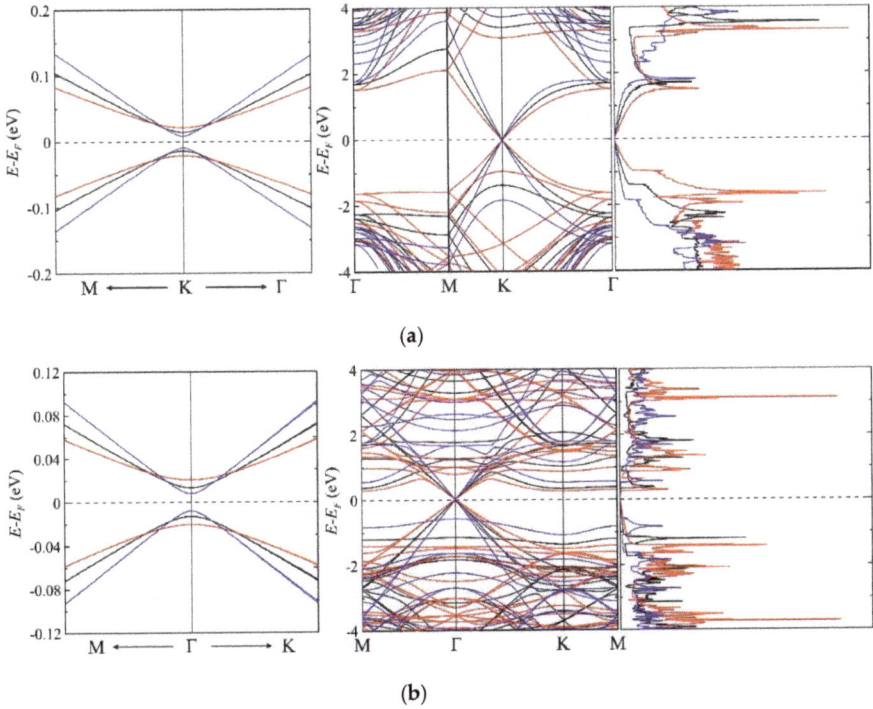

Figure 2. Band structure near the Dirac point (left), band structure (center), and density of electronic states (right) of the graphene deposited on the hBN (**a**) and gC$_3$N$_4$ (**b**) substrates. Blackline corresponds to the unstrained graphene, the blue line corresponds to the 10% compression, and the red line corresponds to the 10% stretching of the graphene sheet on the substrate. Γ, K, and M are the standard notations for the high-symmetry characteristic points in the Brillouin zone, where Γ corresponds to the center of the Brillouin zone. The Fermi level is assigned at zero.

Our subsequent studies are focused on the effects of mechanical stretching and compression on the behavior of the energy gap in graphene on the hBN and gC$_3$N$_4$ substrates. We analyze the impact of uniform biaxial compression and stretching to the electronic characteristics of graphene. These properties differ significantly from those of free graphene. Earlier, it was obtained that the uniaxial (zigzag or armchair) stretching and shearing as well as inhomogeneous deformation opens the energy gap in free-standing graphene [81–83]. On the contrary, isotropic biaxial stretching up to 20% left the graphene gapless [40]. This is due to the fact that, in the case of uniform stretching, the initial symmetry of the graphene lattice is retained, and, therefore, the band gap does not appear. On the contrary, when uniaxial deformation is applied, the graphene lattice symmetry decreases. Such deformations affect the irreducible part of the first Brillouin zone. It varies from its original triangular to the polygonal shape. Thus, the tops of the Dirac cones are no longer located at the high-symmetry points. They move along the Brillouin zone, either for deformations in the armchair or zigzag direction, and the energy gap appears. This is also true in the presence of the other layer or substrate. This effect is observed for the case of twisted bilayer graphene [84]. From a geometrical point of view, the symmetry breaking caused by the presence of the substrate also leads to the appearance of a gap. This effect also conserves during the uniform deformation of the graphene-substrate sample. Therefore, the presence of hBN or gC$_3$N$_4$ substrate makes it possible to tune the bandgap in graphene. In this case, stretching leads to an increase of the gap value, and compression leads to its decrease. For example, uniform stretching of the graphene sheet on the hBN substrate results in the gap of

36.8 meV, and, on the gC_3N_4 substrate, results in the gap of 36.9 meV (see Figure 3). These values are higher than thermal energy at room temperature (k_BT ~26 meV), which indicates the possibility of maintaining the energy gap at normal conditions. Note that, in real straintronic applications, the limits of strain transfer between graphene and substrate should be considered since biaxial deformation is not always completely transferred from the substrate to graphene [85]. We plot the band structures and densities of electronic states of unstrained graphene, graphene stretched by 10%, and graphene compressed by 10% on the hBN and gC_3N_4 substrates (see Figure 2). Note that the bandgap of graphene on gC_3N_4 grows faster than the bandgap of graphene on hBN under stretching (see Figure 3).

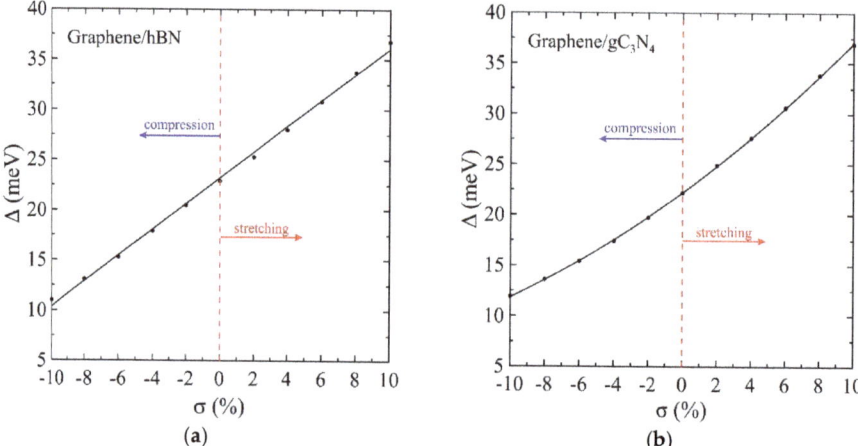

Figure 3. Semiconducting energy gap dependence on the uniform biaxial strain of the graphene deposited on the hBN (**a**) and gC_3N_4 (**b**) substrates. Circles are the results of the calculation, and the solid lines are the least-squares linear (**a**) and quadratic (**b**) fits.

Next, we try to estimate the behavior of the Dirac velocity of unstrained and uniformly stretched/compressed graphene on the hBN and gC_3N_4 substrates. It is already known that, in a gapless free-standing graphene energy dispersion relation has a linear form, i.e., $E(k) = \hbar v_F k$, where v_F is Fermi velocity, k is the modulus of the wave vector in two-dimensional space measured from the Dirac points, and $\hbar$ is the reduced Planck's constant. Thus, it can be said that electrons in pristine graphene have zero effective mass [86]. However, if graphene has a semiconducting gap, as in our case, when it is located on the hBN or gC_3N_4 substrate, the energy dispersion relation can no longer be considered linear near the Dirac points, and the electrons have a nonzero effective mass. In this case, we can approximate the bottom of the conduction band and the top of the valence band ($k = k_0$) by the parabolic functions and estimate the effective mass of particles from the expression $m^* = \hbar^2 \left(\frac{\partial^2 E(k)}{\partial k^2} \right)^{-1} \bigg|_{k=k_0}$, where $p = \hbar k$ is the momentum of the electron. On the other hand, taking into account that the effective mass of charge carriers at the Dirac point given by equation $m^* \approx \frac{\Delta}{2v_D^2}$ [18], where Δ is the value of the energy gap, we can estimate the Dirac velocity v_D. Here, we use the notation for the Dirac velocity to avoid the confusion with the Fermi velocity that is traditionally used for the gapless free-standing graphene. Calculated results for the Dirac velocity for graphene on the hBN and gC_3N_4 substrate under uniform compression and stretching up to 10% are presented in Figure 4. The obtained data for the Dirac velocity depending on mechanical tension σ can be fitted with the following linear equation:

$$v_D(\sigma)[\text{Mm/s}] = -0.0188\sigma + 0.8703 \tag{15}$$

for the graphene on the hBN substrate, using the following linear equation.

$$v_D(\sigma)[\text{Mm/s}] = -0.0157\sigma + 0.8092 \tag{16}$$

for the graphene on the gC$_3$N$_4$ substrate. Thus, uniform stretching can be an effective instrument for the precise tuning of the energy gap of graphene on the considered substrates as well as the Dirac velocity.

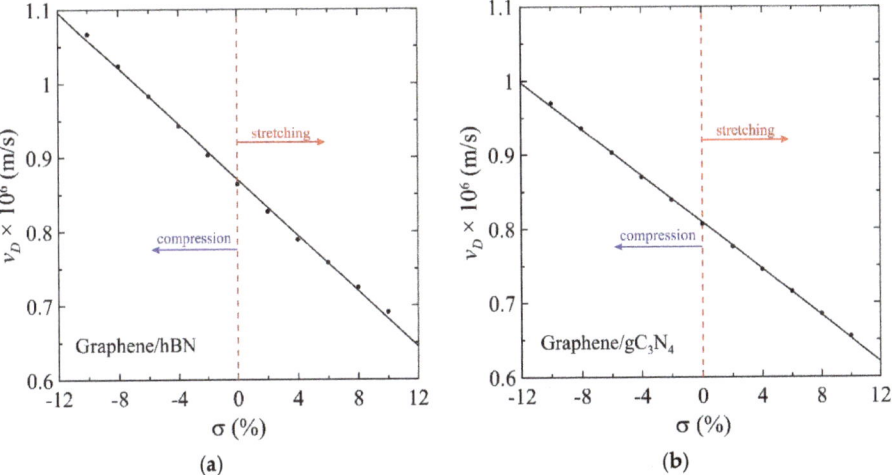

Figure 4. Dirac velocity dependence on the uniform biaxial strain of the graphene deposited on the hBN (a) and gC$_3$N$_4$ (b) substrates. Circles are the results of the calculation and the solid lines are the least-squares linear fits.

As can be seen from Figure 2, the band structure near the Dirac point in the corner of the Brillouin zone for graphene on the hBN substrate is not very different from the graphene on the gC$_3$N$_4$ substrate. However, the Dirac point in the latter case is located in the center of the zone. Therefore, the presented results for the electronic structure described by the two-dimensional radial Dirac equation are limited only to graphene on the hBN substrate.

Figures 3 and 4 show that the gap Δ and Dirac velocity v_D under stretching and compression in the range from $\sigma = -10\%$ to $\sigma = 10\%$ change in opposite directions, which results in much larger regions of variation for the effective mass of charge carriers $m^* = \frac{\Delta}{2v_D^2}$ and the "Compton length" in graphene with a gap in the electronic spectrum $l_D \equiv \frac{\hbar}{m^* v_D} = 2\hbar \frac{v_D}{\Delta}$ (see Figure 5). A continuous transition to the two-dimensional radial Dirac equation near the Dirac point is justified if $a_{CC} \ll r_0$. The value $r_0 = 15\text{Å}$ that is used for calculations in this work, as well as in Reference [46] for the gapless graphene, is very acceptable here since the distance between carbon atoms $a_{CC} \sim 1.5\text{Å}$ at $\sigma = 0\%$ in the considered cases. Stretching and compression by 10% changes these values by no more than 10% so that the dimensionless cutoff radius $R = r_0/l_D$ can vary over a wider region.

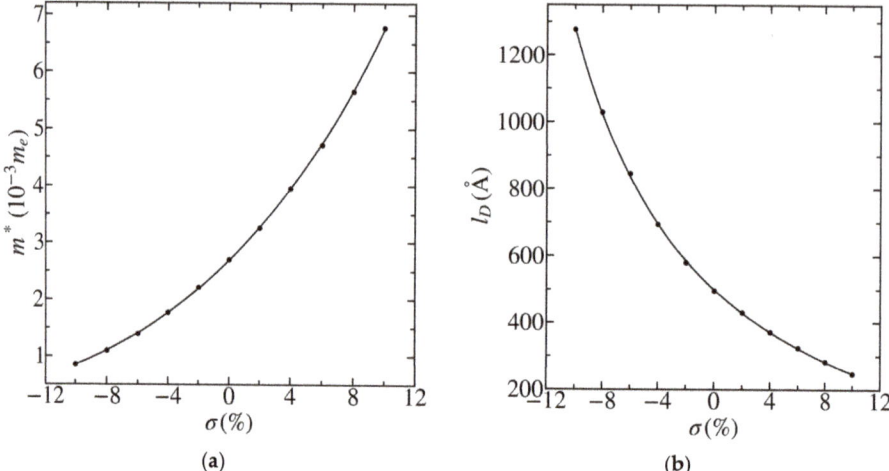

Figure 5. Dependence of the effective mass $m^* = \frac{\Delta}{2v_D^2}$ (a) and the "Compton length" in graphene $l_D \equiv \frac{\hbar}{m^* v_D} = 2\hbar \frac{v_D}{\Delta}$ (b) on mechanical tension σ for graphene deposited on the hBN substrate. The circles correspond to the results calculated for Δ and v_D. The solid lines correspond to their least-squares quadratic fits (see Figures 3 and 4).

Figure 6 shows the critical charge $Z_{cr}(n)$, which is subject to Equation (8), as a function of the radius R for graphene deposited on the hBN substrate. Here, the cutoff function $f(\rho/R)$ for the Coulomb potential (3) matches the uniform distribution of the impurity charge over the ball of radius r_0 and corresponds to the value $f_0 = 2$ in Equations (8) and (11). It is assumed that the effective dielectric constant $\varepsilon = 1$. As one can see, the closest integer value of the impurity charge $Z = 1$ is already supercritical for the first three levels $n = 0, 1, 2$ with the total angular momentum $J = 0$. For the level $n = 3$, it can be both supercritical and subcritical, depending on the mechanical tension σ and the cutoff radius R.

Figure 7 depicts the dependence of the critical charge $Z_{cr}(n)$ on the uniform deformation σ of graphene on the hBN substrate at the cutoff radius $r_0 = 15$Å for the levels $n = 2, 3$ with $J = 0$. It can be seen that the uniform stretching/compression results in a decrease/increase in the critical charge Z_{cr} both for the rectangular cutoff of the Coulomb potential (3) $f_0 = 1$, and for the cutoff function corresponding to the distribution of the impurity charge over the ball $f_0 = 2$. This statement also holds for the first two levels, except for the ground level when $f_0 = 1$ (see Table 1 and Figure 6). The impurity charge $Z = 1$ is supercritical for $f_0 = 1$ and the level $n = 2$ in the entire region of the considered deformations, and for the third excited level, when passing from $\sigma = -10\%$ to $\sigma = 10\%$, the value of $Z = 1$ from subcritical becomes supercritical. At the same time, the charge $Z = 1$ is supercritical in the entire range of σ for the cutoff function with $f_0 = 2$ and the level $n = 3$.

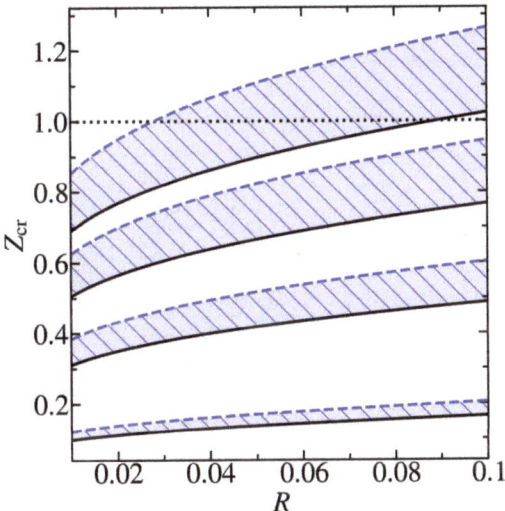

Figure 6. Dependence of the critical charge $Z_{cr}(n)$ on the dimensionless cutoff radius $R = r_0/l_D$ with the cutoff function corresponding to the value $f_0 = 2$, i.e., to the uniform distribution of the impurity charge over the ball of radius r_0 for graphene on the hBN substrate. The shaded areas correspond to the first four levels $n = 0, 1, 2, 3$ with the total angular momentum $J = 0$. The solid black lines correspond to the absence of deformation $\sigma = 0\%$. The dashed blue lines correspond to the compression $\sigma = -10\%$. The dotted line marks the closest integer value of the impurity charge $Z = 1$.

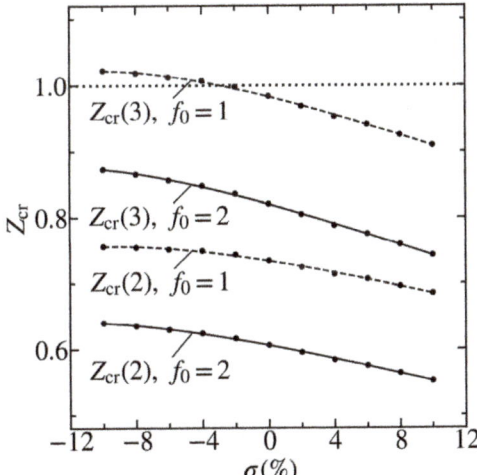

Figure 7. Dependence of the critical charge $Z_{cr}(n)$ on the mechanical tension σ for graphene on the hBN substrate. The lines are shown for the levels $n = 2, 3$ with the total angular momentum $J = 0$, and for the cutoff function corresponding to the value $f_0 = 2$ (solid lines) and the rectangular cutoff with $f_0 = 1$ (dashed lines). The cutoff radius is equal to $r_0 = 15$Å. The circles correspond to the results calculated for Δ and v_D. The lines correspond to their least-squares quadratic fits (see Figures 3 and 4). The dotted line marks the closest integer value of the impurity charge $Z = 1$.

Table 1. Values of the critical charge $Z_{cr}(n)$ versus the mechanical tension σ for graphene on the hBN substrate. The data is given for the first four levels $n = 0, 1, 2, 3$ with the total angular momentum $J = 0$, and for two kinds of cutoff function corresponding to the values $f_0 = 2$ and $f_0 = 1$, with a cutoff radius $r_0 = Rl_D = 15\text{Å}$.

	$f_0 = 2$			$f_0 = 1$		
$\sigma, \%$	10	0	−10	10	0	−10
$R = r_0/l_D$	0.061	0.030	0.012	0.061	0.030	0.012
$Z_{cr}(0)$	0.115	0.122	0.125	0.155	0.156	0.153
$Z_{cr}(1)$	0.349	0.378	0.394	0.442	0.466	0.472
$Z_{cr}(2)$	0.553	0.606	0.640	0.685	0.734	0.756
$Z_{cr}(3)$	0.743	0.820	0.873	0.909	0.984	1.023

Figure 8 shows the phase $\delta_0(\overline{E}; Z; r_0)$ (see Equation (9)) of hole scattering by the impurity with the charge $Z = 1$ in graphene on the hBN substrate for the states with the total angular momentum $J = 0$ for those values of uniform deformation σ for which $Z - Z_{cr} \ll 1$, i.e., when δ_0 changes sharply in the region of the resonance E_0^* over the width Γ^* (see Equations (13) and (14)). Figure 8a corresponds to the resonant scattering of holes at energies close to the quasi-discrete level with $n = 2$ in the case of the rectangular cutoff $f_0 = 1$ of the Coulomb potential (3), and Figure 8b corresponds to the resonant scattering of holes near the quasi-discrete level with $n = 3$ and with the cutoff function matching the uniform distribution of the impurity charge Z over the ball $f_0 = 2$. These quasistationary states with complex energies (13) are described by Equation (12) and correspond to the poles of the partial elastic scattering matrix $S_0(k)$ (9).

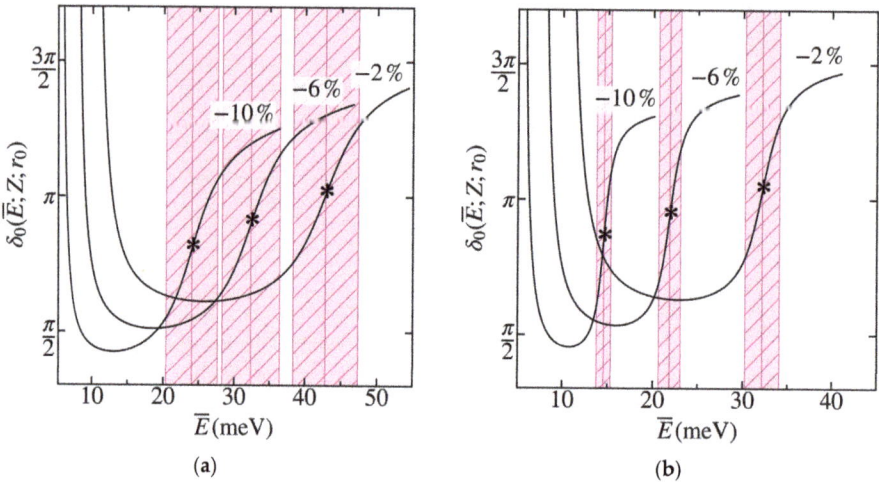

Figure 8. The scattering phase $\delta_0(\overline{E}; Z; r_0)$ as a function of the hole energy $\overline{E}$ in states with the total angular momentum $J = 0$ for graphene on the hBN substrate. The impurity charge $Z = 1$, the cutoff radius $r_0 = 15\text{Å}$, and the cutoff function corresponds to the uniform charge distribution over the sphere of the radius r_0 (**a**) and over the ball of the same radius (**b**). The asterisks mark positions of the resonances E_0^* corresponding to the quasi-stationary states with $n = 2$ (**a**) and $n = 3$ (**b**), the shaded areas mark their widths Γ^*. The numbers indicate the compression values σ.

It can be seen that the resonance E_0^* shifts under the uniform compression toward lower values and becomes sharper, i.e., its width Γ^* shrinks, which is due to a decrease in the supercriticality $Z - Z_{cr}$ (see Figure 7). In addition, when replacing the cutoff corresponding to $f_0 = 2$ with the rectangular one $f_0 = 1$, the resonance becomes sharper if it does not disappear completely, i.e., if the quasidiscrete level does not become the discrete one with $\bar{E} = -E < m^*v_D^2$. In contrast to the behavior of the phase at resonance corresponding to the quasidiscrete level with $n = 0$ (see Figure 7 in Reference [39]), where an abrupt change occurs from the value $\delta_0 = 0$ to $\delta_0 = \pi$, in this case, the phase sharply changes from the value $\delta_0 \approx \pi/2$ to the value $\delta_0 \approx 3\pi/2$, which, apparently, is associated with the accumulation of the background phase δ_{bg} when the level n rises. This means that the resonance should correspond not to the "pure" Breit–Wigner one (14), but to a more general case. In particular, it should correspond to a sharp minimum in the hole scattering cross-section σ_0 (see Reference [44] and Figure 13.3 therein for details).

5. Conclusions

In the presented study, we tried to answer the question of how the choice of the substrate and uniform mechanical stresses can affect the electronic properties in graphene, and, in particular, opens the energy gap and contributes to its fine-tuning. It was found that, when choosing the hexagonal boron nitride (hBN) and graphitic carbon nitride (gC$_3$N$_4$) substrates, the behavior of the electronic properties of deposited graphene is fundamentally different from the free-standing one. In this case, graphene has an energy gap, which makes it possible to classify it as a narrow-gap semiconductor, whereas the uniform stretching increases the value of the gap. These findings can help to overcome the main barrier to using graphene as an element of nano-electronic and straintronic devices. Moreover, uniform stretching affects the Dirac velocity in graphene on the substrate. Dirac velocity is linear with mechanical tension. It increases with uniform compression and decreases with uniform stretching. The Dirac point, near which the dispersion law is similar to the Einstein dependence of the particle energy on its momentum, corresponds, as usual, to the corner of the Brillouin zone for graphene deposited on the hBN substrate, while, for the gC$_3$N$_4$ substrate, it is located in the center of the zone. However, in both cases, the dynamics of charge carriers near the Dirac points is described by the effective two-dimensional radial Dirac equation for states with the total angular momentum $J = M + 1/2$. The orbital angular momentum M can be either integer or half-integer $M = 0, \pm 1/2, \pm 1, \pm 3/2, \ldots$, which is possible in two dimensions. We have limited ourselves to a detailed discussion of states with $J = 0$, i.e., with the half-integer orbital angular momentum $M = -1/2$, which is the first to descend to the top of the valence band in the presence of a supercharged impurity in graphene with a gap.

The dependences of the critical charges Z_{cr} on the homogeneous deformations of graphene on the indicated substrates as well as on the radius of the cutoff of the Coulomb potential at small distances, required for such values of the impurity charge, were calculated. The phases of the scattering of holes by supercritical $Z > Z_{cr}$ impurities were also calculated, and their behavior under the uniform stretching and compression was studied. The same applies to the poles of the elastic scattering matrix, which determine both the position and the width of the quasi-discrete levels.

In addition, the resonant scattering of the holes by the Coulomb potential modified at small distances in states with $J = 0$ was considered in detail for $Z - Z_{cr} \ll 1$, i.e., near the top of the valence band. Such scattering is completely analogous to the resonant scattering in the nonrelativistic single-channel problem on the potential with a low-permeable barrier. In both cases, the partial scattering cross-section has a resonant (Breit–Winger) shape, and there are no other inelastic resonance channels. This is an additional proof that the resonances in scattering at $Z > Z_{cr}$ cannot serve as evidence in favor of the spontaneous production of particle-antiparticle pairs. Thus, the one-particle description is valid not only for $Z < Z_{cr}$, but also in the supercritical region $Z > Z_{cr}$. Therefore, it is possible to compare theoretical calculations with experimental data on the current-voltage characteristics obtained by the scanning tunneling microscopy.

Based on such experimental data, it would be possible to conclude whether half-integer orbital angular momenta are realized in the two-dimensional considered systems. In addition, the effective two-dimensional radial Dirac equation with the integer total angular momenta $J = \pm 1, \pm 2, \ldots$, i.e., with half-integer orbital angular momenta $M = 1/2, \pm 3/2, \ldots$, is exactly the same as the three-dimensional one with the Dirac quantum number $= -J = \mp 1, \mp 2, \ldots$. Thus, one could experimentally establish whether spontaneous production of e^+e^--pairs exist in the supercritical Coulomb problem. However, calculation of the effective dielectric constant of the considered systems is necessary for the direct comparison of experimental data on the spectra of current-voltage characteristics with the theoretical predictions.

Author Contributions: Conceptualization, M.M.M. and V.D.M. Methodology, K.P.K., M.M.M., K.S.K., and V.D.M. Software, K.P.K. and M.M.M. Validation, K.P.K., M.M.M., K.S.K, and V.D.M. Formal analysis, K.P.K., M.M.M., K.S.K., and V.D.M. Investigation, K.P.K., M.M.M., K.S.K., and V.D.M. Resources, K.S.K. and M.M.M. Data curation, M.M.M. and V.D.M. Writing—original draft preparation, M.M.M., K.S.K., and V.D.M. Writing—review and editing, K.P.K. and K.S.K. Visualization, K.P.K. Supervision, V.D.M. Project administration, M.M.M. and V.D.M. Funding acquisition, M.M.M. and V.D.M. All authors have read and agreed to the published version of the manuscript.

Funding: The reported study was funded by RFBR, according to the research projects Nos. 19-02-00643_A and 18-02-00278_A.

Acknowledgments: The Authors are grateful to M.I. Vysotsky, S.I. Godunov, S.V. Ivliev, V.M. Kuleshov, and Yu.E. Lozovik for fruitful discussions.

Conflicts of Interest: The authors declare no conflict of interest. The funders had no role in the design of the study, in the collection, analyses, or interpretation of data, in the writing of the manuscript, or in the decision to publish the results.

References

1. Novoselov, K.S.; Geim, A.K.; Morozov, S.V.; Jiang, D.; Zhang, Y.; Dubonos, S.V.; Grigorieva, I.V.; Firsov, A.A. Electric Field Effect in Atomically Thin Carbon Films. *Science* **2004**, *306*, 666–669. [CrossRef] [PubMed]
2. Wallace, P.R. The Band Theory of Graphite. *Phys. Rev.* **1947**, *71*, 622–634. [CrossRef]
3. Castro Neto, A.H.; Guinea, F.; Peres, N.M.R.; Novoselov, K.S.; Geim, A.K. The electronic properties of graphene. *Rev. Mod. Phys.* **2009**, *81*, 109–162. [CrossRef]
4. Bolotin, K.I.; Sikes, K.J.; Jiang, Z.; Klima, M.; Fudenberg, G.; Hone, J.; Kim, P.; Stormer, H.L. Ultrahigh electron mobility in suspended graphene. *Solid State Commun.* **2008**, *146*, 351–355. [CrossRef]
5. Lee, C.; Wei, X.; Kysar, J.W.; Hone, J. Measurement of the Elastic Properties and Intrinsic Strength of Monolayer Graphene. *Science* **2008**, *321*, 385–388. [CrossRef]
6. Cao, K.; Feng, S.; Han, Y.; Gao, L.; Hue Ly, T.; Xu, Z., Lu, Y. Elastic straining of free-standing monolayer graphene. *Nat. Commun.* **2020**, *11*, 284. [CrossRef]
7. Balandin, A.A.; Ghosh, S.; Bao, W.; Calizo, I.; Teweldebrhan, D.; Miao, F.; Lau, C.N. Superior Thermal Conductivity of Single-Layer Graphene. *Nano Lett.* **2008**, *8*, 902–907. [CrossRef]
8. Cheng, S.-H.; Zou, K.; Okino, F.; Gutierrez, H.R.; Gupta, A.; Shen, N.; Eklund, P.C.; Sofo, J.O.; Zhu, J. Reversible fluorination of graphene: Evidence of a two-dimensional wide bandgap semiconductor. *Phys. Rev. B* **2010**, *81*, 205435. [CrossRef]
9. Robinson, J.T.; Burgess, J.S.; Junkermeier, C.E.; Badescu, S.C.; Reinecke, T.L.; Perkins, F.K.; Zalalutdniov, M.K.; Baldwin, J.W.; Culbertson, J.C.; Sheehan, P.E.; et al. Properties of Fluorinated Graphene Films. *Nano Lett.* **2010**, *10*, 3001–3005. [CrossRef]
10. Nair, R.R.; Ren, W.; Jalil, R.; Riaz, I.; Kravets, V.G.; Britnell, L.; Blake, P.; Schedin, F.; Mayorov, A.S.; Yuan, S.; et al. Fluorographene: A Two-Dimensional Counterpart of Teflon. *Small* **2010**, *6*, 2877–2884. [CrossRef]
11. Son, Y.-W.; Cohen, M.L.; Louie, S.G. Half-metallic graphene nanoribbons. *Nature* **2006**, *444*, 347–349. [CrossRef] [PubMed]
12. Barone, V.; Hod, O.; Scuseria, G.E. Electronic Structure and Stability of Semiconducting Graphene Nanoribbons. *Nano Lett.* **2006**, *6*, 2748–2754. [CrossRef]
13. Guinea, F.; Katsnelson, M.I.; Geim, A.K. Energy gaps and a zero-field quantum Hall effect in graphene by strain engineering. *Nat. Phys.* **2009**, *6*, 30–33. [CrossRef]
14. Ma, Y.; Dai, Y.; Guo, M.; Niu, C.; Zhu, Y.; Huang, B. Evidence of the Existence of Magnetism in Pristine VX2 Monolayers (X = S, Se) and Their Strain-Induced Tunable Magnetic Properties. *ACS Nano* **2012**, *6*, 1695–1701. [CrossRef] [PubMed]
15. Kharche, N.; Nayak, S.K. Quasiparticle Band Gap Engineering of Graphene and Graphone on Hexagonal Boron Nitride Substrate. *Nano Lett.* **2011**, *11*, 5274–5278. [CrossRef] [PubMed]

16. Giovannetti, G.; Khomyakov, P.A.; Brocks, G.; Kelly, P.J.; van den Brink, J. Substrate-induced band gap in graphene on hexagonal boron nitride: Ab initio density functional calculations. *Phys. Rev. B* **2007**, *76*, 073103. [CrossRef]
17. Bokdam, M.; Amlaki, T.; Brocks, G.; Kelly, P.J. Band gaps in incommensurable graphene on hexagonal boron nitride. *Phys. Rev. B* **2014**, *89*, 201404. [CrossRef]
18. Fan, Y.; Zhao, M.; Wang, Z.; Zhang, X.; Zhang, H. Tunable electronic structures of graphene/boron nitride heterobilayers. *Appl. Phys. Lett.* **2011**, *98*, 083103. [CrossRef]
19. Kistanov, A.A.; Cai, Y.; Zhou, K.; Dmitriev, S.V.; Zhang, Y.-W. Effects of graphene/BN encapsulation, surface functionalization and molecular adsorption on the electronic properties of layered InSe: A first-principles study. *Phys. Chem. Chem. Phys.* **2018**, *20*, 12939–12947. [CrossRef]
20. Kistanov, A.A.; Cai, Y.; Zhang, Y.-W.; Dmitriev, S.V.; Zhou, K. Strain and water effects on the electronic structure and chemical activity of in-plane graphene/silicene heterostructure. *J. Phys. Condens. Matter* **2017**, *29*, 095302. [CrossRef]
21. Xu, X.; Liu, C.; Sun, Z.; Cao, T.; Zhang, Z.; Wang, E.; Liu, Z.; Liu, K. Interfacial engineering in graphene bandgap. *Chem. Soc. Rev.* **2018**, *47*, 3059–3099. [CrossRef] [PubMed]
22. Forestier, A.; Balima, F.; Bousige, C.; de Sousa Pinheiro, G.; Fulcrand, R.; Kalbáč, M.; Machon, D.; San-Miguel, A. Strain and Piezo-Doping Mismatch between Graphene Layers. *J. Phys. Chem. C* **2020**, *124*, 11193–11199. [CrossRef]
23. Hoat, D.M.; Vu, T.V.; Obeid, M.M.; Jappor, H.R. Tuning the electronic structure of 2D materials by strain and external electric field: Case of GeI$_2$ monolayer. *Chem. Phys.* **2019**, *527*, 110499. [CrossRef]
24. Nguyen, H.T.T.; Vu, T.V.; Binh, N.T.T.; Hoat, D.M.; Hieu, N.V.; Anh, N.T.T.; Nguyen, C.V.; Phuc, H.V.; Jappor, H.R.; Obeid, M.M.; et al. Strain-tunable electronic and optical properties of monolayer GeSe: Promising for photocatalytic water splitting applications. *Chem. Phys.* **2020**, *529*, 110543. [CrossRef]
25. Kistanov, A.A.; Cai, Y.; Zhou, K.; Srikanth, N.; Dmitriev, S.V.; Zhang, Y.-W. Exploring the charge localization and band gap opening of borophene: A first-principles study. *Nanoscale* **2018**, *10*, 1403–1410. [CrossRef]
26. Hoat, D.M.; Vu, T.V.; Obeid, M.M.; Jappor, H.R. Assessing optoelectronic properties of PbI$_2$ monolayer under uniaxial strain from first principles calculations. *Superlattices Microstruct.* **2019**, *130*, 354–360. [CrossRef]
27. Baimova, J.A.; Korznikova, E.A.; Dmitirev, S.V.; Liu, B.; Zhou, K. Wrinkles and Wrinklons in Graphene and Graphene Nanoribbons under Strain. *Curr. Nanosci.* **2016**, *12*, 184–191. [CrossRef]
28. Baimova, J.A.; Liu, B.; Zhou, K. Folding and crumpling of graphene under biaxial compression. *LoM* **2014**, *4*, 96–99. [CrossRef]
29. Akhiezer, A.I.; Berestetskii, V.B. *Quantum Electrodynamics*; Nauka: Moscow, Russia, 1981.
30. DiVincenzo, D.P.; Mele, E.J. Self-consistent effective-mass theory for intralayer screening in graphite intercalation compounds. *Phys. Rev. B* **1984**, *29*, 1685–1694. [CrossRef]
31. Dirac, P.A.M. *The Principles of Quantum Mechanics*; Clarendon Press: Oxford, UK, 1958.
32. Pauli, W. Über ein kriterium für ein- oder zweiwertigkeit der eigenfunktionen in der wellenmechanik. *Helv. Phys. Acta* **1939**, *12*, 147–167.
33. Van Winter, C. Orbital angular momentum and group representations. *Ann. Phys.* **1968**, *47*, 232–274. [CrossRef]
34. Kowalski, K.; Podlaski, K.; Rembieliński, J. Quantum mechanics of a free particle on a plane with an extracted point. *Phys. Rev. A* **2002**, *66*, 032118. [CrossRef]
35. Kuleshov, V.M.; Mur, V.D.; Narozhny, N.B.; Fedotov, A.M.; Lozovik, Y.E. Coulomb problem for graphene with the gapped electron spectrum. *JETP Lett.* **2015**, *101*, 264–270. [CrossRef]
36. Sommerfeld, A. Zur quantentheorie der spektrallinien. *Ann. Phys.* **1916**, *51*, 1–94. [CrossRef]
37. Pomeranchuk, I.; Smorodinsky, Y. On the energy levels of systems with Z > 137. *J. Phys. USSR* **1945**, *9*, 97–100.
38. Khalilov, V.R.; Ho, C.-L. Dirac electron in a coulomb field in (2+1) dimensions. *Mod. Phys. Lett. A* **1998**, *13*, 615–622. [CrossRef]
39. Kuleshov, V.M.; Mur, V.D.; Fedotov, A.M.; Lozovik, Y.E. Coulomb Problem for $Z > Z_{cr}$ in Doped Graphene. *J. Exp. Theor. Phys.* **2017**, *125*, 1144–1162. [CrossRef]
40. Katin, K.P.; Krylov, K.S.; Maslov, M.M.; Mur, V.D. Tuning the supercritical effective charge in gapless graphene via Fermi velocity modifying through the mechanical stretching. *Diam. Relat. Mater.* **2019**, *100*, 107566. [CrossRef]
41. Mur, V.D.; Popov, V.S.; Voskresenskii, D.N. WKB method at Z > 137. *JETP Lett.* **1978**, *28*, 129–134.

42. Mur, V.D.; Popov, V.S. Quasiclassical approximation for the dirac equation in strong fields. *Sov. J. Nucl. Phys.* **1978**, *28*, 429.
43. Popov, V.S. Electron energy levels at $Z > 137$. *JETP Lett.* **1970**, *11*, 162–165.
44. Taylor, J.R. *Scattering Theory: The Quantum Theory on Nonrelativistic Collisions*; Wiley: New York, NY, USA, 1972.
45. Morgenstern, M. Scanning tunneling microscopy and spectroscopy of graphene on insulating substrates. *Phys. Status Solidi B* **2011**, *248*, 2423–2434. [CrossRef]
46. Wang, Y.; Wong, D.; Shytov, A.V.; Brar, V.W.; Choi, S.; Wu, Q.; Tsai, H.-Z.; Regan, W.; Zettl, A.; Kawakami, R.K.; et al. Observing Atomic Collapse Resonances in Artificial Nuclei on Graphene. *Science* **2013**, *340*, 734–737. [CrossRef] [PubMed]
47. Mao, J.; Jiang, Y.; Moldovan, D.; Li, G.; Watanabe, K.; Taniguchi, T.; Masir, M.R.; Peeters, F.M.; Andrei, E.Y. Realization of a tunable artificial atom at a supercritically charged vacancy in graphene. *Nat. Phys.* **2016**, *12*, 545–549. [CrossRef]
48. Mur, V.D.; Narozhny, N.B.; Petrosyan, A.N.; Lozovik, Y.E. Quantum dot version of topological phase: Half-integer orbital angular momenta. *JETP Lett.* **2008**, *88*, 688–692. [CrossRef]
49. Kuleshov, V.M.; Mur, V.D.; Narozhny, N.B.; Lozovik, Y.E. Topological Phase and Half-Integer Orbital Angular Momenta in Circular Quantum Dots. *Few Body Syst.* **2016**, *57*, 1103–1126. [CrossRef]
50. Wang, Y.; Yang, R.; Shi, Z.; Zhang, L.; Shi, D.; Wang, E.; Zhang, G. Super-Elastic Graphene Ripples for Flexible Strain Sensors. *ACS Nano* **2011**, *5*, 3645–3650. [CrossRef]
51. Dmitriev, S.V.; Baimova, Y.A.; Savin, A.V.; Kivshar', Y.S. Stability range for a flat graphene sheet subjected to in-plane deformation. *JETP Lett.* **2011**, *93*, 571–576. [CrossRef]
52. Giannozzi, P.; Baroni, S.; Bonini, N.; Calandra, M.; Car, R.; Cavazzoni, C.; Ceresoli, D.; Chiarotti, G.L.; Cococcioni, M.; Dabo, I.; et al. Quantum espresso: A modular and open-source software project for quantum simulations of materials. *J. Phys. Condens. Matter* **2009**, *21*, 395502. [CrossRef]
53. Giannozzi, P.; Andreussi, O.; Brumme, T.; Bunau, O.; Buongiorno Nardelli, M.; Calandra, M.; Car, R.; Cavazzoni, C.; Ceresoli, D.; Cococcioni, M.; et al. Advanced capabilities for materials modelling with Quantum ESPRESSO. *J. Phys. Condens. Matter* **2017**, *29*, 465901. [CrossRef]
54. Perdew, J.P.; Burke, K.; Ernzerhof, M. Generalized Gradient Approximation Made Simple. *Phys. Rev. Lett.* **1996**, *77*, 3865–3868. [CrossRef] [PubMed]
55. Blöchl, P.E. Projector augmented-wave method. *Phys. Rev. B* **1994**, *50*, 17953–17979. [CrossRef] [PubMed]
56. Kresse, G.; Joubert, D. From ultrasoft pseudopotentials to the projector augmented-wave method. *Phys. Rev. B* **1999**, *59*, 1758–1775. [CrossRef]
57. Grimme, S.; Antony, J.; Ehrlich, S.; Krieg, H. A consistent and accurate ab initio parametrization of density functional dispersion correction (DFT-D) for the 94 elements H-Pu. *J. Chem. Phys.* **2010**, *132*, 154104. [CrossRef] [PubMed]
58. Monkhorst, H.J.; Pack, J.D. Special points for Brillouin-zone integrations. *Phys. Rev. B* **1976**, *13*, 5188–5192. [CrossRef]
59. Methfessel, M.; Paxton, A.T. High-precision sampling for Brillouin-zone integration in metals. *Phys. Rev. B* **1989**, *40*, 3616–3621. [CrossRef]
60. Blöchl, P.E.; Jepsen, O.; Andersen, O.K. Improved tetrahedron method for Brillouin-zone integrations. *Phys. Rev. B* **1994**, *49*, 16223–16233. [CrossRef]
61. Kang, Y.-J.; Kang, J.; Chang, K.J. Electronic structure of graphene and doping effect on SiO_2. *Phys. Rev. B* **2008**, *78*, 115404. [CrossRef]
62. Nevius, M.S.; Conrad, M.; Wang, F.; Celis, A.; Nair, M.N.; Taleb-Ibrahimi, A.; Tejeda, A.; Conrad, E.H. Semiconducting Graphene from Highly Ordered Substrate Interactions. *Phys. Rev. Lett.* **2015**, *115*, 136802. [CrossRef]
63. Zhou, S.Y.; Gweon, G.-H.; Fedorov, A.V.; First, P.N.; de Heer, W.A.; Lee, D.-H.; Guinea, F.; Castro Neto, A.H.; Lanzara, A. Substrate-induced bandgap opening in epitaxial graphene. *Nat. Mater.* **2007**, *6*, 770–775. [CrossRef]
64. Pereira, V.M.; Kotov, V.N.; Castro Neto, A.H. Supercritical Coulomb impurities in gapped graphene. *Phys. Rev. B* **2008**, *78*, 085101. [CrossRef]
65. Pieper, W.; Greiner, W. Interior electron shells in superheavy nuclei. *Z. Phys.* **1969**, *218*, 327–340. [CrossRef]
66. Gershteĭn, S.S.; Zel'dovich, Y.B. Positron production during the mutual approach of heavy nuclei and the polarization of the vacuum. *JETP* **1970**, *30*, 358–361.

67. Zel'dovich, Y.B.; Popov, V.S. Electronic structure of superheavy atoms. *Sov. Phys. Usp.* **1972**, *14*, 673–694. [CrossRef]
68. Okun, L.B. Superbound electrons. *Comments Nucl. Part. Phys.* **1974**, *6*, 25–27.
69. Voronov, B.L.; Gitman, D.M.; Tyutin, I.V. The Dirac Hamiltonian with a superstrong Coulomb field. *Theor. Math. Phys.* **2007**, *150*, 34–72. [CrossRef]
70. Kuleshov, V.M.; Mur, V.D.; Narozhny, N.B.; Fedotov, A.M.; Lozovik, Y.E.; Popov, V.S. Coulomb problem for a $Z > Z_{cr}$ nucleus. *Phys. Usp.* **2015**, *58*, 785–791. [CrossRef]
71. Krylov, K.S.; Mur, V.D.; Fedotov, A.M. On the resonances near the continua boundaries of the Dirac equation with a short-range interaction. *Eur. Phys. J. C* **2020**, *80*, 270. [CrossRef]
72. Schwinger, J. On Gauge Invariance and Vacuum Polarization. *Phys. Rev.* **1951**, *82*, 664–679. [CrossRef]
73. Godunov, S.I.; Machet, B.; Vysotsky, M.I. Resonances in positron scattering on a supercritical nucleus and spontaneous production of e^+e^- pairs. *Eur. Phys. J. C* **2017**, *77*, 782. [CrossRef]
74. Behera, H.; Mukhopadhyay, G. Strain-tunable band gap in graphene/h-BN hetero-bilayer. *J. Phys. Chem. Solids* **2012**, *73*, 818–821. [CrossRef]
75. Li, X.; Dai, Y.; Ma, Y.; Han, S.; Huang, B. Graphene/g-C3N4 bilayer: Considerable band gap opening and effective band structure engineering. *Phys. Chem. Chem. Phys.* **2014**, *16*, 4230. [CrossRef] [PubMed]
76. Zuo, Z.; Xu, Z.; Zheng, R.; Khanaki, A.; Zheng, J.-G.; Liu, J. In-situ epitaxial growth of graphene/h-BN van der Waals heterostructures by molecular beam epitaxy. *Sci. Rep.* **2015**, *5*, 14760. [CrossRef]
77. Yu, Q.; Guo, S.; Li, X.; Zhang, M. Template free fabrication of porous g-C$_3$N$_4$/graphene hybrid with enhanced photocatalytic capability under visible light. *Mater. Technol.* **2014**, *29*, 172–178. [CrossRef]
78. Elias, D.C.; Nair, R.R.; Mohiuddin, T.M.G.; Morozov, S.V.; Blake, P.; Halsall, M.P.; Ferrari, A.C.; Boukhvalov, D.W.; Katsnelson, M.I.; Geim, A.K.; et al. Control of Graphene's Properties by Reversible Hydrogenation: Evidence for Graphane. *Science* **2009**, *323*, 610–613. [CrossRef] [PubMed]
79. Girit, Ç.Ö.; Meyer, J.C.; Erni, R.; Rossell, M.D.; Kisielowski, C.; Yang, L.; Park, C.-H.; Crommie, M.F.; Cohen, M.L.; Louie, S.G.; et al. Graphene at the Edge: Stability and Dynamics. *Science* **2009**, *323*, 1705–1708. [CrossRef] [PubMed]
80. Kan, E.; Ren, H.; Wu, F.; Li, Z.; Lu, R.; Xiao, C.; Deng, K.; Yang, J. Why the Band Gap of Graphene Is Tunable on Hexagonal Boron Nitride. *J. Phys. Chem. C* **2012**, *116*, 3142–3146. [CrossRef]
81. Sahalianov, I.Y.; Radchenko, T.M.; Tatarenko, V.A.; Cuniberti, G.; Prylutskyy, Y.I. Straintronics in graphene: Extra large electronic band gap induced by tensile and shear strains. *J. Appl. Phys.* **2019**, *126*, 054302. [CrossRef]
82. Ni, Z.H.; Yu, T.; Lu, Y.H.; Wang, Y.Y.; Feng, Y.P.; Shen, Z.X. Uniaxial Strain on Graphene: Raman Spectroscopy Study and Band-Gap Opening. *ACS Nano* **2008**, *2*, 2301–2305. [CrossRef]
83. Naumov, I.I.; Bratkovsky, A.M. Gap opening in graphene by simple periodic inhomogeneous strain. *Phys. Rev. B* **2011**, *84*, 245444. [CrossRef]
84. Chen, S.; He, M.; Zhang, Y.-H.; Hsieh, V.; Fei, Z.; Watanabe, K.; Taniguchi, T.; Cobden, D.H.; Xu, X.; Dean, C.R.; et al. Electrically tunable correlated and topological states in twisted monolayer–bilayer graphene. *Nat. Phys.* **2020**. [CrossRef]
85. Bousige, C.; Balima, F.; Machon, D.; Pinheiro, G.S.; Torres-Dias, A.; Nicolle, J.; Kalita, D.; Bendiab, N.; Marty, L.; Bouchiat, V.; et al. Biaxial Strain Transfer in Supported Graphene. *Nano Lett.* **2017**, *17*, 21–27. [CrossRef] [PubMed]
86. Novoselov, K.S.; Geim, A.K.; Morozov, S.V.; Jiang, D.; Katsnelson, M.I.; Grigorieva, I.V.; Dubonos, S.V.; Firsov, A.A. Two-dimensional gas of massless Dirac fermions in graphene. *Nature* **2005**, *438*, 197–200. [CrossRef] [PubMed]

Publisher's Note: MDPI stays neutral with regard to jurisdictional claims in published maps and institutional affiliations.

© 2020 by the authors. Licensee MDPI, Basel, Switzerland. This article is an open access article distributed under the terms and conditions of the Creative Commons Attribution (CC BY) license (http://creativecommons.org/licenses/by/4.0/).

Article

A Facile Method for the Generation of Fe₃C Nanoparticle and Fe–N$_x$ Active Site in Carbon Matrix to Achieve Good Oxygen Reduction Reaction Electrochemical Performances

Yuzhe Wu, Yuntong Li, Conghui Yuan * and Lizong Dai *

Fujian Provincial Key Laboratory of Fire Retardant Materials, College of Materials, Xiamen University, Xiamen 361005, China; 20720150150066@stu.xmu.edu.cn (Y.W.); lyt@stu.xmu.edu.cn (Y.L.)
* Correspondence: yuanch@xmu.edu.cn (C.Y.); lzdai@xmu.edu.cn (L.D.); Tel./Fax: +86-592-2186178 (C.Y. & L.D.)

Received: 14 September 2020; Accepted: 23 October 2020; Published: 26 October 2020

Abstract: Introduction of both nitrogen and transition metal elements into the carbon materials has demonstrated to be a promising strategy to construct highly active electrode materials for energy shortage. In this work, through the coordination reaction between Fe^{3+} and 1,3,5–tris(4–aminophenyl)benzene, metallosupramolecular polymer precursors are designed for the preparation of carbon flakes co-doped with both Fe and N elements. The as-prepared carbon flakes display wrinkled edges and comprise Fe₃C nanoparticle and active site of Fe–N$_x$. These carbon materials exhibit excellent electrocatalytic performance. Towards oxygen reduction reaction (ORR), the optimized sample has E_{onset} and $E_{half-wave}$ of 0.93 V and 0.83 V in alkaline system, respectively, which are very close to that of Pt/C. This approach may offer a new way to high performance and low-cost electrochemical catalysts.

Keywords: coordination; metallosupramolecular polymer; active site; carbon materials; oxygen reduction reaction

1. Introduction

In recent years, much attention has been focused on the development of facile and applicable methods to fabricate high-activity, low-cost oxygen reduction reaction (ORR) catalysts. This is significant to overcome the challenges in the commercialization and industrialization of hydrogen fuel cells. Nevertheless, Pt/C still acts as a main role in the commercialized ORR catalysts, even though it is relatively expensive [1–4]. Therefore, non-precious metal-based materials with high catalytic performance have attracted great research interest [5–8].

Indeed, transition metal elements have been widely introduced into the carbon materials to achieve high electrochemical performances [9–12]. Incorporation of transition metal elements into the carbon matrix generally relies on the design of composite precursors. For example, by using the reaction between salts (like iron(III) nitrate nonahydrate, nickel(II) nitrate hexahydrate, manganese(II) acetate tetrahydrate, and cobalt(II) nitrate hexahydrate) and graphite oxide, graphite oxide-metal-based precursors can be generated. Subsequently, a thermal procedure leads to the formation of Mn, Fe, Co, and Ni-coped graphene, which exhibited improved ORR performance [13]. Design of metal-organic precursors has been demonstrated to be a controllable method to generate transition metal elements doped carbon materials. It has been reported that precursors derived from the coordination between transition metals salts (MCl$_x$, M = Cu, Ni, Co, Fe and Mn) and melamine/aniline, can be used to prepare transition metal doped carbon materials with enhanced ORR properties [14]. More interesting, utilization of metal-organic frameworks as precursors for the fabrication of transition metal-doped carbon materials has attracted great attention, because of the designable composition, pore structure, and tunable metal species [15]. Notably, co-doping of both iron and N elements in the carbon materials is of great advantage for improving the electrochemical properties, due to the generation of iron-nitrogen-carbon (Fe–N–C) active sites (Fe–N_x and Fe–C_x) [16–18]. Usually, Fe–N–C catalysts were prepared by thermal cracking of iron macrocyclic polymers, iron-organic salts, or N-containing compounds [19,20]. The complicated synthetic process and high cost greatly limited their practical application [21,22]. So, it is still desirable to explore simple method to prepare Fe–N–C catalysts from commercial resources.

In this report, we show that a simple ligand-Fe^{3+} coordination strategy using commercial resources as starting materials, can create metallosupramolecular polymer precursors for the fabrication of Fe and N elements co-doped carbon materials with high ORR activity. The organic ligand adopted in this research is 1,3,5–tris(4–aminophenyl)benzene (denoted as TA) possessing three amine groups and conjugated structure. Because of the rigid structure of TA, the as-formed TA–Fe coordinative networks display uniform layer structure. After carbonization, carbon flakes with wrinkled edges (denoted as CNSs) can be easily generated. We have found that Fe_3C nanoparticle and active site of Fe–N_x are formed in the carbon matrix, which can promote the ORR activity of the carbon materials. In addition, the feature of this research lies in the simple synthesis of CNSs. It can introduce Fe_3C nanoparticle and Fe–N_x into the carbon matrix by one step. This synthetic route has certain universality and representativeness. It is also found that the appropriate molar ratio between amino ligand and Fe^{3+} was the most important factor that determines the activity of CNSs.

2. Experimental Section

2.1. Materials

1,3,5–Tris(4–aminophenyl)benzene, iron(III) chloride hexahydrate and dichloromethane were supplied by Aladdin Company (Shanghai, China) and directly used. Anhydrous ethanol, anhydrous methanol, KOH, and hydrochloric acid were purchased from Shanghai Chemical Reagent Industry (Shanghai, China). Nafion (5 wt%) was supplied by Sigma-Aldrich (Shanghai, China).

2.2. Catalysts Preparation

TA (0.3 g, 0.85 mmol) was firstly dissolved in 100 mL of dichloromethane with vigorous stirring for 3 h. Then, 1.38, 1.84, and 2.30 mL methanol solutions of Iron(III) chloride hexahydrate (100 mg/mL, 0.37 mmol/mL) were added into the solution drop by drop under vigorous stirring under N_2 atmosphere. Thus, the corresponding molar ratios between TA and Fe^{3+} were 1.0:0.6, 1.0:0.8, and 1.0:1.0, respectively. The mixtures became yellow, and suspensions of metallosupramolecular polymers were formed after 12 h of reaction. The yellow powders were collected by centrifugation and washing with a mixed solvent of 30 mL dichloromethane and 30 mL anhydrous methanol three times. After drying in vacuum at 50 °C overnight, TA–Fe precursors were obtained (denoted as TA–0.6Fe, TA–0.8Fe and TA–1.0Fe). These precursors were then carbonized at 850 °C for 2 h in argon gas with a heating rate of 5 °C/min to generate the carbon materials. The carbon materials were washed by 6 M HCl for 5 h and 100 mL of ultrapure water three times at room temperature, then dried by freeze-drying overnight. A second carbonization was performed at 850 °C for 2 h in argon gas with a heating rate of 10 °C/min to get the target carbon materials (denoted as CNS–0.6Fe, CNS–0.8Fe and CNS–1.0Fe).

2.3. Characterization

The scanning electron microscopy (SEM) images were observed through an SU-70 microscopy instrument (HITACHI, Tokyo, Japan). The FTIR measurements were tested by an AVATAR 360 FTIR (Nicolet Instrument, Tokyo, Japan) at room temperature. The powder X-ray diffraction (XRD) patterns were measured through a Desktop X-ray Diffractometer ((Rigaku, Tokyo, Japan) using Cu (600 W) Kα radiation. Raman spectra were tested by a Labram HR800 Evolution (Horiba, Lille, France). The X-ray photoelectron spectroscopy (XPS) were tested by PHI Quantum-2000 photoelectron spectrometer

(Physical Electronics, Inc., Chanhassen, MN, USA). The pore volume and Brunauer–Emmett–Teller (BET) were taken through an ASAP 2460 system (Norcross, GA, USA). Electron paramagnetic resonance (EPR) experiments were conducted on an electron spin resonance spectrometer (Bruker EMX-10/12, Bruker UK, Coventry, UK) at 90 K. Transmission electron microscopic (TEM) measurements were performed using a JEM-2100 microscope (JEOL, Tokyo, Japan). The elemental energy-dispersive X-ray spectroscopy (EDX) were obtained by using a FEI TECNAI F20 microscope (Hillsboro, OR, USA).

2.4. Electrochemical Measurements

The electrochemical experiments were conducted on an electrochemical workstation (CHI 760E, Chenhua, Shanghai, China), by using the typical three-electrode system. A standard rotating disk electrode with a glassy carbon disk (5 mm in diameter) was applied as working electrode. Before test, CNS–0.6Fe, CNS–0.8Fe, and CNS–1.0Fe (5.0 mg) were dispersed in 1.0 mL of homogeneous solvent with 500 μL of anhydrous ethanol, 450 μL of H_2O and 50 μL of 5 wt% Nafion. The above newly made slurry (4.5 μL) was carefully dropped onto a glassy carbon electrode as working electrode. The ORR performance was tested in newly made KOH aqueous solution (0.1 mol/L) at room temperature. Pt foil and Ag/AgCl (KCl saturation) electrode were separately applied as counter electrode and reference electrode. The potential in this study was relative to the Ag/AgCl electrode.

3. Results and Discussion

3.1. Characterization of Metallosupramolecular Polymer Precursors

The synthetic process of precursor is very simple. As shown in Scheme 1, metallosupramolecular polymer precursors are generated from the direct reaction between commercially available resources. In a dichloromethane solution, the high coordination affinity between TA and Fe^{3+} can easily induce the formation precipitates. After a pyrolysis treatment of the precursors, carbon materials comprising both Fe_3C nanoparticle and $Fe–N_x$ active site can be fabricated easily.

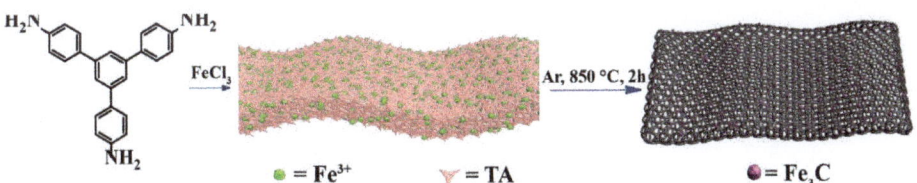

Scheme 1. Synthetic process of the CNSs.

The typical SEM image of TA–0.8Fe is displayed in Figure 1a, from which sheet-like morphology can be observed. The coordination reaction between TA and Fe^{3+} was verified by FTIR as shown in Figure 1b. The peaks located at 3350–3343 cm^{-1} correspond to the characteristic signals of amino groups of TA, and the peaks at 845, 685, and 600 cm^{-1} are derived from iron(III) chloride hexahydrate. Comparing the spectra of precursors with TA, the characteristic peak of amino groups shifts and becomes broaden. Also, the characteristic peaks of Fe^{3+} in the precursors were evidently weakened. These results indicate the coordination between Fe^{3+} and TA [23].

In the EPR spectra (Figure 2a), all samples show a small radical signal of amino group at $g' = 2.00$. For TA–0.6Fe, TA–0.8Fe, and TA–1.0Fe, the representative signal of Fe^{3+} at $g' = 4.25$ can be observed, indicating the presence of Fe^{3+} in the precursor [24]. The XPS survey spectra of TA, TA–0.6Fe, TA–0.8Fe, and TA–1.0Fe are displayed in Figure 2b. The representative signals of Fe 2p locate at 714.4 ± 0.1 and 725.4 ± 0.1 eV, which can be attributed to the binding energies of $2p_{3/2}$ and $2p_{1/2}$ orbitals of Fe^{3+}, respectively. Figure 2c–f shows the high-resolution XPS spectra of N 1s of TA and the precursors. A signal at 399.4 ± 0.1 eV is attributed to the amino group. In the case of TA–0.6Fe, TA–0.8Fe, and TA–1.0Fe, a peak at 401.6 ± 0.1 eV is attributed to amino group perturbed by Fe^{3+} [25], which is helpful for the formation of Fe–N–C active site during carbonization. As showed in Table S1, TA–0.8Fe has higher content of Fe–NH_2 than TA–0.6Fe and TA–1.0Fe, which may result in more content of Fe–N_x active site after carbonization.

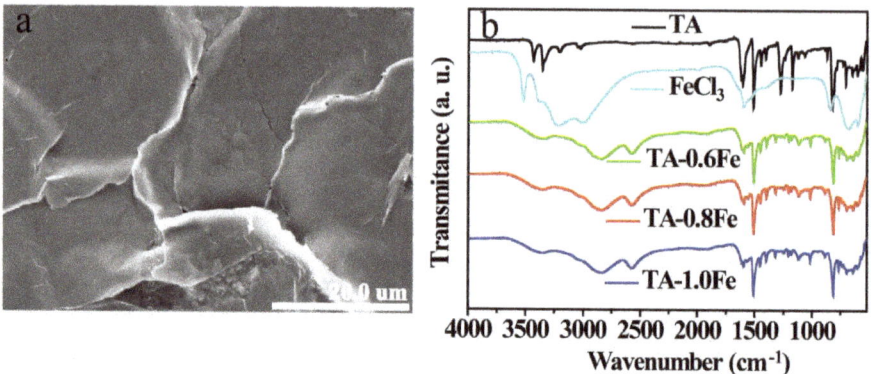

Figure 1. (a) SEM image of TA–0.8Fe. (b) FTIR spectra of TA, iron(III) chloride hexahydrate and TA–0.6Fe, TA–0.8Fe, and TA–1.0Fe.

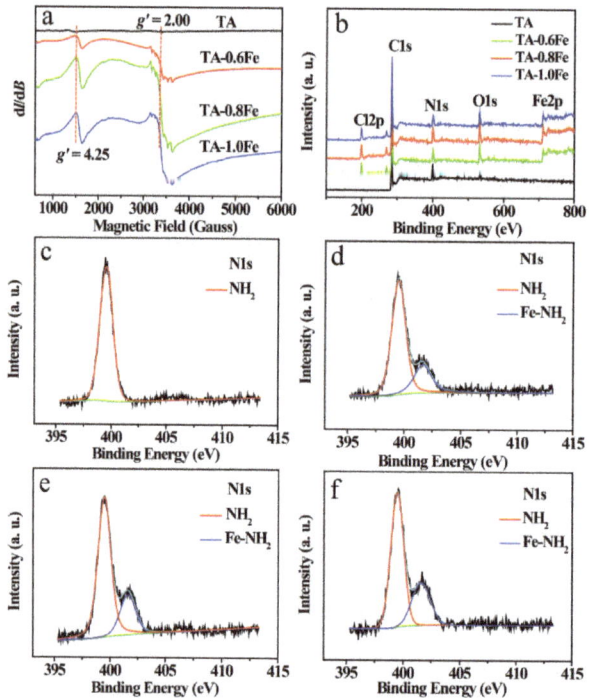

Figure 2. (a) EPR spectra of TA, TA–0.6Fe, TA–0.8Fe and TA–1.0Fe. (b) XPS survey spectra of TA, TA–0.6Fe, TA–0.8Fe and TA–1.0Fe. High-resolution XPS spectra of N 1s of TA (c), TA–0.6Fe (d), TA–0.8Fe (e) and TA–1.0Fe (f).

3.2. Structure and Composition of CNSs

After twice carbonization at 850 °C, the as obtained CNSs can maintain the lamellar structure, but the edges become wrinkled (Figure 3a,b and Figure S1a,b). This structure may be helpful for the direct contact between the active sites with the oxygen, thus improving the electrocatalytic activity of carbon materials. Notably, CNS–0.8Fe possesses the most uniform lamellar morphology (Figure 3b). The high-resolution TEM images of CNS–0.8Fe show clear inter-planar distance of 0.201 nm derived from the (031) plan of Fe_3C nanoparticle (Figure 3c,d). The outer carbon coating on the Fe_3C nanoparticle has a good lattice structure with a spacing of 0.34 nm, corresponding to the (002) plan of graphitic carbon (Figure 3c,d). The outer layer of graphitized carbon on Fe_3C nanoparticles has good electrical conductivity. During the ORR catalytic process, Fe_3C nanoparticles may not contact with electrolyte directly, but can play the catalytic role indirectly through the outer layer of graphitized carbon to improve the catalytic activity. The Fe_3C nanoparticle generated in CNSs can improve the ORR activity of carbon materials, which was confirmed already [26]. Figure 3e–i gives the dark-field TEM image and EDX mapping of CNS–0.8Fe. Obviously, elements of C, N, Fe, and O are homogeneously dispersed all through the carbon materials.

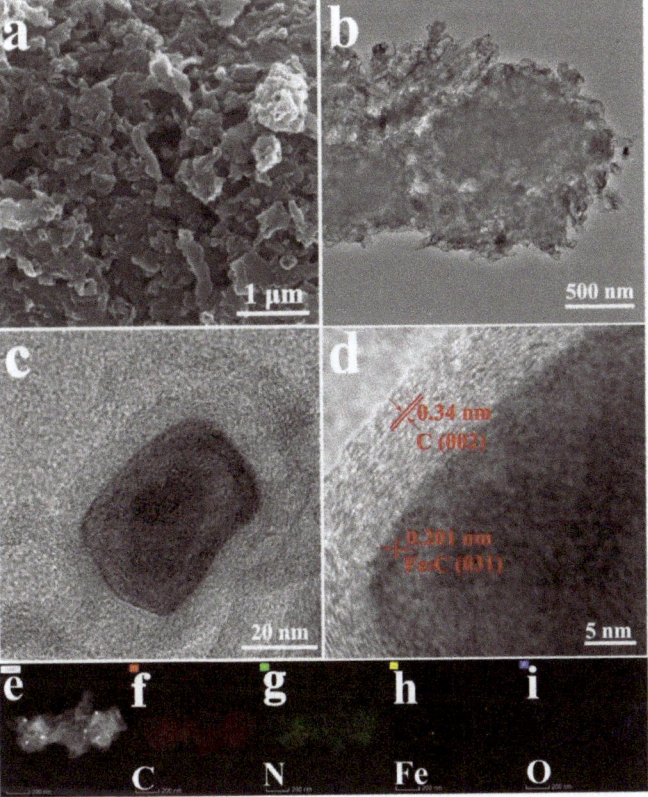

Figure 3. SEM image of CNS–0.8Fe (**a**). TEM image of CNS–0.8Fe (**b**). High-resolution TEM images of CNS–0.8Fe (**c,d**). Dark-field image and EDX mappings C, N, Fe, and O elements (**e–i**) of CNS–0.8Fe.

Figure 4a illustrates the Raman spectra of the CNSs. For CNS–0.6Fe, CNS–0.8Fe and CNS–1.0Fe, the intensity ratios between D band (1340 cm^{-1}) derived from disordered graphitic structure and G band (1571 cm^{-1}) derived from ordered carbon structure are calculated to be 0.97, 0.95, and 1.04, respectively. This result indicates that the graphitic degree of CNS–0.8Fe is higher than that of the other samples. The crystalline structures of CNSs were evaluated by the XRD. As displayed in Figure 4b, a broad diffraction peak at about 25° is attributed to the (002) plane of ordered graphitic structure. Moreover, the XRD results clearly confirm that the iron element is remained in the carbon matrix as a Fe_3C form. All the diffraction peaks are in good agreement with that of the Fe_3C (JCPDS Card No.65–2413). These results, in combination with the high-resolution TEM images, prove the presence of Fe_3C nanoparticle in the CNSs catalysts, which may promote the ORR activity of carbon materials [27].

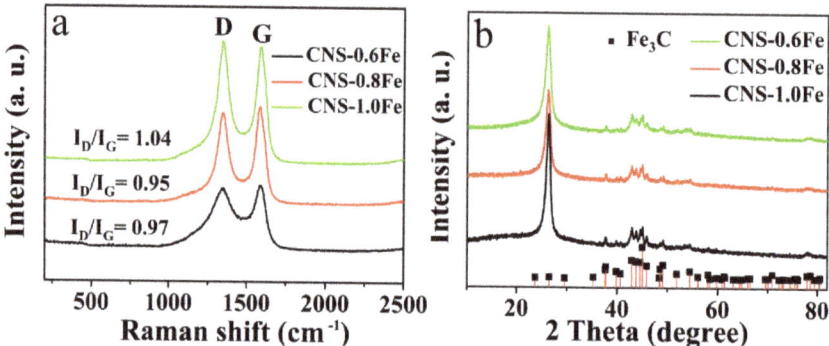

Figure 4. (a) Raman spectra of CNS–0.6Fe, CNS–0.8Fe and CNS–1.0Fe catalysts. (b) XRD patterns of CNS–0.6Fe, CNS–0.8Fe, and CNS–1.0Fe catalysts.

The pore character of CNSs was characterized through the physisorption of nitrogen at 77 K. As shown in Figure 5, all samples show well-developed micro-pore and mesoporous-pore structures. Table 1 shows the corresponding information about BET surface area as well as total pore volumes of CNSs. The surface areas of CNS–0.6Fe, CNS–0.8Fe, and CNS–1.0Fe are 167.86, 196.20, and 145.20 $m^2 \cdot g^{-1}$, with relevant pore volumes of 0.16, 0.17, and 0.17 $cm^3 \cdot g^{-1}$, respectively. Obviously, surface areas of CNSs are resulted from both micro-pore and mesoporous-pore structures. The CNS–0.8Fe has a higher BET surface area than CNS–0.6Fe and CNS–1.0Fe. We consider that the large surface area of CNS–0.8Fe is attributed to the moderate crosslinking degree of the metallosupramolecular polymer networks.

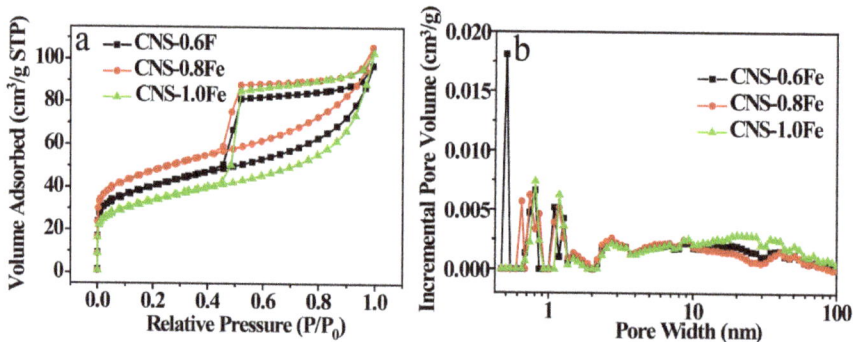

Figure 5. (a) N_2 adsorption and desorption isotherms of CNS–0.6Fe, CNS–0.8Fe and CNS–1.0Fe. (b) Pore size distribution of CNS–0.6Fe, CNS–0.8Fe and CNS–1.0Fe.

Table 1. Surface area, porosity of CNSs.

Samples	S_{BET} [a] [$m^2 \cdot g^{-1}$]	S_{micro} [b] [$m^2 \cdot g^{-1}$]	$S_{meso+macro}$ [c] [$m^2 \cdot g^{-1}$]	V_{total} [d] [$cm^3 \cdot g^{-1}$]
CNS–0.6Fe	167.86	73.36	94.49	0.16
CNS–0.8Fe	196.20	95.39	100.81	0.17
CNS–1.0Fe	145.20	62.43	82.77	0.17

[a] Specific surface area derived from BET. [b] Surface area about micropores calculated through the t-plot method. [c] Surface area of mesopores and macropores calculated through the t-plot method. [d] Total pore volume.

The XPS survey spectra of the CNSs are shown in Figure 6a. Also, the high-resolution XPS spectra of C 1s, N 1s, and Fe 2p of CNS–0.8Fe are displayed Figure 6b–d. The C 1s signal is split into three representative peaks at 284.4 ± 0.1 (C=C, C–C), 285.4 ± 0.1 (C–O, C–N), and 288.2 ± 0.1 eV (C=O). Four peaks of N 1s signal at 398.5 ± 0.1, 399.5 ± 0.1, 401.2 ± 0.1, and 404.5 ± 0.1 eV are respectively belong to the pyridinic N, Fe–N_x, graphitic N, and oxidized N. Notably, Fe–N_x and pyridinic N are recognized to be promising for the improvement of ORR activity [28,29]. The N 1s spectra of CNS–0.6Fe and CNS–1.0Fe are also showed in Figure S2. For the Fe 2p spectrum, the peak located at 725.4 ± 0.1 eV is assigned to the binding energy of Fe^{3+} for the $2p_{1/2}$ band, and the peak of Fe^{2+} is detected at 723.2 ± 0.1 eV for the $2p_{1/2}$ band. Another two peaks at 714.4 ± 0.1 and 710.5 ± 0.1 eV can be respectively attributed to the binding energies of $2p_{3/2}$ orbitals of Fe^{3+} as well as Fe^{2+} species. The last signal at 719.6 ± 0.1 eV is the satellite peak. These XPS results, in combination with the XRD results, clearly confirm that the presence of Fe_3C nanoparticle, and active sites of Fe–N_x and pyridinic N in the carbon matrix of CNSs. Moreover, as listed in Table 2, the pyridinic N and Fe–N_x contents of CNS–0.8Fe are 0.28 and 0.43 at.%, respectively, which are much higher than that of CNS–0.6Fe (0.17 and 0.20 at.%) and CNS–1.0Fe (0.25 and 0.22 at.%). This result indicates that CNS–0.8Fe may have more active sites towards ORR. In summary, the Fe and N elements co-doping effect leads to the generation of both Fe–N_x active site and Fe_3C nanoparticle when carbonization, which can greatly improve the catalytic activity towards ORR. That is the main mechanism and contribution of the co-doping effect of Fe and N elements in CNSs.

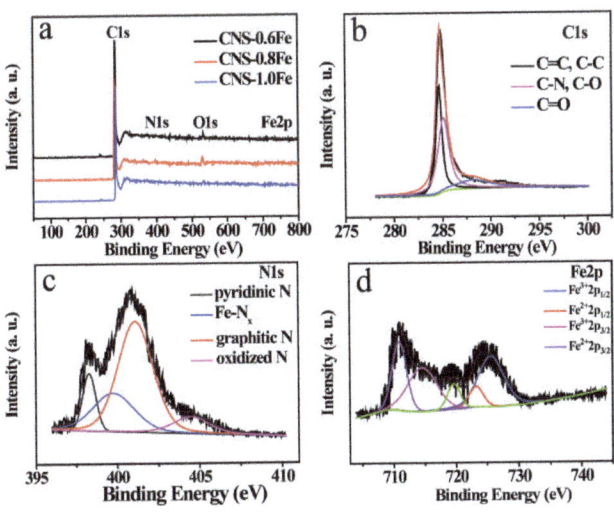

Figure 6. (a) XPS survey spectra of CNS–0.6Fe, CNS–0.8Fe and CNS–1.0Fe. High-resolution XPS spectra of (b) C 1s, (c) N 1s, (d) Fe 2p of CNS–0.8Fe.

Table 2. Contents (at.%) of nitrogen with different chemical environments calculated from the N1s XPS spectra.

Samples	N (at.%) [a]			
	Pyridinic N	Fe–N$_x$	Graphitic N	Oxidized N
CNS-0.6Fe	0.17	0.20	0.37	0.19
CNS-0.8Fe	0.28	0.43	0.89	0.21
CNS-1.0Fe	0.25	0.22	0.35	0.16

[a] The different contents of nitrogen (at.%) calculated by the analysis of the peak area as for pyridinic N, Fe–N$_x$, graphitic N and oxidized N.

3.3. ORR Performance of CNSs

The CV curves of CNSs were tested in both Ar or O$_2$-saturated 0.1 M KOH solution (Figure 7a). The samples show no reduction peak in Ar-saturated solution but show a typical reduction peak when changed into O$_2$-saturated solution. The double-layer capacitance of the three samples was researched, which are showed in Figure S3. The CV area of CNS–0.8Fe in Ar is larger than CNS–0.6Fe and CNS–1.0Fe. This indicates that CNS–0.8Fe owns larger electrochemically active surface area than the other two samples, which is helpful for improving the catalytic activity of ORR. Figure 7b list the LSV curves of CNSs. The onset potential (E$_{onset}$) of CNS–0.6Fe, CNS–0.8Fe, and CNS–1.0Fe for ORR are 0.89, 0.93, and 0.88 V vs. RHE, separately. The half-wave (E$_{half-wave}$) of the CNS–0.6Fe, CNS–0.8Fe and CNS–1.0Fe are 0.78, 0.83, and 0.80 V vs. RHE, separately. The above results were confirmed by LSV tests in the newly made O$_2$-saturated solution at the scanned rate of 10 mV/s with the fixed rotation speed of 1600 rpm. As a control experiment, the ORR activity of Pt/C catalyst was also tested with the same experimental condition (Figure 7b). Apparently, the E$_{onset}$ (0.93 V) and E$_{half-wave}$ (0.83 V) values of CNS–0.8Fe are very close to Pt/C catalyst (E$_{onset}$ = 0.95 V and E$_{half-wave}$ = 0.85 V). The Tafel plots of CNSs were tested (Figure S4). The Tafel slopes of CNS–0.6Fe and CNS–1.0Fe are 89 and 84 mV·dec^{-1}, respectively. However, a Tafel slope of 73 mV·dec^{-1} is detected for CNS–0.8Fe, which is lower than that of Pt/C (78 mV·dec^{-1}) and directly indicating that CNS–0.8Fe owns faster ORR kinetics.

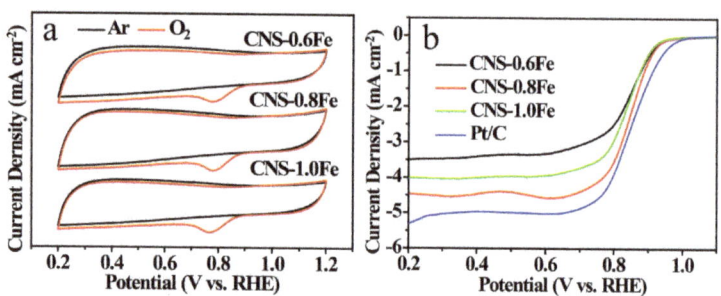

Figure 7. (a) CV curves of CNS–0.6Fe, CNS–0.8Fe and CNS–1.0Fe in Ar or O$_2$, scan rate: 50 V/s. (b) LSV curves of CNS–0.6Fe, CNS–0.8Fe and CNS–1.0Fe at 1600 rpm with a scan rate of 10 mV/s.

The evidently improved electrocatalytic performance of CNS–0.8Fe among the three samples can be explained by the following four reasons. First, CNS–0.8Fe has relative higher specific surface area in comparison with CNS–0.6Fe and CNS–1.0Fe, thus resulting in the expose of more active sites. Second, CNS–0.8Fe possesses a better developed lamellar structure than CNS–0.6Fe and CNS–1.0Fe as indicated by the TEM images, which is beneficial for the contact between active sites and oxygen molecules during ORR process. Third, the calculated I$_D$/I$_G$ value testified that the graphitic degree of CNS–0.8Fe is higher than the others, thus endowing this sample with a better electrical conductivity. Fourth and most importantly, CNS–0.8Fe possesses higher pyridinic N and Fe–N$_x$ contents than

CNS–0.6Fe and CNS–1.0Fe, which can provide more active sites towards ORR, thus greatly enhancing the catalytic activity.

The LSV curves of CNSs were also collected with different rotation rates. The current density of CNS–0.6Fe, CNS–0.8Fe, and CNS–1.0Fe increase gradually when consecutively changing the rotation speeds from 400 to 1600 rpm, as listed in Figure S5. Probably, the shortening of the diffusion distance directly leads to this regular phenomenon. To further explore the reaction kinetics of the ORR process, rotating ring disk electrode (RRDE) experiments were performed to calculate the generation of HO_2^- also with electron transfer numbers (n) values. CNS–0.6Fe, CNS–0.8Fe, CNS–1.0Fe, and the commercial Pt/C displayed higher disk current but minor ring current, as shown in Figure 8a. With the increase of potential from 0.2 to 0.5 V, the corresponding HO_2^- yield ranges of CNS–0.6Fe, CNS–0.8Fe, CNS–1.0Fe, and Pt/C catalyst are 10.57 to 11.90%, 3.99 to 6.22%, 4.66 to 5.93%, and 1.81 to 2.71%, as shown in Figure 8b. Also, the corresponding n values of CNS–0.6Fe, CNS–0.8Fe, CNS–1.0Fe, and Pt/C catalyst are 3.67 to 3.78, 3.87 to 3.92, 3.88 to 3.90, and 3.94 to 3.96 as shown in Figure 8c. So, these results just could indicate that CNSs catalyze ORR by the typical dominant four-electron transfer pathway [30–33].

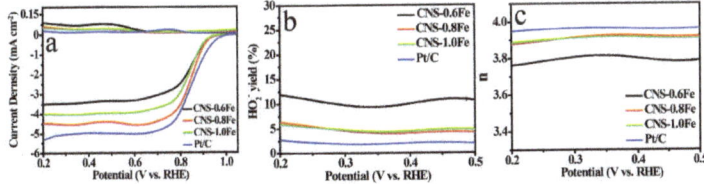

Figure 8. (a) Rotating ring disk electrode (RRDE) tests of CNS–0.6Fe, CNS–0.8Fe, CNS–1.0Fe and Pt/C at 1600 rpm. (b) The HO_2^- yields, (c) electron transfer number.

Taking CNS–0.8Fe as an example, the durability of CNSs was evaluated at 1600 rpm with consecutive 1000 cycles of CV scan in O_2-saturated 0.1 M KOH solution (Figure S6). The decrease of the onset and half-wave potentials is not evident. The LSV curves recorded before and after 1000 cycles reveal negative shifts of $E_{half-wave}$ of 7 mV for CNS–0.8Fe, which is lower than that of Pt/C (12 mV) as reported [21]. This result indicates that CNS–0.8Fe is relatively stable for ORR. The relevant crossover effects tests were conducted by taking CNS–0.8Fe as an example through chronoamperometric measurement to evaluate the catalytic selectivity of the catalysts. The methanol oxidation reaction resulted that the current density of Pt/C catalyst directly decreased at once when scrupulously adding 3.0 M methanol, as shown in Figure 9. However, this was not the case for CNS–0.8Fe, as the current density of that did not show evident change (Figure 9). These results confirmed that CNSs have good catalytic selectivity for ORR [34–36].

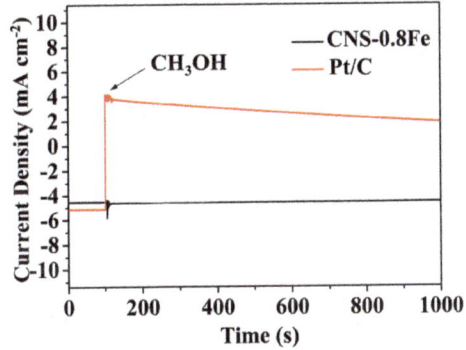

Figure 9. Methanol crossover tests of CNS-0.8Fe and Pt/C at 1600 rpm.

4. Conclusions

In summary, we prepared a new type of Fe and N co-doped carbon materials through a simple and effective method in one step. Direct coordination between amino ligand and Fe^{3+} could easily afford metallosupramolecular polymer precursors. After two carbonization processes, carbon flakes with wrinkled edges and active site of Fe–N_x and Fe_3C nanoparticle were fabricated. The catalytic activity of the carbon materials towards ORR were detailed investigated. The carbon material of CNS–0.8Fe possessed E_{onset} = 0.93 V and $E_{half-wave}$ = 0.83 V vs. RHE in alkaline system, which were comparable to Pt/C catalyst. The ligand TA and Fe^{3+} could generate more content of Fe–NH_2 in the precursor at a proper proportion through the coordination reaction and further led to the generation of more content of Fe–N_x active site when carbonization. So, the appropriate molar ratio between amino ligand and Fe^{3+} was the most important factor that determined the activity of CNSs. We considered that this simple method and conclusion might be of practical interest for the exploration of electrocatalysts with excellent ORR activity.

Supplementary Materials: The following are available online at http://www.mdpi.com/1996-1944/13/21/4779/s1, Figure S1: TEM images of CNS–0.6Fe (**a**) and CNS–1.0Fe (**b**), Figure S2: High-resolution XPS spectra of N 1s for CNS–0.6Fe (**a**) and CNS–1.0Fe (**b**), Figure S3: CV curves of CNS–0.6Fe, CNS–0.8Fe and CNS–1.0Fe in Ar-saturated 0.1 M KOH solution, scan rate: 50 V/s, Figure S4: Tafel slopes derived from the LSV curves of CNS–0.6Fe, CNS–0.8Fe, CNS–1.0Fe and Pt/C, Figure S5: LSV curves of (**a**) CNS–0.6Fe, (**b**) CNS–0.8Fe and (**c**) CNS–1.0Fe at 400–1600 rpm with a scan rate of 10 mV/s in O_2-saturated 0.1 M KOH solution, Figure S6: LSV curves of CNS–0.8Fe at 1600 rpm before and after durability test in O_2-saturated 0.1 M KOH solution, scan rate: 50 mV/s, Table S1: Contents (atomic %) of N element with different chemical environments calculated from the N 1s XPS spectrum.

Author Contributions: The experiments were designed by Y.W.; the experiment was carried out and the manuscript was written by Y.W., Y.L., C.Y. and L.D. All authors have read and agreed to the published version of the manuscript.

Funding: The National Natural Science Foundation of China (51673161, 51773172), Scientific and Technological Innovation Platform of Fujian Province (2014H2006), National Science and Technology Ministry (2014BAF08B03) supported our work.

Conflicts of Interest: There are no conflicts to declare.

References

1. Shao, M.; Chang, Q.; Dodelet, J.-P.; Chenitz, R. Recent Advances in Electrocatalysts for Oxygen Reduction Reaction. *Chem. Rev.* **2016**, *116*, 3594–3657. [CrossRef]
2. Lu, X.; Wang, D.; Ge, L.; Xiao, L.; Zhang, H.; Liu, L.; Zhang, J.; An, M.; Yang, P. Enriched graphitic N in nitrogen-doped graphene as a superior metal-free electrocatalyst for the oxygen reduction reaction. *New J. Chem.* **2018**, *42*, 19665–19670. [CrossRef]
3. Asahi, M.; Yamazaki, S.-I.; Morimoto, Y.; Itoh, S.; Ioroi, T. Crystal structure and oxygen reduction reaction (ORR) activity of copper(II) complexes of pyridylmethylamine ligands containing a carboxy group. *Inorganica Chim. Acta* **2018**, *471*, 91–98. [CrossRef]
4. Bukka, S.; Badam, R.; Vedarajan, R.; Matsumi, N. Photo-generation of ultra-small Pt nanoparticles on carbon-titanium dioxide nanotube composites: A novel strategy for efficient ORR activity with low Pt content. *Int. J. Hydrogen Energy* **2019**, *44*, 4745–4753. [CrossRef]
5. Banhamd, D.; Ye, S.; Pei, K.; Ozaki, J.-I.; Kishimoto, T.; Imashiro, Y. A review of the stability and durability of non-precious metal catalysts for the oxygen reduction reaction in proton exchange membrane fuel cells. *J. Power Sources* **2015**, *285*, 334–348. [CrossRef]
6. Wu, J.; Wang, J.; Lv, X.; Wang, X. A thin slice-like Co_3O_4/N-doped graphene hybrid as an efficient catalyst for oxygen reduction reaction. *Inorg. Chem. Commun.* **2019**, *106*, 128–134. [CrossRef]
7. Yan, Z.; Dai, C.; Zhang, M.; Lv, X.; Zhao, X.; Xie, J. Nitrogen doped porous carbon with iron promotion for oxygen reduction reaction in alkaline and acidic media. *Int. J. Hydrogen Energy* **2019**, *44*, 4090–4101. [CrossRef]
8. Chu, Y.; Gu, L.; Du, H.; Qu, K.; Zhang, Y.; Zhao, J.; Xie, Y. The synthesis of phenanthroline and bipyridine based ligand for the preparation of Fe–N_x/C type electrocatalyst for oxygen reduction. *Int. J. Hydrogen Energy* **2018**, *43*, 21810–21823. [CrossRef]

9. Hassan, D.; El-Safty, S.A.; Khalil, K.A.; Dewidar, M.; Abu El-Magd, G. Carbon Supported Engineering NiCo$_2$O$_4$ Hybrid Nanofibers with Enhanced Electrocatalytic Activity for Oxygen Reduction Reaction. *Materials* **2016**, *9*, 759. [CrossRef]
10. Abdelwahab, A.; Carrasco-Marín, F.; Pérez-Cadenas, A.F. Binary and Ternary 3D Nanobundles Metal Oxides Functionalized Carbon Xerogels as Electrocatalysts toward Oxygen Reduction Reaction. *Materials* **2020**, *13*, 3531. [CrossRef]
11. Mahammed, A.; Gross, Z. Metallocorroles as Electrocatalysts for the Oxygen Reduction Reaction (ORR). *Isr. J. Chem.* **2016**, *56*, 756–762. [CrossRef]
12. Dou, S.; Wang, X.; Wang, S. Rational Design of Transition Metal-Based Materials for Highly Effcient Electrocatalysis. *Small Methods* **2019**, *3*, 1800211. [CrossRef]
13. Toh, R.J.; Poh, H.L.; Sofer, Z.; Pumera, M. Transition Metal (Mn, Fe, Co, Ni)-Doped Graphene Hybrids for Electrocatalysis. *Chem. Asian J.* **2013**, *8*, 1295–1300. [CrossRef] [PubMed]
14. Peng, H.; Liu, F.; Liu, X.; Liao, S.; You, C.; Tian, X.; Nan, H.; Luo, F.; Song, H.; Fu, Z.; et al. Effect of Transition Metals on the Structure and Performance of the Doped Carbon Catalysts Derived From Polyaniline and Melamine for ORR Application. *ACS Catal.* **2014**, *4*, 3797–3805. [CrossRef]
15. Eisenberg, D.; Slot, T.K.; Rothenberg, G. Understanding Oxygen Activation on Metal- and Nitrogen-Codoped Carbon Catalysts. *ACS Catal.* **2018**, *8*, 8618–8629. [CrossRef]
16. Guo, J.; Cheng, Y.; Xiang, Z. Confined-Space-Assisted Preparation of Fe$_3$O$_4$-Nanoparticle-Modified Fe–N–C Catalysts Derived from a Covalent Organic Polymer for Oxygen Reduction. *ACS Sustain. Chem. Eng.* **2017**, *5*, 7871–7877. [CrossRef]
17. Wang, Q.; Zhou, Z.-Y.; Lai, Y.-J.; You, Y.; Liu, J.-G.; Wu, X.-L.; Terefe, E.; Chen, C.; Song, L.; Rauf, M.; et al. Phenylenediamine-Based FeN$_x$/C Catalyst with High Activity for Oxygen Reduction in Acid Medium and Its Active-Site Probing. *J. Am. Chem. Soc.* **2014**, *136*, 10882–10885. [CrossRef]
18. Wang, M.-Q.; Ye, C.; Wang, M.; Li, T.-H.; Yu, Y.-N.; Bao, S.-J. Synthesis of M (Fe$_3$C, Co, Ni)-porous carbon frameworks as high-efficient ORR catalysts. *Energy Storage Mater.* **2018**, *11*, 112–117. [CrossRef]
19. Tei, G.; Tamaki, T.; Hayashi, T.; Nakajima, K.; Sakai, A.; Yotsuhashi, S.; Ogawa, T. Oxygen Reduction Reaction (ORR) Activity of a Phenol-Substituted Linear FeIII-Porphyrin Dimer. *Eur. J. Inorg. Chem.* **2017**, *2017*, 3229–3232. [CrossRef]
20. Li, W.; Sun, L.; Hu, R.; Liao, W.; Li, Z.; Li, Y.; Guo, C. Surface Modification of Multi-Walled Carbon Nanotubes via Hemoglobin-Derived Iron and Nitrogen-Rich Carbon Nanolayers for the Electrocatalysis of Oxygen Reduction. *Materials* **2017**, *10*, 564. [CrossRef]
21. Li, Y.; Li, Z.; Wu, Y.; Wu, H.; Zhang, H.; Wu, T.; Yuan, C.; Xu, Y.; Zeng, B.; Dai, L. Carbon particles co-doped with N, B and Fe from metal-organic supramolecular polymers for boosted oxygen reduction performance. *J. Power Sources* **2019**, *412*, 623–630. [CrossRef]
22. Gu, W.; Hu, L.; Li, J.; Wang, E. Recent Advancements in Transition Metal-Nitrogen-Carbon Catalysts for Oxygen Reduction Reaction. *Electroanalysis* **2018**, *30*, 1217–1228. [CrossRef]
23. Zhang, Z.; Li, H.; Liu, H. Insight into the adsorption of tetracycline onto amino and amino–Fe^{3+} gunctionalized mesoporous silica: Effect of functionalized groups. *J. Environ. Sci.* **2018**, *65*, 171–178. [CrossRef]
24. Li, L.; Yuan, C.; Zhou, D.; Ribbe, A.E.; Kittilstved, K.R.; Thayumanavan, S. Utilizing Reversible Interactions in Polymeric Nanoparticles To Generate Hollow Metal-Organic Nanoparticles. *Angew. Chem.* **2015**, *127*, 13183–13187. [CrossRef]
25. Zuo, J.-C.; Tong, S.-R.; Yu, X.-L.; Wu, L.-Y.; Cao, C.-Y.; Ge, M.; Song, W.-G. Fe^{3+} and amino functioned mesoporous silica: Preparation, structural analysis and arsenic adsorption. *J. Hazard. Mater.* **2012**, *235*, 336–342. [CrossRef]
26. Jiang, W.-J.; Gu, L.; Li, L.; Zhang, Y.; Zhang, X.; Zhang, L.-J.; Wang, J.; Hu, J.-S.; Wei, Z.; Wan, L.-J. Understanding the High Activity of Fe–N–C Electrocatalysts in Oxygen Reduction: Fe/Fe$_3$C Nanoparticles Boost the Activity of Fe–N$_x$. *J. Am. Chem. Soc.* **2016**, *138*, 3570–3578. [CrossRef]
27. Wei, J.; Liang, Y.; Hu, Y.; Kong, B.; Simon, G.P.; Zhang, J.; Jiang, S.P.; Wang, H. A Versatile Iron-Tannin-Framework Ink Coating Strategy to Fabricate Biomass-Derived Iron Carbide/Fe–N-Carbon Catalysts for Efficient Oxygen Reduction. *Angew. Chem. Int. Ed.* **2015**, *55*, 1355–1359. [CrossRef]
28. Shen, H.; Gracia-Espino, E.; Xamxikamar, M.; Zang, K.; Luo, J.; Wang, L.; Gao, S.; Mamat, X.; Sanshuang, G.; Wagberg, T.; et al. Synergistic Effects between Atomically Dispersed Fe–N–C and C–S–C for the Oxygen Reduction Reaction in Acidic Media. *Angew. Chem. Int. Ed.* **2017**, *56*, 13800–13804. [CrossRef]

29. Chang, Y.; Yuan, C.; Li, Y.; Liu, C.; Wu, T.; Zeng, B.; Xu, Y.; Thayumanavan, S. Controllable fabrication of a N and B co-doped carbon shell on the surface of TiO_2 as a support for boosting the electrochemical performances. *J. Mater. Chem. A* **2017**, *5*, 1672–1678. [CrossRef]
30. Hou, Y.; Huang, T.; Wen, Z.; Mao, S.; Cui, S.; Chen, J. Metal–Organic Framework-Derived Nitrogen-Doped Core-Shell-Structured Porous Fe/Fe_3C@C Nanoboxes Supported on Graphene Sheets for Efficient Oxygen Reduction Reactions. *Adv. Energy Mater.* **2014**, *4*, 1400337. [CrossRef]
31. Jiao, L.; Wan, G.; Zhang, R.; Zhou, H.; Yu, S.-H.; Jiang, H.-L. From Metal-Organic Frameworks to Single-Atom Fe Implanted N doped Porous Carbons: Efficient Oxygen Reduction in Both Alkaline and Acidic Media. *Angew. Chem. Int. Ed.* **2018**, *57*, 1–6. [CrossRef] [PubMed]
32. Chang, Y.; Yuan, C.; Liu, C.; Mao, J.; Li, Y.; Wu, H.; Wu, Y.; Xu, Y.; Zeng, B.; Dai, L. B, N co-doped carbon from cross-linking induced self-organization of boronate polymer for supercapacitor and oxygen reduction reaction. *J. Power Sources* **2017**, *365*, 354–361. [CrossRef]
33. Rouhet, M.; Bozdech, S.; Bonnefont, A.; Savinova, E.R. Influence of the proton transport on the ORR kinetics and on the H_2O_2 escape in three-dimensionally ordered electrodes. *Electrochem. Commun.* **2013**, *33*, 111–114. [CrossRef]
34. Liu, S.-H.; Wu, J.-R.; Pan, C.-J.; Hwang, B.-J. Synthesis and characterization of carbon incorporated Fe–N/carbonsfor methanol-tolerant oxygen reduction reaction of polymer electrolyte fuel cells. *J. Power Sources* **2014**, *250*, 279–285. [CrossRef]
35. Kosmala, T.; Bibent, N.; Sougrati, M.T.; Dražić, G.; Agnoli, S.; Jaouen, F.; Granozzi, G. Stable, Active, and Methanol-Tolerant PGM-Free Surfaces in an Acidic Medium: Electron Tunneling at Play in Pt/FeNC Hybrid Catalysts for Direct Methanol Fuel Cell Cathodes. *ACS Catal.* **2020**, *10*, 7475–7485. [CrossRef]
36. Vecchio, C.L.; Sebastián, D.; Lázaro, M.J.; Aricò, A.S.; Baglio, V. Methanol-Tolerant M–N–C Catalysts for Oxygen Reduction Reactions in Acidic Media and Their Application in Direct Methanol Fuel Cells. *Catalysts* **2018**, *8*, 650. [CrossRef]

Publisher's Note: MDPI stays neutral with regard to jurisdictional claims in published maps and institutional affiliations.

© 2020 by the authors. Licensee MDPI, Basel, Switzerland. This article is an open access article distributed under the terms and conditions of the Creative Commons Attribution (CC BY) license (http://creativecommons.org/licenses/by/4.0/).

Article

Effects of Boron Carbide on Coking Behavior and Chemical Structure of High Volatile Coking Coal during Carbonization

Qiang Wu, Can Sun, Zi-Zong Zhu *, Ying-Dong Wang and Chong-Yuan Zhang

College of Materials Science and Engineering, Chongqing University, Chongqing 400044, China; qiangwucqu@163.com (Q.W.); suncan2230@163.com (C.S.); 18243934868@163.com (Y.-D.W.); zhangchongyuan0811@163.com (C.-Y.Z.)
* Correspondence: zzzhu666@cqu.edu.cn; Tel.: +86-139-8327-0208

Abstract: Modified cokes with improved resistance to CO_2 reaction were produced from a high volatile coking coal (HVC) and different concentrations of boron carbide (B_4C) in a laboratory scale coking furnace. This paper focuses on modification mechanism about the influence of B_4C on coking behavior and chemical structure during HVC carbonization. The former was studied by using a thermo-gravimetric analyzer. For the latter, four semi-cokes prepared from carbonization tests for HVC with or without B_4C at 450 °C and 750 °C, respectively, were analyzed by using Fourier transform infrared spectrum and high-resolution transmission electron microscopy technologies. It was found that B_4C will retard extensive condensation and crosslinking reactions by reducing the amount of active oxygen obtained from thermally produced free radicals and increase secondary cracking reactions, resulting in increasing size of aromatic layer and anisotropic degree in coke structure, which eventually improves the coke quality.

Keywords: high volatile coking coal; boron carbide; coking behavior; chemical structure; coke quality

Citation: Wu, Q.; Sun, C.; Zhu, Z.-Z.; Wang, Y.-D.; Zhang, C.-Y. Effects of Boron Carbide on Coking Behavior and Chemical Structure of High Volatile Coking Coal during Carbonization. *Materials* **2021**, *14*, 302. https://doi.org/10.3390/ma14020302

Received: 8 December 2020
Accepted: 5 January 2021
Published: 8 January 2021

Publisher's Note: MDPI stays neutral with regard to jurisdictional claims in published maps and institutional affiliations.

Copyright: © 2021 by the authors. Licensee MDPI, Basel, Switzerland. This article is an open access article distributed under the terms and conditions of the Creative Commons Attribution (CC BY) license (https://creativecommons.org/licenses/by/4.0/).

1. Introduction

Adding cheap materials into coal blends to produce metallurgical coke has been extensively researched, due to the gradual rise of coking coal price and inadequate supply of the prime-coking coals with medium volatility. The most studied method is harnessing non-coking coal to replace part of coking coals, but the caking property of coal blends will deteriorate in such a situation. To improve the caking property, on the one hand, various high bond-ability substances, such as pitches, coal tar pitches, coal extracts, and solvent-refined coals, have been used in the carbonization process of coal blending with low bond-ability [1–7]. On the other hand, non-coking coal was pretreated by thermal treatments [8], hydrothermal treatments [9,10], and steam treatments [11,12]. Additionally, to reduce the costs of coal blends and the amount of CO_2 emission, adding small amount of biomass material into coal blends to produce metallurgical coke has also been suggested [13–15]. Although these methods can broaden coking coal resources, few industrial applications about the above studies are successful because of factors such as coke quality, cost, production conditions, etc.

It is widely known that the reserve of high volatile coking coal (HVC) occupies about half of all coking coal reserves, and the price of HVC is relatively cheap. Therefore, increasing the usage amount of HVC in coal blending to produce metallurgical coke is also an effective solution to reduce the cost of coke and maintain the sustainable development of the coke industry. However, the usage amount of HVC in coal blending is usually limited to 20–30% in the producing process of metallurgical coke [5,12], which is extremely disproportionate to the reserve of HVC. This is because HVC is a low proportion of metamorphism coking coal and possesses plenty of oxygen-containing groups as well as aliphatic side chains, which results in the rapid formation of large quantities of gases, free radicals, and plastic mass with high fluidity during HVC carbonization, subsequently leaving a weak coke with a thin-walled porous structure [13,14,16–19]. In such a case, Qian et al found that the addition of pitches could contribute to developing intermediate texture during HVC carbonization [20]. This improves the coke quality, but the complex operation process and plugging problem limit its industrial application. Vega et.al., asserted that mild oxidation was an effective method to improve the coking performance of HVC with low oxygen content (less than 5%) [19]. However, HVC usually contains high oxygen content (more than 10%) so that mild oxidation will lead to excessive oxygen in HVC, which worsens its thermoplasticity and coking performance.

Admittedly, exploring a new method which is able to improve the coking performance of HVC, finally increasing its usage in coal blends, is urgent and essential. Previous studies have indicated that during resin pyrolysis, B_4C reacts with oxygen-containing fragments released from resin to generate boric oxide, whose formation further affects the viscosity of resin at high temperature [21,22]. The compositions of the oxygen-containing fragments obtained from the pyrolysis process of resin and HVC are partly similar; that B_4C may react with the oxygen-containing fragments derived from the plastic zone of HVC. In such a case, there are more indigenous donor hydrogen stabilizing free radicals due to the decrease in reaction between oxygen-containing fragments and transferable hydrogen. Consequently, B_4C may be a promising additive to improve HVC's coking performance and coke quality.

Initially, this work aims to investigate whether adding B_4C can improve the quality of coke obtained from HVC. Five carbonization tests were carried out in a laboratory scale coking furnace, and the quality of resulting coke was reflected by coke reactivity towards CO_2 (CRI) and coke strength after reaction (CSR) indexes. Secondly, one of the important tasks of this work was to acquire the mechanism of improvement of coke quality in detail. The influence of B_4C on coking behavior during HVC carbonization was evaluated by using a thermo-gravimetric analyzer (TG). Based on the coke quality data and TG analysis, four semi-cokes with or without B_4C were manufactured under two characteristic temperatures and were analyzed by Fourier transform infrared spectrum (FTIR) and high-resolution transmission electron microscopy (HRTEM) techniques to investigate the effects of B_4C on chemical structure during HVC carbonization. Having a better understanding of the interactional mechanism between HVC and B_4C contributes to increasing the proportion of low-rank coking coal (such as HVC) in coal blending, especially in the case of adding B_4C, to produce low-cost metallurgical coke. Simultaneously, this method is easy to apply in industry.

2. Materials and Methods

2.1. Samples and Carbonization Experiments of Coke

The HVC used in the current study was collected from Kubai coal field of Xinjiang province, located in northwest China. The boron carbide (B_4C) powder with particle diameter <58 μm had a purity of 95%. The coke carbonization experiments between HVC and B_4C were carried out in a laboratory scale coking furnace, and experimental schemes are shown in Table 1.

The coking furnace temperature was controlled automatically by a programmable controller and was heated by resistance wire. The density and moisture of coal blends were limited to 0.95 g/cm^{-3} and 10 wt.%, respectively. Approximately 2 kg of the coal blend (particle size of HVC less than 3 mm) was mixed evenly and placed in a coking retort with an internal diameter of 100 mm and a length of 500 mm. Next, this coking retort was put into the coking furnace after the experimental temperature reached 700 °C, combining with a heating rate 10 °C/min. After that, it was consecutively heated to 1000 °C at the rate of 5 °C/min, and the target temperature was maintained for 5 h, finally cooling the retort to room temperature in the atmosphere. The GB1997-89 standard was applied to produce coke samples whose CRI and CSR indexes were measured using GB/T4000-1996 standard. These indexes are shown as the average value of three trials in a later context.

Table 1. Carbonization schemes for coke and semi-coke.

Samples	High Volatile Coking Coal (HVC) (wt.%)	B_4C (wt.%)	Temperature (°C)
Scheme of Cokes			
Coke-0	100.00	0	1000
Coke-1	99.75	0.25	1000
Coke-2	99.50	0.50	1000
Coke-3	99.25	0.75	1000
Coke-4	99.00	1.00	1000
Scheme of Semi-Cokes			
C450	100.00	0	450
C450M	99.50	0.50	450
C750	100.00	0	750
C750M	99.50	0.50	750

2.2. TG Measurements

Generally, the carbonization process of coal is similar to its pyrolysis process under an inert gas. In order to simulate and evaluate the influence of adding B_4C on coking behavior in the carbonization process of HVC, two groups of thermo-gravimetric analysis experiments for HVC (particle size of <74 μm in diameter) with and without 0.5 wt.% B_4C were carried out on a NETZSCH STA 449 C analyzer (Nestal, Selbu, Germany). About 10 mg of coal blend was placed in an alumina cell and heated from ambient temperature to 1000 °C at a rate of 10 °C/min under a continuous argon atmosphere with flow rate of 50 mL/min.

2.3. Carbonization Experiments of Semi-Coke

To investigate the influence of B4C on the chemical structure in the carbonization process of HVC, the semi-coke carbonization experiment was carried out in an electrically heated oven using a 200 mL crucible and corresponding schemes, based on the coke quality analysis and characteristic temperatures analysis from TG, are listed in Table 1. Approximately 100 g coal blending (particle size of HVC less than 1 mm) was loaded into the crucible. The crucible was placed in the oven's chamber filled with inert atmosphere, heated at the rate of 10 °C/min to 450 °C and 750 °C, respectively, held at the target temperature for 5 min, and then removed and cooled to room temperature under an N_2 atmosphere.

2.4. Preparation of Demineralized Samples

Raw coal samples and semi-coke samples were ground and sieved to obtain particles of <74 μm in diameter. To eliminate the potential effects of minerals on the FTIR and HRTEM analyses, these samples were acid-washed using HCl and HF solution at room temperature as described in a previous study [23]. Generally, acid treatment under such conditions does not cause significant structural changes [24]. These demineralized samples were dried for 12 h in vacuum room at 60 °C and stored under nitrogen atmosphere. Ultimate analyses of these samples were determined according to GB/T 476-208 criterion and the analytical results are listed in Table 2.

Table 2. Elemental analyses of studied samples (wt.%, daf).

Sample	H	C	N	S	O [a]	H/C
Raw coal	4.83	76.93	1.32	0.51	16.41	0.753
C450	4.00	81.41	1.35	0.54	12.69	0.590
C450M	4.12	80.32	1.31	0.55	13.70	0.616
C750	1.75	86.47	1.53	0.47	9.78	0.243
C750M	1.71	86.83	1.51	0.48	9.47	0.236

[a] By difference; wt.%, daf: weight percentage of various elements on a dry and ash-free basis.

2.5. FTIR Measurements

The FTIR spectra of the demineralized samples were recorded on a Thermo-Nicolet iS5 FTIR spectrometer (Thermo Fisher Scientific, MMAS, Waltham, MA, USA). All samples for the FTIR measurement were prepared by mixing the investigated sample with dried KBr powder, and the mixture was pressed to form a pellet under 12 MPa for 2 min. All spectra were obtained within the 400–4000 cm^{-1} wave number range at a resolution of 4 cm^{-1}, and 32 scans per spectrum were performed.

2.6. HRTEM Measurements

The influence of adding B_4C on lattice fringes in the carbonization process of HVC was investigated by using HRTEM. The demineralized samples were grounded in ethanol and sonicated in an ultrasonic washer for 10 min, and then sprayed on a copper microgrid as HRTEM specimens. The HRTEM images of the samples were acquired from a 200 kV transmission electron microscope (JEM-2100F, JEOL, Tokyo, Japan). Detailed procedures and conditions of HRTEM analysis were derived from a previous thesis [25].

3. Results and Discussion

3.1. Coke Quality Analysis

In order to easily observe the changing trend of strength, two smooth curves in Figure 1 were automatically performed by linking data points based on the B-spline mode of software Origin 9.1. As shown in Figure 1, the CRI index of coke decreased by 26.3%, while CSR index of coke increased by 18.5% when 0.25 wt.% B_4C was added into HVC, indicating that the coke quality is improved distinctly through the addition of B_4C. In addition, these thermal strength indexes were further promoted when B_4C content increased to 0.50 wt.%. However, continuing to increase the content of B_4C to 0.75 wt.% even 1 wt.% led to little change of these indexes. These results indicate that adding 0.5 wt.% B_4C into HVC to improve the coke quality is sufficient. Therefore, the addition amount of B_4C was set as 0.5 wt.% in the following research on the modification mechanism.

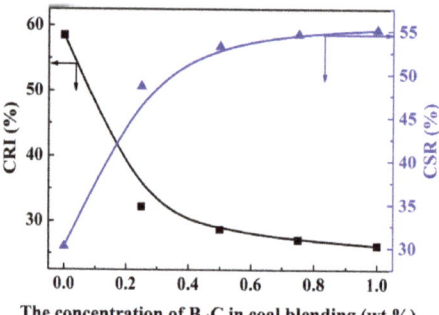

Figure 1. The relationship between thermal strength indexes of coke and the concentration of B_4C in coal blending. CRI: coke reactivity towards CO_2; CSR: coke strength after reaction.

3.2. Ultimate Analyses

As listed in Table 2, raw coal has a high oxygen content (16.41 wt.%, daf) and low carbon content (76.93 wt.%, daf), suggesting that raw coal is a low coalification coking coal with abundant oxygen-containing groups. In the carbonization process, the contents of hydrogen and oxygen decrease in semi-cokes while the content of carbon increases with the increase in temperature (Table 2), which is largely caused by releasing small molecular weight gases, such as CO_2, CO, H_2O, H_2, etc. It is worth noting that some differences on element content and H/C ratio were observed in semi-cokes under the same temperature, indicating that B_4C causes a change of carbonization process of HVC.

3.3. TG Analysis

Figure 2 shows the curves of mass loss (TG) and their derivatives (DTG) for raw coal and modified coal, respectively. According to previous research [26], seven characteristic temperatures derived from the TG and DTG curves are defined and the two curves are divided into five stages as shown in Figure 2. For the drying stage (room temperature—T_i) and slow pyrolysis stage (T_i–T_m; the process with a low reaction rate—Figure 2), there is little difference in both coals, indicating that the B_4C has little effect on the dehydration process, the releasing process of gas soaked in the pores, and the decomposition process of unstable functional groups below 400 °C.

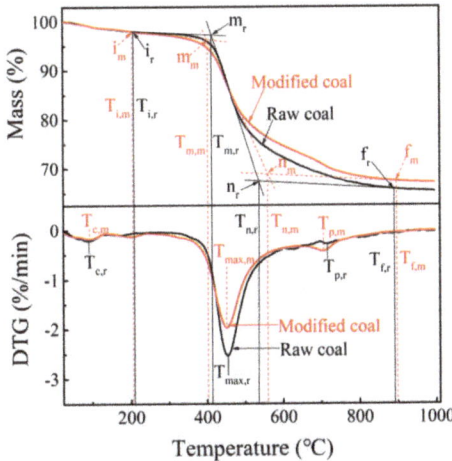

Figure 2. Thermo-gravimetric (TG) and DTG curves of raw coal and modified coal (addition of 0.5 wt.% B_4C into raw coal) at a heating rate of 10 °C/min. $T_{*,r}$: represents the characteristic temperatures of raw coal; $T_{*,m}$: represents the characteristic temperatures of modified coal; T_c: moisture loss peak temperature; T_i: pyrolysis initial temperature; T_m: initial temperature of the pyrolysis process with a fast reaction rate; T_{max}: highest pyrolysis peak temperature; T_n: finish temperature of the pyrolysis process with a fast reaction rate; T_c: second cracking peak temperature; T_f: pyrolysis finish temperature.

For the fast pyrolysis stage (T_m–T_n), the process with a fast reaction rate, a weight loss peak with a high mass loss rate (2.53 %/min) was observed in the DTG curve of raw coal, indicating that many considerable reactions occurred, and these reactions caused the formation of abundant volatile matters, free radicals, and molecular fragments in this stage. Compared with the weight loss peak, however, the corresponding peak in modified coal showed a lower mass loss rate (1.98 %/min), indicating that the reaction rate of pyrolysis reactions in the fast pyrolysis stage are slowed down by adding B_4C. This may be because partial active oxygen obtained from thermally produced molecular fragments are combined with B_4C to form boron oxide [21,27–29], which decreases the consumption of transferable hydrogen used to stabilize reactive oxygen substances so that more transferable hydrogen can be used to stabilize the free radicals. In such a case, on the one hand, more stable free radicals can be arranged in an ordered structure, resulting in the development of anisotropic mesophase structures in semi-coke [30]. On the other hand, the reaction rate of condensation and crosslinking reactions will be reduced in modified coal, which is consistent with a lower mass loss rate peak in the fast pyrolysis stage.

For the fast polycondensation stage (T_n–T_f), the peak at $T_{p,m}$ is more intensive than the peak at $T_{p,r}$, indicating that the intensity of secondary cracking reactions in modified coal is higher than that in raw coal. This is because, with the development of mesophase structures in modified coal, more macromolecular weight polymers will form and participate in secondary cracking reactions [26].

For the slow polycondensation stage (T_f–1000 °C), there were no obvious differences in the transformation process from char to coke for both coals, but the weight of pyrolytic residues of modified coal was higher than that of raw coal after the pyrolysis finished. This is attributed not only to the residue of boron oxide, but also the lower release of small molecular weight gases in the fast pyrolysis stage.

3.4. FTIR Analysis

Figure 3a shows the FTIR spectra of raw coal and the four semi-coke samples, which exhibit similar absorption bands primarily consisting of oxygen-containing groups, aliphatic C–H groups, aromatic nucleus C=C, and substituted aromatic rings, while their intensities of absorption bands vary considerably. For further observing the changes of significant functional groups in the FTIR spectra, the baselines of zone representing aliphatic C–H groups from 3000 to 2800 cm^{-1}, zones representing oxygen-containing functional groups from 1800 to 1500 cm^{-1} and 1350 to 1000 cm^{-1}, and zone representing aromatic C–H groups from 900 to 700 cm^{-1} were corrected, as shown in Figure 3b–e, respectively. According to the literature [23,24,26,31–33], band assignments are shown in Table 3.

Table 3. Band assignments derived from FTIR spectra.

Band Position (cm^{-1})	Assignments
3415–3350	–OH stretching vibration
2975–2955	Aliphatic CH$_3$ asymmetric stretching vibration
2925–2919	Aliphatic CH$_2$ asymmetric stretching vibration
2855–2850	Aliphatic CH$_2$ symmetric stretching vibration
1705–1695	Aromatic (carbonyl/carboxyl groups) (C=O)
1640–1605	Aromatic ring stretching C=O or C=C
1470–1450	aliphatic chains CH$_3$–, CH$_2$–
1274–1260	C–O stretching vibration in aryl ethers
1165–1155	C–O stretching vibration in phenols, ethers
1098–1095	C–O stretching vibration in alcohols or aromatic ring C–H bending vibration
1035–1030	Aromatic ring stretching vibration or C–O stretching vibration
1010	C–O stretching vibration
876–872	Aromatic nucleus CH, one adjacent H deformation
810–801	Aromatic nucleus CH, two adjacent H deformation
750	Aromatic nucleus CH, four adjacent H deformation

As shown in Figure 3a,b, the intensities of peaks at 2955 cm^{-1}, 2920 cm^{-1}, 2852 cm^{-1}, and 1465 cm^{-1} for aliphatic C–H functional groups steadily decrease with increasing temperatures due to the thermal cleavage of aliphatic chains. Besides, these fractured aliphatic segments will participate in polycondensation reactions during aromatization to form larger condensed aromatic nucleus polymers. It is worth noting that the peak intensity at 3000–2800 cm^{-1} of C450M is greater than that of C450, indicating that C450M has a higher concentration of aliphatic C–H groups as well as a lower condensation degree of aromatic nuclei. Conversely, the content of aliphatic C–H groups of C750M are slightly lower than that of C750, suggesting that the condensation degree of aromatic nuclei in C750M increases with the addition of B$_4$C.

As can be seen in Figure 3a,c,d, the intensities of peaks at 3400 cm^{-1}, 1702 cm^{-1}, 1262 cm^{-1}, 1097 cm^{-1}, 1032 cm^{-1}, and 1010 cm^{-1} for oxygen-containing functional groups monotonously decrease with the increasing temperatures, even the four peaks at 1702 cm^{-1}, 1165 cm^{-1}, 1032 cm^{-1}, and 1010 cm^{-1} disappear in semi-cokes at 750 °C. This is because these –COOH, O–H, and C–O bands are gradually broken or destroyed in the HVC carbonization process and released through small molecules gases [17,18]. It is interesting that the intensity of peak at 1610 cm^{-1} for C=C stretching band also declines with the increase in temperature. Ideally, the C=C stretching band representing the condensation degree of aromatic nuclei will increase with increasing carbonization temperature, but the presence of plenty of phenolic groups and COO– groups in low-rank HVC (C 76.93 wt.% in Table 1) is likely to increase intensity of the 1610 cm^{-1} band [34]. Therefore, the decrease in intensity of the peak at 1610 cm^{-1} is mainly due to the decomposition of oxygen-containing groups in the carbonization process of HVC.

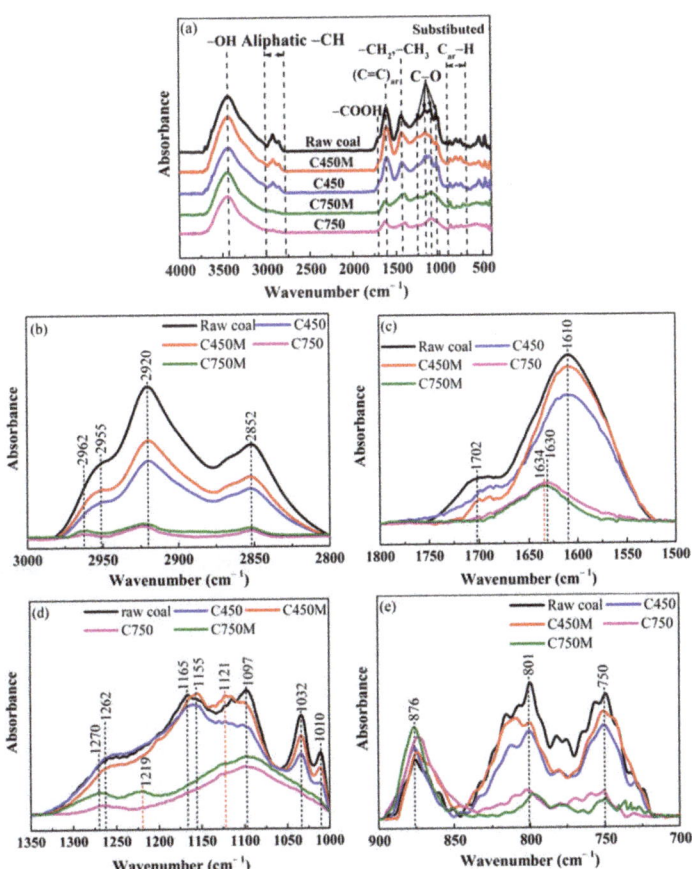

Figure 3. FTIR spectra of raw coal and four semi-cokes: (**a**) FTIR spectra from the 4000–400 cm^{-1} zone; (**b**) FTIR spectra after subtracting the baseline from the 3000–2800 cm^{-1} zone; (**c**) FTIR spectra after subtracting the baseline from the 1800–1500 cm^{-1} zone; (**d**) FTIR spectra after subtracting the baseline from the 1350–1000 cm^{-1} zone; (**e**) FTIR spectra after subtracting the baseline from the 900–700 cm^{-1} zone.

It is worth noting that a new peak at 1121 cm^{-1} (in Figure 3d) attributed to a B–O bond is observed in C450M [29,35,36], indicating that partial active oxygen obtained from oxygen-containing fragments were consumed by reaction with B$_4$C to form the boron oxide in the plastic zone of HVC. In addition, by comparing the peak strength of oxygen-containing functional groups at the same position in C450 and C450M, the strength of all oxygen-containing peaks was increased by adding B$_4$C. One of hypotheses was that substitution reactions between boron compound and oxygen-containing functional groups occur, which leads to the formation of organically bound B–O groups with higher bond energies. Simultaneously, a shift of the C=C stretching band from 1630 cm^{-1} (in C750) to 1634 cm^{-1} (in C750M) and a new band at around 1219 cm^{-1} representing B–C stretching vibration are observed. Generally, an increase in wavelength of the C=C stretching band is attributed to the generation of the B–C band. Meanwhile, boron atoms should be incorporated in the sp^2 C networks in coke structure. These two explanations are clarified in related articles [29,36]. However, compared with conditions in these articles, the experimental temperature in this work was lower. Therefore, whether the explanations introduced in the above articles are suitable to support the phenomena in this work still needs to be illustrated.

As shown in Figure 3e, these fingerprint absorption peaks at 900–700 cm^{-1} are caused by aromatic C–H out-of-plane bending vibrations. It is generally accepted that the degree of aromatic substitution and condensation of aromatic nuclei rely on the number of adjacent hydrogens per ring [24]. The intensities of peaks at 812 cm^{-1} and 750 cm^{-1} decrease gradually and the intensity of peaks at 870 cm^{-1} increases gradually with increasing temperature, indicating that the concentrations of highly-substituted aromatic rings and aromatic structures with 1–2 rings decrease in the carbonization process and the size of condensed aromatic nuclei becomes larger. It is noteworthy that the peak intensity at 876 cm^{-1} of C450M is lower than that of C450, but the peak intensity at 876 cm^{-1} of C750M is higher than that of C750 when the temperature arrives at 750 °C. Simultaneously, both peak intensities at 801 cm^{-1} and 750 cm^{-1} of C450M are higher than those of C450, but both peak intensities of C750M are lower than those of C750. These results show that the addition of B_4C can retard condensation reactions in the plastic zone and increase the secondary cracking reactions in the fast polycondensation zone of HVC, which will lead to the formation of larger sized condensed aromatic nuclei in C750M.

3.5. HRTEM Analysis

The lattice fringes cannot be observed directly from HRTEM images, so these original images were organized to acquire the lattice fringe images and every step is briefly described as follows: firstly, using Fourier transformation to cope with original images obtains frequency domain images; secondly, using rounded filtering eliminates the disordered part in these images; thirdly, using inverse Fourier transformation transfers the images obtained after the second step to new images; then, using threshold segmentation to handle the new images obtains black-and-white binary images that can present microcrystalline fringe; next, initially etch, then expend, and finally skeleton process the black-and-white binary images to obtain lattice fringe images [37]. HRTEM images and the corresponding lattice fringe images of raw coal and four semi-coke samples are shown in Figure 4. According to Figure 4, these lattice fringe images show a striking difference in the shape, size, and orientation of the layers. For the raw coal, the majority of layers are small, twisted, and lack orientation; that few stacks can be observed in Figure 4b, which is consistent with its low coalification. With the elevation of carbonization temperature, aromatic layers' stacking-number and their size increase while they become better-orientated in semi-cokes, as shown in Figure 4d,f. In addition, these changes of aromatic layers become more evident in semi-cokes at 750 °C (in Figure 4h,j). These results show that the crystallinity of semi-coke increases with the increase in carbonization temperature.

According to the literature [25,37], the number of aromatic carbon atoms can be determined by the lattice fringe length of HRTEM image; therefore, the extracted lattice fringe images were analyzed by image processing software (ImageJ V1.47) to calculate the aromatic fringe size. The relationships between lattice fringe length and aromatic sheet assignments are listed in Table 4, and the distribution frequency of lattice fringe length in the samples is also shown in Table 4 through the form of the average value based on the examination of three different regions containing 1000 aromatic fringes.

Table 4 shows a summary of the classification of the aromatic fringes by fringe lengths and their frequency of occurrence in the aromatic fringe population. For raw coal, the concentrations of benzene and parallelogram-shape aromatic structures of (<4 × 4) occupy 17.16% and 80.09%, respectively, while the ratio of parallelogram-shape aromatic structures of (>3 × 3) are only 2.75%. This result indicates that HVC contains large quantities of small size condensed aromatic structure, which are responsible for its weak thermal stability and the formation of plenty of free radicals in the plastic zone [13,16,26,38]. With increasing carbonization temperature, the concentrations of parallelogram-shape aromatic structures of (>3 × 3) increase significantly in semi-cokes, as shown in Table 4, suggesting that large quantities of condensation and repolymerization reactions occur in semi-cokes at 450 °C and 750 °C. It is noticeable that the concentration of parallelogram-shape aromatic structures of (>3 × 3) in C450M is lower than that in C450 whereas the concentration of the above structures in C750M is higher than that in C750, indicating that adding B_4C contribute to constraining condensation reactions in the plastic zone and promoting repolymerization reactions in the fast polycondensation zone.

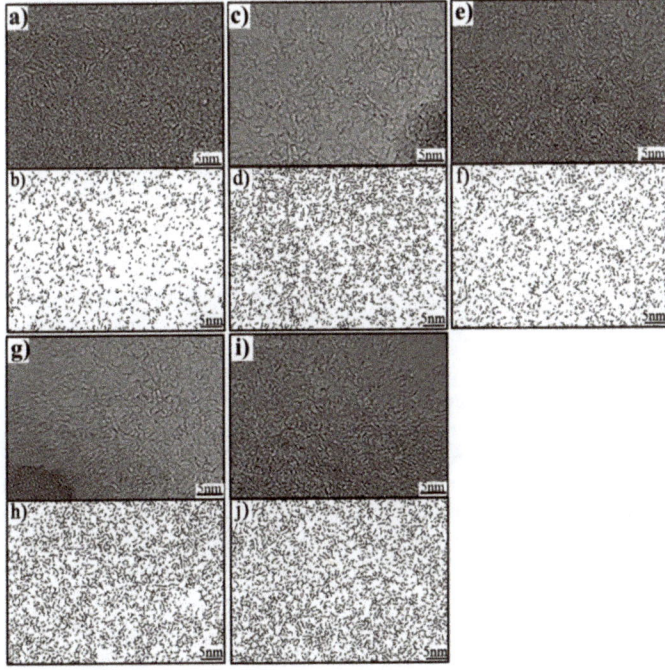

Figure 4. High-resolution transmission electron microscopy (HRTEM) images and the corresponding extracted lattice fringe images of coal and semi-coke samples: (**a,b**) raw coal; (**c,d**) C450; (**e,f**) C450M; (**g,h**) C750; (**i,j**) C750M.

Table 4. Aromatic fringe assignments and distribution frequency based on the analysis of HRTEM fringe images.

Aromatic Fringe Assignments		Distribution Frequency of Aromatic Fringes (%)				
Aromatic Sheet	Grouping (Å)	Raw Coal	C450	C450M	C750	C750M
Benzene	2.5–2.9	17.16	13.57	16.05	14.20	13.06
1 × 1	3.0–5.4	52.65	41.92	50.16	40.16	39.27
2 × 2	5.5–7.4	14.66	16.32	16.01	15.45	15.84
3 × 3	7.5–11.4	12.78	15.11	13.34	14.51	13.14
4 × 4	11.5–14.4	2.14	4.44	2.08	5.49	6.39
5 × 5	14.5–17.4	0.31	2.58	0.88	3.45	3.59
6 × 6	17.5–20.4	0.20	1.62	0.88	1.57	2.37
7 × 7	20.5–24.4	0.10	1.94	0.24	1.73	1.55
8 × 8	24.5–28.4	0	0.73	0.16	1.10	1.55
>8 × 8	>28.5	0	1.78	0.16	2.35	3.35

3.6. Role of the B_4C in the Improvement of Coke Quality

B_4C reacts with active oxygen obtained from oxygen-containing compounds in the plastic zone of HVC, which leads to reducing the reactions between active oxygen and transferable hydrogen. Therefore, more transferable hydrogens are available to stabilize free radicals, reducing the condensation and crosslinking reactions of free radicals. In this case, the anisotropic mesophase structures develop in the plastic zone of modified coal, which contributes to increasing the degree of anisotropy in coke. Simultaneously, more stable free radical fragments will be involved in the polymerization reaction under higher temperature, increasing the size of the aromatic sheet and condensed degree in semi-coke, as shown in C750M. Consequently, with further elevating temperature, the size of aromatic layer will increase in coke structures by adding B_4C. In summary, adding B_4C into HVC to produce coke has the result of increasing the size of the aromatic layer and anisotropic degree in the coke structure, which contributes to enhancing the coke resistance to CO_2 reaction, ultimately resulting in the significant improvement of coke quality.

4. Conclusions

In this work, in addition to a new additive (B_4C) that is primarily introduced, the modification mechanisms of coke quality were analyzed by using TG, FTIR, and HRTEM techniques. The main conclusions are summarized as follows:

(1) HVC contains large quantities of oxygen-containing functional groups, aliphatic side chains, and small molecular weight aromatic molecules, resulting in the coke derived from HVC with a high CRI index and low CSR index.
(2) B_4C can considerably improve the quality of low-strength coke prepared from HVC. Regarding optimal thermal strength indexes, the CRI index of coke decreases by 29.8%, while the CSR index of coke enhances by 23.1% when adding 0.5 wt.% B_4C into HVC.
(3) The reaction between B_4C and active oxygen derived from oxygen-containing compounds during HVC carbonization leads to reduced condensation and crosslinking reactions and increased secondary cracking reactions, which result in phenomena—increasing size of the aromatic layer and anisotropic degree in the modified coke structure—that are responsible for significantly improvements of the coke quality.

Author Contributions: Conceptualization, Q.W. and Z.-Z.Z.; software, Q.W. and C.-Y.Z.; validation, Q.W., C.S. and Y.-D.W.; formal analysis, Q.W. and C.S.; investigation, Q.W. and C.S.; resources, Z.-Z.Z.; data curation, Q.W.; writing—original draft preparation, Q.W.; writing—review and editing, Q.W. and C.S.; visualization, Q.W.; supervision, Z.-Z.Z. and C.-Y.Z.; project administration, Q.W.; funding acquisition, Z.-Z.Z. All authors have read and agreed to the published version of the manuscript.

Funding: This research was funded by the Key Project of Science and Technology of Chongqing, grant number CSTS.2010AB4084.

Institutional Review Board Statement: Not applicable.

Informed Consent Statement: Not applicable.

Data Availability Statement: The data presented in this study are available on request from the corresponding author. The data are not publicly available due to privacy issues.

Acknowledgments: The authors gratefully appreciate the National Natural Science Foundation of China, grant number 51044005 for financial support.

Conflicts of Interest: The authors declare no conflict of interest.

References

1. Diez, M.A.; Alvarez, R.; Melendi, S.; Barriocanal, C. Feedstock recycling of plastic wastes/oil mixtures in cokemaking. *Fuel* **2009**, *88*, 1937–1944. [CrossRef]
2. Fernández, A.M.; Barriocanal, C.; Díez, M.A.; Alvarez, R. Influence of additives of various origins on thermoplastic properties of coal. *Fuel* **2009**, *88*, 2365–2372. [CrossRef]

3. Melendi, S.; Diez, M.A.; Alvarez, R.; Barriocanal, C. Plastic wastes, lube oils and carbochemical products as secondary feedstocks for blast-furnace coke production. *Fuel Process. Technol.* **2011**, *92*, 471–478. [CrossRef]
4. Yan, J.C.; Bai, Z.Q.; Bai, J.; Guo, Z.X.; Li, W. Effects of organic solvent treatment on the chemical structure and pyrolysis reactivity of brown coal. *Fuel* **2014**, *128*, 39–45. [CrossRef]
5. Mollah, M.M.; Marshall, M.; Qi, Y.; Knowles, G.P.; Taghavimoghaddam, J.; Jackson, W.R.; Chaffee, A.L. Attempts to produce blast furnace coke from Victorian brown coal. 4. Low surface area char from alkali treated brown coal. *Fuel* **2016**, *186*, 320–327. [CrossRef]
6. Nomura, S.; Arima, T. Influence of binder (coal tar and pitch) addition on coal caking property and coke strength. *Fuel Process. Technol.* **2017**, *159*, 369–375. [CrossRef]
7. Mochizuki, Y.K.; Kubota, Y.; Uebo, K.; Tsubouchi, N. Influence of tarry material deposition on low-strength cokes or pyrolyzed chars of low rank coals on the strength. *Fuel* **2018**, *232*, 780–790. [CrossRef]
8. Krzesińska, M.; Szeluga, U.; Czajkowska, S.; Muszyński, J.; Zachariasz, J.; Pusz, S.; Kwiecińska, B.; Koszorek, A.; Pilawa, B. The thermal decomposition studies of three Polish bituminous coking coals and their blends. *Int. J. Coal Geol.* **2009**, *77*, 350–355. [CrossRef]
9. Mukherjee, D.K.; Sengupta, A.N.; Choudhury, D.P.; Sanyal, P.K.; Rudra, S.R. Effect of hydrothermal treatment on caking propensity of coal. *Fuel* **1996**, *75*, 477–482. [CrossRef]
10. Sarkar, N.B.; Sarkar, P.; Choudhury, A. Effect of hydrothermal treatment of coal on the oxidation susceptibility and electrical resistivity of HTT coke. *Fuel Process. Technol.* **2005**, *86*, 487–497. [CrossRef]
11. Graff, R.A.; Brandes, S.D. Modification of Coal by Subcritical Steam—Pyrolysis and Extraction Yields. *Energy Fuels* **1987**, *1*, 84–88. [CrossRef]
12. Shui, H.F.; Li, H.P.; Chang, H.T.; Wang, Z.C.; Gao, Z.; Lei, Z.P.; Ren, S.B. Modification of sub-bituminous coal by steam treatment: Caking and coking properties. *Fuel Process. Technol.* **2011**, *92*, 2299–2304. [CrossRef]
13. Soncini, R.M.; Means, N.C.; Weiland, N.T. Co-pyrolysis of low rank coals and biomass: Product distributions. *Fuel* **2013**, *112*, 74–82. [CrossRef]
14. Flores, B.D.; Flores, I.V.; Guerrero, A.; Orellana, D.R.; Pohlmann, J.G.; Diez, M.A.; Borrego, A.G.; Osorio, E.; Vilela, A.C.F. Effect of charcoal blending with a vitrinite rich coking coal on coke reactivity. *Fuel Process. Technol.* **2017**, *155*, 97–105. [CrossRef]
15. Wu, Z.Q.; Zhang, J.; Zhang, B.; Guo, W.; Yang, G.D.; Yang, B.L. Synergistic effects from co-pyrolysis of lignocellulosic biomass main component with low-rank coal: Online and offline analysis on products distribution and kinetic characteristics. *Appl. Energy* **2020**, *276*, 115461. [CrossRef]
16. Shi, L.; Liu, Q.Y.; Guo, X.J.; Wu, W.Z.; Liu, Z.Y. Pyrolysis behavior and bonding information of coal—A TGA study. *Fuel Process. Technol.* **2013**, *108*, 125–132. [CrossRef]
17. Xu, Y.; Zhang, Y.F.; Wang, Y.; Zhang, G.J.; Chen, L. Gas evolution characteristics of lignite during low-temperature pyrolysis. *J. Anal. Appl. Pyrolysis* **2013**, *104*, 625–631. [CrossRef]
18. He, W.J.; Liu, Z.Y.; Liu, Q.Y.; Ci, D.H.; Lievens, C.; Guo, X.F. Behaviors of radical fragments in tar generated from pyrolysis of 4 coals. *Fuel* **2014**, *134*, 375–380. [CrossRef]
19. Vega, M.F.; Fernández, A.M.; Díaz-Faes, E.; Barriocanal, C. Improving the properties of high volatile coking coals by controlled mild oxidation. *Fuel* **2017**, *191*, 574–582. [CrossRef]
20. Qian, Z.; Marsh, H. The Co-Carbonization Behavior of Chinese Coals with Pitch Additives. *J. Mater. Sci.* **1984**, *19*, 3311–3318. [CrossRef]
21. Jiang, H.Y.; Wang, J.G.; Duan, Z.C.; Li, F. Study on the microstructure evolution of phenol-formadehyde resin modified by ceramic additive. *Chin. J. Mater. Res.* **2006**, *20*, 203–207.
22. Wang, J.G.; Jiang, N.; Guo, Q.G.; Liu, L.; Song, J.R. Study on the structural evolution of modified phenol–formaldehyde resin adhesive for the high-temperature bonding of graphite. *J. Nucl. Mater.* **2006**, *348*, 108–113. [CrossRef]
23. Wu, D.; Liu, G.J.; Sun, R.Y.; Fan, X. Investigation of Structural Characteristics of Thermally Metamorphosed Coal by FTIR Spectroscopy and X-ray Diffraction. *Energy Fuels* **2013**, *27*, 5823–5830.
24. Ibarra, J.V.; Munoz, E.; Moliner, R. FTIR study of the evolution of coal structure during the coalification process. *Org. Geochem.* **1996**, *24*, 725–735. [CrossRef]
25. Mathews, J.P.; Fernandez-Also, V.; Jones, A.D.; Schobert, H.H. Determining the molecular weight distribution of Pocahontas No. 3 low-volatile bituminous coal utilizing HRTEM and laser desorption ionization mass spectra data. *Fuel* **2010**, *89*, 1461–1469. [CrossRef]
26. Song, H.J.; Liu, G.R.; Zhang, J.Z.; Wu, J.H. Pyrolysis characteristics and kinetics of low rank coals by TG-FTIR method. *Fuel Process. Technol.* **2017**, *156*, 454–460. [CrossRef]
27. Viricelle, J.P.; Goursat, P.; Bahloul-Hourlier, D. Oxidation behaviour of a boron carbide based material in dry and wet oxygen. *J. Therm. Anal. Calorim.* **2000**, *63*, 507–515. [CrossRef]
28. Werheit, H.; Au, T.; Schmechel, R.; Shalamberidze, S.O.; Kalandadze, G.I.; Eristavi, A.M. IR-Active Phonons and Structure Elements of Isotope-Enriched Boron Carbide. *J. Solid State Chem.* **2000**, *154*, 79–86. [CrossRef]
29. Khai, T.V.; Na, H.G.; Kwak, D.S.; Kwon, Y.J.; Ham, H.; Shim, K.B.; Kim, H.W. Comparison study of structural and optical properties of boron-doped and undoped graphene oxide films. *Chem. Eng. J.* **2012**, *211*, 369–377. [CrossRef]

30. Grint, A.; Mehani, S.; Trewhella, M.; Crook, M.J. Role and Composition of the Mobile Phase in Coal. *Fuel* **1985**, *64*, 1355–1361. [CrossRef]
31. Orrego-Ruiz, J.A.; Cabanzo, R.; Mejia-Ospino, E. Study of Colombian coals using photoacoustic Fourier transform infrared spectroscopy. *Int. J. Coal Geol.* **2011**, *85*, 307–310. [CrossRef]
32. Lin, X.C.; Wang, C.H.; Ideta, K.; Miyawaki, J.; Nishiyama, Y.; Wang, Y.G.; Yoon, S.; Mochida, I. Insights into the functional group transformation of a chinese brown coal during slow pyrolysis by combining various experiments. *Fuel* **2014**, *118*, 257–264. [CrossRef]
33. He, X.Q.; Liu, X.F.; Nie, B.S.; Song, D.Z. FTIR and Raman spectroscopy characterization of functional groups in various rank coals. *Fuel* **2017**, *206*, 555–563. [CrossRef]
34. Painter, P.C.; Snyder, R.W.; Starsinic, M.; Coleman, M.M.; Kuehn, D.W.; Davis, A. Concerning the Application of Ft-Ir to the Study of Coal—A Critical-Assessment of Band Assignments and the Application of Spectral-Analysis Programs. *Appl. Spectrosc.* **1981**, *35*, 475–485. [CrossRef]
35. Zhang, W.G.; Cheng, H.M.; Sano, H.; Uchiyama, Y.; Kobayashi, K.; Zhou, L.J.; Shen, Z.H.; Zhou, B.L. The effects of nanoparticulate SiC upon the oxidation behavior of C-SiC-B4C composites. *Carbon* **1998**, *36*, 1591–1595. [CrossRef]
36. Rodriguez, M.G.; Kharissova, O.V.; Ortiz-Mendez, U. Formation of boron carbide nanofibers and nanobelts from heated by microwave. *Rev. Adv. Mater. Sci.* **2004**, *7*, 55–60.
37. Sharma, A.; Kyotani, T.; Tomita, A. Quantitative evaluation of structural transformations in raw coals on heat-treatment using HRTEM technique. *Fuel* **2001**, *80*, 1467–1473. [CrossRef]
38. Lievens, C.; Ci, D.H.; Bai, Y.; Ma, L.G.; Zhang, R.; Chen, J.Y.; Gai, Q.Q.; Long, Y.H.; Guo, X.F. A study of slow pyrolysis of one low rank coal via pyrolysis-GC/MS. *Fuel Process. Technol.* **2013**, *116*, 85–93. [CrossRef]

Crumpled Graphene-Storage Media for Hydrogen and Metal Nanoclusters

Liliya R. Safina [1,*], Karina A. Krylova [2,3], Ramil T. Murzaev [2], Julia A. Baimova [2,3] and Radik R. Mulyukov [1,2]

1. Ufa State Petroleum Technological University, Kosmonavtov Str. 1, 450062 Ufa, Russia; radik@imsp.ru
2. Institute for Metals Superplasticity Problems of the Russian Academy of Sciences, Khalturina 39, 450001 Ufa, Russia; bukreevakarina@gmail.com (K.A.K.); mur611@mail.ru (R.T.M.); julia.a.baimova@gmail.com (J.A.B.)
3. Bashkir State University, Validy Str. 32, 450076 Ufa, Russia
* Correspondence: saflia@mail.ru

Abstract: Understanding the structural behavior of graphene flake, which is the structural unit of bulk crumpled graphene, is of high importance, especially when it is in contact with the other types of atoms. In the present work, crumpled graphene is considered as storage media for two types of nanoclusters—nickel and hydrogen. Crumpled graphene consists of crumpled graphene flakes bonded by weak van der Waals forces and can be considered an excellent container for different atoms. Molecular dynamics simulation is used to study the behavior of the graphene flake filled with the nickel nanocluster or hydrogen molecules. The simulation results reveal that graphene flake can be considered a perfect container for metal nanocluster since graphene can easily cover it. Hydrogen molecules can be stored on graphene flake at 77 K, however, the amount of hydrogen is low. Thus, additional treatment is required to increase the amount of stored hydrogen. Remarkably, the size dependence of the structural behavior of the graphene flake filled with both nickel and hydrogen atoms is found. The size of the filling cluster should be chosen in comparison with the specific surface area of graphene flake.

Keywords: crumpled graphene; Ni-graphene composite; hydrogen; molecular dynamics; storage media

1. Introduction

The transformation of nanoscale structural elements into three-dimensional (3D) complex architecture is currently an important task of materials science. Since the discovery of fullerenes in 1985, a lot of new carbon structures have been proposed. Carbon polymorphs can be used to obtain nanostructures with unique mechanical and physical properties applicable in nanoelectronics, energy storage devices, sensors, supercapacitors, Li-ion batteries, etc. [1–7].

Graphene is capable of enhancing the performance, functionality as well as durability of many applications. The one-atom thin structure can serve as the platform for other materials especially since the graphene layer can be bent or crumpled. Extensive studies have been carried out in recent decades to investigate the crumpling behavior of thin sheets like graphene, by both theoretical and experimental methods [8–19]. It was shown that crumpled structures can have excellent compression and aggregation-resistant properties [5,6,9,10,20–22]. Crumpled graphene can be used for the production of new supercapacitors [19]. However, one of their new and important applications is that of a natural container for other atoms.

It is known that coupling metal nanoclusters with carbon materials can efficiently promote catalysis and electrocatalytic activities [23,24] or prevent the corrosion of metals [25]. redAmong metals, some can be easily attached to its surface, while others even repulse from the surface. The bigger the carbon solubility of metal, the more amounts of carbon can be attracted to the metal surface. Metals such as Ni, Cu, Pt, or Au are the most preferable as catalysts for the growth of graphene [26–28]. However, nickel is one of the most broadly used metals for a range of graphene applications, including the growth of carbon nanotubes (CNTs) and graphene [29–31]. Lattice mismatch is one of the important factors for choosing the metal for metal–graphene interface and Ni(111) surface is the closest matched interface with respect to the graphene of all transition metals [26]. In [29], a simple and scalable method for the synthesis of hollow graphene balls using a Ni nanocluster template was developed. The interface between graphene and Ni attracts considerable interest due to the possibility of the synthesis of large-area graphene on metal substrate [32–36]. Since the fabrication of graphene layers on the metal substrate was very successful, the idea of the fabrication of metal–graphene layered composites was raised by [37–40]. Graphene added into the copper and nickel matrix by chemical vapor deposition can considerably improve the strength of metal [38]. The overview of graphene–nickel composites can be found in [39]. Both the theoretical and experimental works on Ni-graphene layered composites showed that the mechanical properties of metal-matrix composite reinforced with the graphene layer considerably improved. In the present work, the other approach to obtain the Ni-graphene composite was applied: introducing metal nanoparticles to the crumpled graphene matrix. In [41–44] it was shown that such composites can be obtained through compression combined with high temperatures.

The other important advantages of such structures are a large specific surface area (SSA) and, respectively, a high rate of gas adsorption, which makes it possible to predict their use in hydrogen storage [45–53]. Various structural parameters can affect the degree of hydrogen accumulation [45]. For example, the amount of adsorbed hydrogen increases with an increase in the CNT diameter, since this increases the surface area on which hydrogen can be adsorbed. An increase in the distance between graphene sheets in the structure also leads to an increase in the gravimetric hydrogen absorption density. Another promising carbon structure is two layers of graphene connected by short CNTs with graphene cones added on top of the structure. The idea of such a structure appeared when it was shown that the hydrogenation of graphene can lead to the buckling of the graphene sheet [54]: the amount of accumulated hydrogen increased from only 3 wt% for the simple graphene plane to 20 wt% for graphene plane with graphene cone. Carbon nanostructures doped with alkali metals (Li and K) can adsorb even more hydrogen [48–53]: 14 wt% (Li) and 20 wt% (K) of hydrogen at moderate conditions, in contradiction with the lower values reported later [50]. The Li-doped activated carbon [55] can store from 2.1 to 2.6 wt% of H_2 at 77 K and at 2 MPa, which shows that at given pressure–temperature conditions the amount of stored hydrogen can be increased. It should be mentioned that the required amount of stored hydrogen for carbon nanostructures was still not achieved. Thus, an active search for new materials and structures for hydrogen storage and transportation is of high importance.

In the present work, two types of nanofillers for crumpled graphene flake were considered—Ni and hydrogen nanoclusters. Thus, the idea of using crumpled graphene flake for the storage of different atoms using Ni and hydrogen clusters as the example was realized. It is important to understand the dynamics of interaction between single graphene flake and a metal or non-metal nanocluster before the study of storage in three-dimensional crumpled graphene. The behavior of the hybrid structure is studied by atomistic simulation at different temperatures. For the carbon–nickel system, room temperature (300 K) and temperature close to the melting point of Ni (1000 K) were chosen. However, for the carbon–hydrogen system, the temperature of liquid nitrogen (77 K) and room temperature were chosen. The dynamics of the structure were studied during exposure for 20 ps.

2. Materials and Methods

To study the interaction between graphene flake (GF) and atomic nanocluster, an initial structure composed of the GF of one size (N_C = 252) and atomic clusters of different sizes (N = 21; 38; 47; 66; 78) inside the flake was considered (see Figure 1a). One graphene flake was a small CNT(11,11) 1.3 nm-long with two atomic rows deleted along the axis of the nanotube. Since the size of this small sample of graphene along three dimensions is very small (1.3 nm along the z axis, 3.6 nm along the y axis and one-atom-thick along the x axis) it is called "flake" rather than "nanoribbon". This GF was filled with nanoclusters of two types: (i) nickel nanocluster and (ii) hydrogen nanocluster.

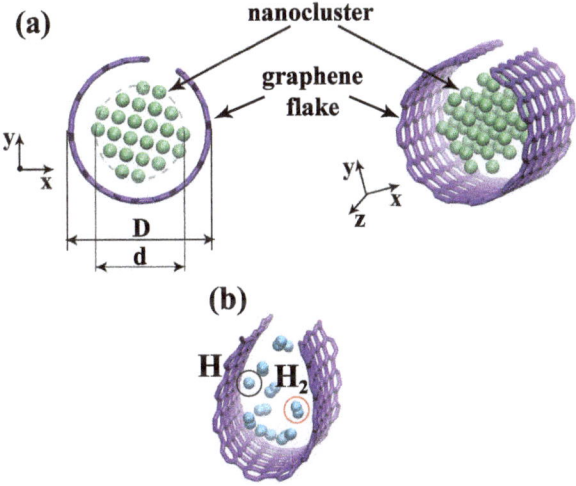

Figure 1. (a) Initial structure: graphene flake with the diameter D filled with nanocluster with the diameter d; (b) atomic and molecular hydrogen inside the graphene flake.

The maximum size of the nanocluster was chosen to almost completely fill the cavity of the graphene flake. The size of the structural elements is shown in Table 1. The distance between the edge of the nanocluster and the side of the flake can be defined as $a = (D-d)/2$. However, these parameters affect the structural transformation just in the case of metal nanocluster since it is a solid particle. a hydrogen cluster is gas and small H atoms can easily spread inside GF.

Table 1. The size of the structural elements.

D, Å	N	d, Å	a, Å
	21	6.2	4.25
	38	8.1	3.30
14.7	47	9.0	2.85
	66	9.4	2.65
	78	10.4	2.15

It should be mentioned, that nanocluster Ni_{78} was considered here as nanocluster of critical size for the chosen size of the GF. However, for the metal cluster, this size was too big and the sides of GF were closer than it should have been in a real system. To eliminate the negative effect of the overly sized nanocluster, nanocluster Ni_{66} was considered the characteristic for the case of metal–graphene interaction. While for the case of hydrogen–graphene interaction, an even bigger size of the nanocluster could be considered since the size of the hydrogen atoms was much smaller than the size of Ni atoms.

The simulation was conducted using a large-scale atomic/molecular massively parallel simulator (LAMMPS). Equations of motion for the atoms were integrated numerically using the fourth-order Verlet method with the time step of 0.2 fs. The Nose–Hoover thermostat was used to control the system temperature. The periodic boundary conditions were applied in all directions, however, the simulation box is much bigger than the size of GF filled with nanocluster. The adaptive intermolecular reactive empirical bond order potential (AIREBO) [56] was used to describe the interatomic interactions between carbon atoms, including both covalent bonds in the basal plane of graphene and van der Waals interactions between GF and nanocluster. The simulation configurations were visualized by Visual Molecular Dynamics (VMD) Software [57].

2.1. Graphene with Nickel Nanocluster

Graphene flake with the nanocluster was exposed at 300 K and 1000 K for 20 ps to study the dynamics of nanocluster inside graphene flake. Previously [41], it was shown that the melting temperature of the Ni nanocluster was about 1300 K. Thus, in the present work, the highest considered temperature was 1000 K. At this temperature, Ni nanocluster was close to melting but not melted yet, even for the smallest nanocluster, which consisted of 21 atoms. Initially, Ni atoms were packed into the face-centered cubic lattice.

To describe the interatomic interaction between Ni–Ni and Ni–C, Morse interatomic potential was used with the parameters D_e = 0.4205 eV, R_e = 2.78 Å and β = 1.4199 1/Å for Ni–Ni [58]; and D_e = 0.433 eV, R_e = 2.316 Å, β = 3.244 1/Å for Ni–C. The parameters of the Morse potential for describing the interaction of nickel and carbon atoms were recently proposed using *ab-initio* simulation [59,60].

2.2. Graphene with Hydrogen Nanocluster

Graphene flake with hydrogen nanocluster was exposed at 77 K and 300 K for 20 ps. The temperature of 77 K was chosen since it was previously shown that better sorption of hydrogen can be found at 77 K [61]. The temperature of 300 K was chosen to study the dynamics at room temperature, however, it is known that this temperature is too high for hydrogen storage [46].

To describe the interatomic interaction between C and H, AIREBO interatomic potential was used.

It should be mentioned that the initially obtained structure contains atomic hydrogen, however, H atoms transform to H_2 molecules and just several H atoms remain single (see Figure 1b). As it was shown, the lowest binding energy between graphene and H_2 was observed when the distance between C and H_2 in the range from 2.9 to 3.2 Å [61]. Hydrogen molecules can be bonded by van der Waals interaction when they move close enough to the side of GF. Single hydrogen atoms can be chemically adsorbed on graphene by covalent bonding.

3. Results

3.1. Graphene with Ni Nanocluster

At first, the evolution of the potential energy of the system during the crumpling process for graphene flake filled within spherical the Ni nanocluster was analyzed. In Figure 2, potential energy as the function of exposure time at 300 K (a) and 1000 K (b) for five types of nanoclusters was shown. It was found that the total potential energy of the system was saturated to a practically constant value at the end of the equilibration process, indicating that the system reached equilibrium and a stable state. All the changes in the energy curves correspond to some structural changes.

The longest time of stabilization at 300 K was found for Ni_{21} (5 ps). Curves for nanoclusters of close diameter are almost the same. Graphene flake with Ni_{38} and Ni_{47} reach the equilibrium state at about 4 ps of exposure, while the GF with Ni_{66} and Ni_{78} reach equilibrium state at 3.2 ps. The bigger the diameter of the nanocluster, the less the equilibration time. This can be explained by the mutual arrangement of the nanocluster and GF. For a small nanocluster, the distance a is two times higher than for the biggest one which means that the time required to attach the nanocluster by the graphene flake is longer. For nanocluster Ni_{78}, 3 ps is enough for GF to fully cover the nanocluster, while for Ni_{21}, not only coverage took place, but also the further crumpling of the flake with the changing of the round shape of nanocluster.

At higher temperature (1000 K), again a strong correlation between equilibration time and the size of the nanocluster was found. For GF with Ni_{21} nanocluster, the transformation was fast since the temperature was close to the melting point and the nanocluster can be easily destroyed. Temperature fluctuations facilitate the crumpling process and the appearance of new bonds between the edges of GF. For bigger nanoclusters, the temperature slightly affects the time of equilibration. Again, the same values of equilibration time were obtained for Ni_{38} and Ni_{47} (about 4 ps) and for Ni_{66} and Ni_{78} (about 3.2 ps). However, temperature decreases the total potential energy of the system.

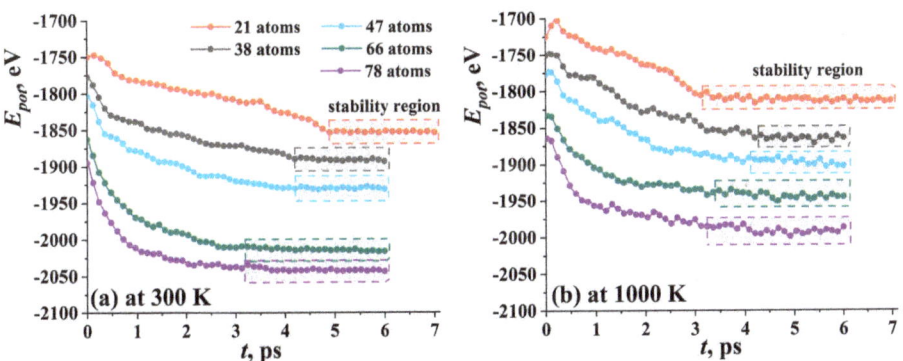

Figure 2. Potential energy as the function of exposure time at 300 K (**a**) and 1000 K (**b**) for five types of nanoclusters.

In Figure 3, the process of crumpling for nanoclusters Ni_{21} (a), Ni_{47} (b) and Ni_{66} (c) was shown in details after exposure at 300 K (blue line) and 1000 K (red line). At the beginning of the crumpling process, GF startED to change its round shape, and the fcc crystalline order of Ni nanocluster WAs destroyed. Metal atoms are attracted by the graphene surface and tend to occupy equilibrium positions above the center of the carbon hexagon, which was also mentioned in [41]. Ni atoms WEre interacting with graphene hexagons by van der Waals forces. The edges of GF can be bonded during exposure.

At first, consider GF with Ni_{21} (see Figure 3a, lower line of snapshots). At 300 K, GF lost its round shape at about 0.5 ps, edges of GF move towards each other, and at $t = 1.5$ ps one covalent bond appeared between two edges. Nanocluster disturbed by the temperature and several atoms attach to the graphene plane (state II). This leads to the spreading of the Ni atoms over the graphene plane (state III). Simultaneously, such a small nanocluster allowed the graphene flake to easily bend and the structural unit transforms to the bi-layer graphene with Ni atoms spread inside (state IV). Several more covalent bonds appeared on the side edges of GF. At 300 K, not many covalent bonds between the edges of GF can appear, since the temperature is far from the melting temperature of graphene [62,63].

Increase in the exposure temperature to 1000 K facilitates the transformation. At 300 K, equilibration time is equal to 5 ps, while at 1000 K, it is equal to 3.5 ps. Subsquently, the structure came to a stable state and no further changes were observed despite slight thermal fluctuations. At 1000 K, more carbon atoms on the edges of GF found neighbors since the structure is disturbed by thermal fluctuations.

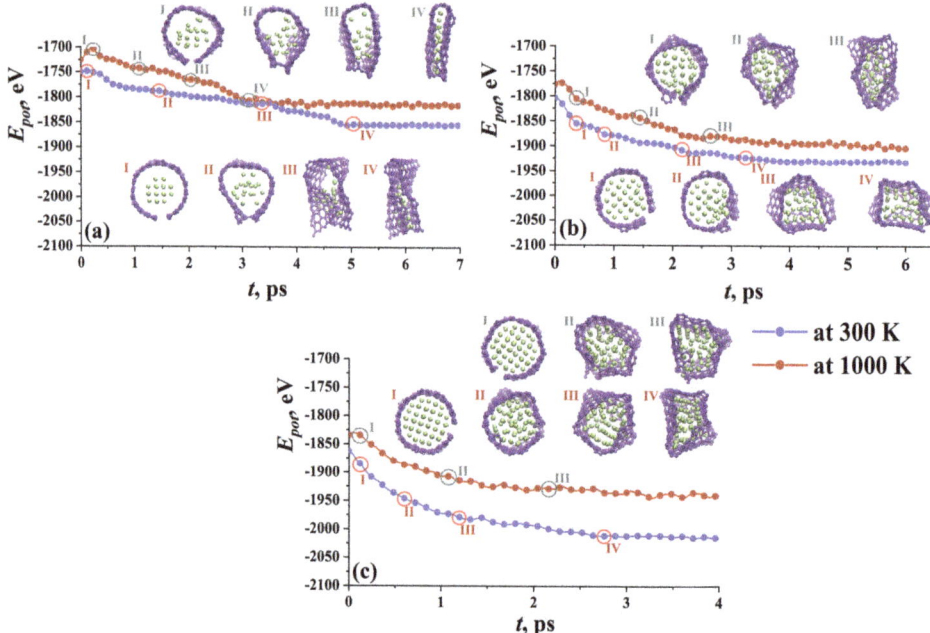

Figure 3. Potential energy as the function of exposure time for Ni_{21} (**a**), Ni_{47} (**b**) and Ni_{66} (**c**). Corresponding snapshots of the structure of crumpled graphene flake filled with nanocluster during exposure at 300 K (bottom line of snapshots) and 1000 K (upper line of snapshots). Carbon atoms are shown by violet and nickel atoms are shown by green.

For structures with Ni_{47} and Ni_{66}, the behavior is qualitatively close. An almost round Ni nanocluster inside the graphene flake was observed at the initial state at both 300 K and 1000 K (as can be seen in Figure 3b,c). The edges of GF can attach to each other with the formation of new covalent bonds. During exposure, graphene flake transforms into a capsule containing nickel nanocluster. The nickel nanocluster almost completely fills the graphene flake, in comparison with Ni_{21}. As a result, it is difficult to deform such a structure. However, at 1000 K, the nanocluster is more disturbed and the stable state (state III, the top line of snapshots) reached at a lower equilibration time. The nickel nanocluster became more planar since GF is rigid and can bend the soft metal nanocluster. In the case of Ni_{47}, GF transforms the "bag" for nickel, while Ni_{66} nanocluster is too big and edges on both sides cannot be attached. In the case of Ni_{66}, GF just covers the Ni nanocluster as much is possible.

A graphene flake with Ni_{38} behaves in a similar way to Ni_{47}, while GF with Ni_{78} behaves similarly to Ni_{66}. Note that GF always tends to wrap the metal cluster.

One of the important applications of such structural elements is the fabrication of composites. In Figure 4, the initial structure (a, a') was composed of graphene flake filled with Ni nanocluster Ni_{21} and Ni_{78} correspondingly. The initial structure is compressed at 1000 K to obtain graphene–nickel composite material (c, c'). Composite can be fabricated under hydrostatic compression at high temperatures [41–44].

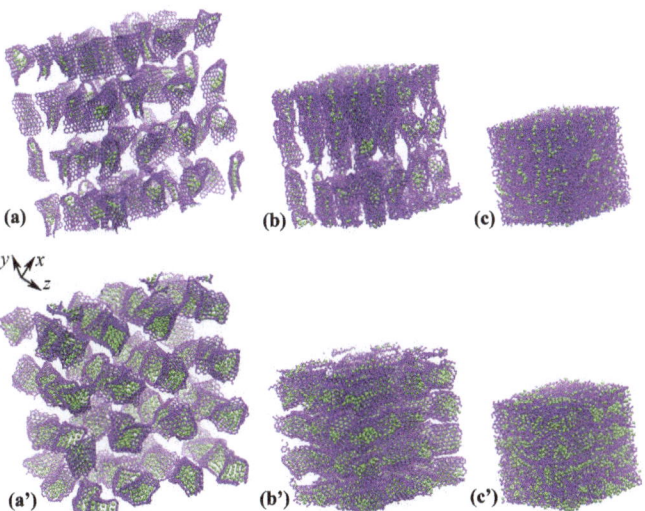

Figure 4. Snapshots of the structure of crumpled graphene filled with (**a**–**c**) Ni_{21} and (**a'**–**c'**) Ni_{78} nanocluster. Initial structure was obtained by annealing at 300 K, while composite was obtained by hydrostatic compression at 1000 K. **Ni** atoms are shown by purple and **C** atoms are shown by green color.

3.2. Graphene with Hydrogen Nanocluster

In Figure 5, the potential energy of GF filled with hydrogen cluster during exposure at 77 K (a) and 300 K (b) is presented as the function of equilibration time. Similar to GF with the Ni nanocluster, five structural units were divided into three groups—GF with 21 hydrogen atoms; GF with 38 and 47 H atoms; and GF with 66 and 78 H atoms. Despite that the hydrogen atoms transform into hydrogen molecules, the initial number of atoms was also used, since the number of H_2 molecules and single hydrogen atoms can change from one simulation run to another which depends on thermal fluctuations.

At 77 K (Figure 5a), the equilibrium state is the state when hydrogen atoms found their sites and GF change the cylinder shape to the one with minimal energy. At 300 K (Figure 5a), the equilibrium state is the state when all hydrogen atoms disrobed from the side of GF or even fully leave the cavity of GF. This would be discussed later together with the description of corresponding snapshots. The biggest drop of the energy value took place during the first picosecond and connected with the change of initial nonequilibrium configuration of the cluster and slightly with the change of shape of the GF.

At 77 K, the time of equilibration was the longest (3 ps) for hydrogen clusters consist of 77, 66, and 38 atoms, and the shortest (1.5 ps) is for 21 and 47 atoms. At that temperature, the size of the cluster plays quite an important role which is connected with the number of sites for hydrogen on the side of GF. All hydrogen molecules and atoms can easily find sites on graphene for an initial number of atoms less than 47, while for bigger clusters, the number of sites is not enough and some molecules and atoms will move outside GF or seek better sits near GF.

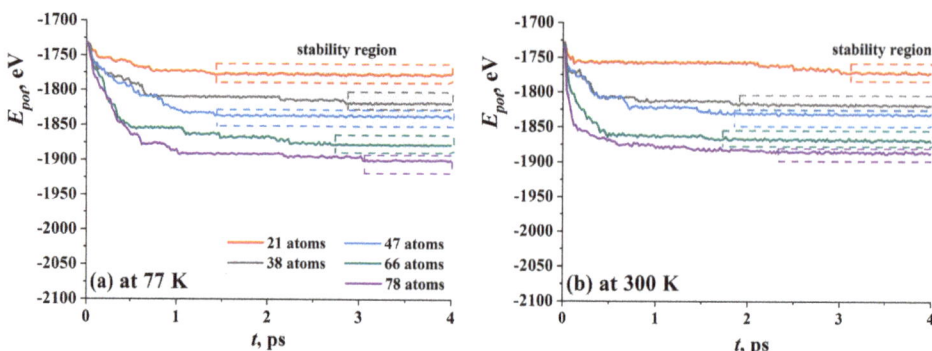

Figure 5. Potential energy as the function of exposure time at 77 K (**a**) and 300 K (**b**) for five types of hydrogen nanoclusters.

At 300 K, the main point is the process of dehydrogenation which is quite quick at such a high temperature, and for clusters with 38–78 atoms, about 1.5 ps is enough for all molecules to detach the side of GF and move outside. The long time of equilibration for cluster H_{21} can be explained simply: there are about nine molecules and three atoms and all the time spent for the slow movement of this hydrogen. A large number of atoms in the cluster pushes the sides of GF and opens it much faster.

To understand the dynamics of the hydrogen sorption/desorption, snapshots of the structural units altogether with their potential energy are presented in Figure 6 for three groups. If the hydrogen cluster is quite small (21 atoms), there are a lot of vacant places on the graphene plane to attach hydrogen. Only several atoms or molecules can move outside GF at a low temperature equal to 77 K. Most of the hydrogen molecules attached by van der Waals force to graphene and slightly moving along graphene flake.

At 300 K, hydrogen molecules and atoms have no chance to form even a weak bond to graphene. At 300 K, all hydrogen atoms move outside GF which is quite understandable. Numerous theoretical and experimental works confirm the sorption of molecular hydrogen on carbon nanostructures [50,64–66] in a specific temperature range (50–200 K). Here, such a high temperature was used to analyze how hydrogen will leave the cavity of GF. As it can be seen from Figure 6a, this process is very fast and at 4 ps, only two molecules stay inside GF (stage IV, upper line).

A large number of hydrogen atoms almost completely fills the cavity of the GF. At 77 K, hydrogen atoms are easily converted into hydrogen molecules and under the action of chemical attraction or van der Waals forces, attach to the walls of the graphene flake. Mainly, hydrogen, which is located in the center of the flake under the influence of temperature, tends to leave the graphene cavity. Therefore, in structures with 66 and 78 hydrogen atoms, the stabilization of the structure takes longer (3 ps), in contrast to small hydrogen clusters, such as 21 and 47 (1.5 ps).

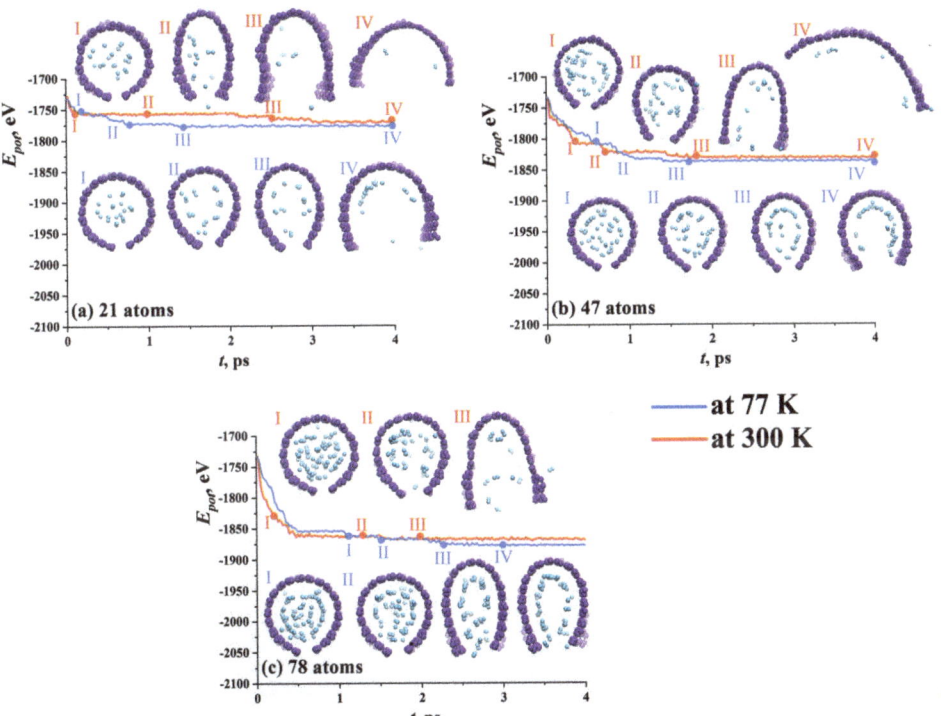

Figure 6. Potential energy as the function of exposure time for three types of hydrogen nanoclusters: (**a**) 21 atoms; (**b**) 47 atoms; and (**c**) 78 atoms. Corresponding snapshots of graphene flake filled with nanocluster during exposure at 77 K (bottom line of snapshots) and 300 K (upper line of snapshots). Carbon atoms are shown by violet and hydrogen atoms are shown by blue.

For a bigger nanocluster, energy curves almost coincide for two temperatures, since hydrogen molecules at 77 K can find the places for sorption just after the first steps. In this case, hydrogen atoms are placed near graphene and link it as far as the van der Waals radius of the molecule reaches. Thus, the hydrogen adsorption capacities depend considerably on the initial size of the hydrogen cluster. The SSA of the GF should be big enough for a chosen number of H atoms and H_2 molecules. Here, SSA is equal to 1153.9 m/g^2, which is enough to settle down 21–38 H atoms, but not enough for a bigger number of H atoms.

In Figure 7, snapshots of GF with 78 hydrogen atoms during exposure at 77 K for 20 ps are presented. As it can be seen, some hydrogen molecules attach the opposite side of GF since there were no sites inside the cavity. During exposure, even at 77 K, GF opens and then closes again. If the exposure time would be increased to even 100 ps, GF will move like the wings of a flying bird with adsorbed hydrogen atoms. This state is equilibrium and can be preserved for a long time.

As well as for a metal nanocluster, the three-dimensional structure of crumpled graphene filled with hydrogen is presented in Figure 8. The corresponding structure was considered in [46,47], where the effect of hydrostatic compression on the possibility of hydrogen storage was studied. In Figure 8a, the initial structure of crumpled graphene was presented. As it can be seen, initially there are too many channels for hydrogen to move out of the structure. Thus, additional treatment was required to save hydrogen inside the pores of crumple graphene. In Figure 8b, the structure after 40% of hydrostatic compression was presented. In such a compressed structure, hydrogen can be stored much more effectively than in undeformed crumpled graphene [46,47]. It can be concluded that crumpled graphene is an effective storage media for hydrogen. However, a search for the improvement of the quantity of stored hydrogen should be found, for instance, in the introduction of metal atoms. From this point of view, understanding the interaction between hydrogen and metal nanoclusters and GF is of high importance.

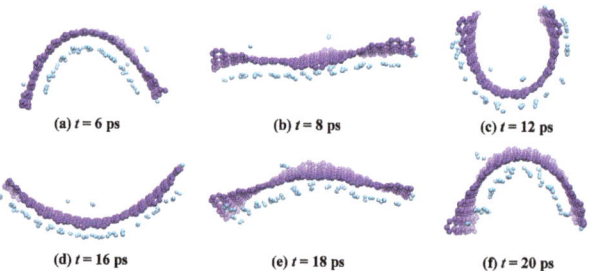

Figure 7. Snapshots of graphene flake with 78 hydrogen atoms during exposure at 77 K for 20 ps. Colors are as in Figure 6.

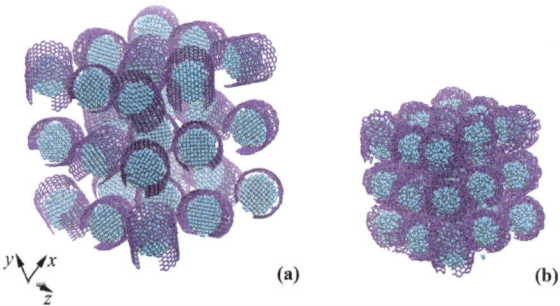

Figure 8. Snapshots of crumpled graphene filled with hydrogen atoms before (a) and after (b) hydrostatic compression. Colors are as in Figure 6.

4. Conclusions

Molecular dynamics simulation is used to study the dynamics of graphene flakes filled with the nanoclusters of two different types: metal and non-metal. The obtained results can shed the light on understanding the possibility of using crumpled graphene as a storage media from the point of single flake behavior.

It was found that cavities of crumpled graphene (the structure consists of crumpled graphene flakes) can be used as containers for metal nanoclusters—for instance, for Ni. A nanocluster consisting of 21 to 78 Ni atoms was considered. A graphene flake can easily cover nanocluster, however, the dynamics of the interaction strongly depend on the nanocluster size. Small nanoclusters can be easily bent by rigid graphene flake, while the biggest conserve the shape. Such a structure, composed of Ni nanoclusters and graphene flakes, can be further used to obtain composite material with improved mechanical properties.

The problem of hydrogen storage has been of high importance for decades and the application of carbon nanostructures from this point of view also looks promising. Crumpled graphene has a high specific surface area, light weight and a lot of pores and cavities which can be filled with hydrogen. It was shown that at 77 K, hydrogen molecules and atoms can be absorbed by both sides of graphene flakes. However, a single flake cannot store enough hydrogen for practical application. However, when graphene flakes composed in another structure and with additional treatment like hydrostatic compression [46,47], it can be successfully used for hydrogen storage and transportation.

Author Contributions: Software and initial simulation setup were conducted by L.R.S. and R.T.M. The molecular dynamics study and drafting the manuscript were conducted by J.A.B., K.A.K. and R.R.M. All authors have read and agreed to the published version of the manuscript.

Funding: Part of this research was supported by the grant of Russian Science Foundation (No. 20-72-10112). Work of R.R.M. and R.T.M. supported by the program of fundamental researches of Government Academy of Sciences of IMSP RAS.

Institutional Review Board Statement: Not applicable.

Informed Consent Statement: Not applicable.

Data Availability Statement: Not applicable.

Conflicts of Interest: The authors declare no conflict of interest.

Abbreviations

The following abbreviations are used in this manuscript:

MD	Molecular Dynamics
CNT	Carbon Nanotubes
GF	Graphene Flake
SSA	Specific Surface Area

References

1. Iwan, A.; Malinowski, M.; Pasciak, G. Polymer fuel cell components modified by graphene: Electrodes, electrolytes and bipolar plates. *Renew. Sustain. Energy Rev.* **2015**, *49*, 954–967. [CrossRef]
2. Baimova, J.A.; Rysaeva, L.K.; Liu, B.; Dmitriev, S.V.; Zhou, K. From flat graphene to bulk carbon nanostructures. *Phys. Status Solidi (b)* **2015**, *252*, 1502–1507. [CrossRef]
3. Xiao, J.; Mei, D.; Li, X.; Xu, W.; Wang, D.; Graff, G.L.; Bennett, W.D.; Nie, Z.; Saraf, L.V.; Aksay, I.A.; et al. Hierarchically Porous Graphene as a Lithium–Air Battery Electrode. *Nano Lett.* **2011**, *11*, 5071–5078. [CrossRef] [PubMed]
4. Liu, T.; Zhang, L.; Cheng, B.; Hu, X.; Yu, J. Holey Graphene for Electrochemical Energy Storage. *Cell Rep. Phys. Sci.* **2020**, *1*, 100215. [CrossRef]
5. Baimova, Y.A.; Murzaev, R.T.; Dmitriev, S.V. Mechanical properties of bulk carbon nanomaterials. *Phys. Solid State* **2014**, *56*, 2010–2016. [CrossRef]
6. Niu, L.; Xie, J.; Chen, P.; Li, G.; Zhang, X. Quasi-static compression properties of graphene aerogel. *Diam. Relat. Mater.* **2021**, *111*, 108225. [CrossRef]
7. Yola, M.L. Development of Novel Nanocomposites Based on Graphene/Graphene Oxide and Electrochemical Sensor Applications. *Curr. Anal. Chem.* **2019**, *15*, 159–165. [CrossRef]
8. Meyer, J.C.; Geim, A.K.; Katsnelson, M.I.; Novoselov, K.S.; Booth, T.J.; Roth, S. The structure of suspended graphene sheets. *Nature* **2007**, *446*, 60–63. [CrossRef]
9. Tallinen, T.; Åström, J.A.; Timonen, J. The effect of plasticity in crumpling of thin sheets. *Nat. Mater.* **2008**, *8*, 25–29. [CrossRef]
10. Matan, K.; Williams, R.B.; Witten, T.A.; Nagel, S.R. Crumpling a Thin Sheet. *Phys. Rev. Lett.* **2002**, *88*, [CrossRef]
11. Zhang, L.; Zhang, F.; Yang, X.; Long, G.; Wu, Y.; Zhang, T.; Leng, K.; Huang, Y.; Ma, Y.; Yu, A.; et al. Porous 3D graphene-based bulk materials with exceptional high surface area and excellent conductivity for supercapacitors. *Sci. Rep.* **2013**, *3*. [CrossRef]
12. Balankin, A.S.; Ochoa, D.S.; León, E.P.; de Oca, R.C.M.; Rangel, A.H.; Cruz, M.Á.M. Power law scaling of lateral deformations with universal Poisson's index for randomly folded thin sheets. *Phys. Rev. B* **2008**, *77*. [CrossRef]
13. Baimova, J.A.; Liu, B.; Zhou, K. Folding and crumpling of graphene under biaxial compression. *Lett. Mater.* **2014**, *4*, 96–99. [CrossRef]
14. Krainyukova, N.V.; Zubarev, E.N. Carbon Honeycomb High Capacity Storage for Gaseous and Liquid Species. *Phys. Rev. Lett.* **2016**, *116*. [CrossRef]
15. Semat, J.M.; Rashmi, W.; Mahesh, V.; Khalid, M.; Priyanka, J. Synthesis of crumpled graphene by fast cooling method. In *Proceedings of the International Engineering Research Conference-12th Eureca 2019*; AIP Publishing: Melville, NY, USA, 2019 [CrossRef]
16. Savin, A.; Savina, O.I. The Effect of Layers Interaction on the Stiffness of Bending Deformations of Multilayered Carbon Nanoribbons. *Phys. Solid State* **2019**, *61*, 686–692. [CrossRef]
17. van Bruggen, E.; van der Linden, E.; Habibi, M. Tailoring relaxation dynamics and mechanical memory of crumpled materials by friction and ductility. *Soft Matter* **2019**, *15*, 1633–1639. [CrossRef]
18. Liao, Y.; Li, Z.; Xia, W. Size-dependent structural behaviors of crumpled graphene sheets. *Carbon* **2021**, *174*, 148–157. [CrossRef]
19. Liu, S.; Wang, A.; Li, Q.; Wu, J.; Chiou, K.; Huang, J.; Luo, J. Crumpled Graphene Balls Stabilized Dendrite-free Lithium Metal Anodes. *Joule* **2018**, *2*, 184–193. [CrossRef]

20. Luo, J.; Jang, H.D.; Sun, T.; Xiao, L.; He, Z.; Katsoulidis, A.P.; Kanatzidis, M.G.; Gibson, J.M.; Huang, J. Compression and Aggregation-Resistant Particles of Crumpled Soft Sheets. *ACS Nano* **2011**, *5*, 8943–8949. [CrossRef]
21. Baimova, J.A.; Liu, B.; Dmitriev, S.V.; Zhou, K. Mechanical properties of crumpled graphene under hydrostatic and uniaxial compression. *J. Phys. D Appl. Phys.* **2015**, *48*, 095302. [CrossRef]
22. Khan, M.B.; Wang, C.; Wang, S.; Fang, D.; Chen, S. The mechanical property and microscopic deformation mechanism of nanoparticle-contained graphene foam materials under uniaxial compression. *Nanotechnology* **2020**, *32*, 115701. [CrossRef] [PubMed]
23. Cui, X.; Ren, P.; Deng, D.; Deng, J.; Bao, X. Single layer graphene encapsulating non-precious metals as high-performance electrocatalysts for water oxidation. *Energy Environ. Sci.* **2016**, *9*, 123–129. [CrossRef]
24. Wang, X.X.; Tan, Z.H.; Zeng, M.; Wang, J.N. Carbon nanocages: A new support material for Pt catalyst with remarkably high durability. *Sci. Rep.* **2014**, *4*. [CrossRef] [PubMed]
25. Deng, J.; Ren, P.; Deng, D.; Bao, X. Enhanced Electron Penetration through an Ultrathin Graphene Layer for Highly Efficient Catalysis of the Hydrogen Evolution Reaction. *Angew. Chem. Int. Ed.* **2015**, *54*, 2100–2104. [CrossRef] [PubMed]
26. Dahal, A.; Batzill, M. Graphene–nickel interfaces: A review. *Nanoscale* **2014**, *6*, 2548. [CrossRef] [PubMed]
27. Sutter, P.; Sadowski, J.T.; Sutter, E. Graphene on Pt(111): Growth and substrate interaction. *Phys. Rev. B* **2009**, *80*, 245411. [CrossRef]
28. Gao, L.; Guest, J.R.; Guisinger, N.P. Epitaxial Graphene on Cu(111). *Nano Lett.* **2010**, *10*, 3512–3516. [CrossRef]
29. Yoon, S.M.; Choi, W.M.; Baik, H.; Shin, H.J.; Song, I.; Kwon, M.S.; Bae, J.J.; Kim, H.; Lee, Y.H.; Choi, J.Y. Synthesis of Multilayer Graphene Balls by Carbon Segregation from Nickel Nanoparticles. *ACS Nano* **2012**, *6*, 6803–6811. [CrossRef]
30. Ji, Z.; Wang, Y.; Yu, Q.; Shen, X.; Li, N.; Ma, H.; Yang, J.; Wang, J. One-step thermal synthesis of nickel nanoparticles modified graphene sheets for enzymeless glucose detection. *J. Colloid Interface Sci.* **2017**, *506*, 678–684. [CrossRef]
31. Ai, L.; Tian, T.; Jiang, J. Ultrathin Graphene Layers Encapsulating Nickel Nanoparticles Derived Metal–Organic Frameworks for Highly Efficient Electrocatalytic Hydrogen and Oxygen Evolution Reactions. *ACS Sustain. Chem. Eng.* **2017**, *5*, 4771–4777. [CrossRef]
32. Batzill, M. The surface science of graphene: Metal interfaces, CVD synthesis, nanoribbons, chemical modifications, and defects. *Surf. Sci. Rep.* **2012**, *67*, 83–115. [CrossRef]
33. Omukai, A.; Yoshimura, A.; Watanabe, F.; Tachibana, M. Graphene films synthesized by chemical vapor deposition with ethanol. *Trans. Mater. Res. Soc. Jpn.* **2011**, *36*, 359–362. [CrossRef]
34. Li, X.; Cai, W.; Colombo, L.; Ruoff, R.S. Evolution of Graphene Growth on Ni and Cu by Carbon Isotope Labeling. *Nano Lett.* **2009**, *9*, 4268–4272. [CrossRef]
35. Chae, S.J.; Güneş, F.; Kim, K.K.; Kim, E.S.; Han, G.H.; Kim, S.M.; Shin, H.J.; Yoon, S.M.; Choi, J.Y.; Park, M.H.; et al. Synthesis of Large-Area Graphene Layers on Poly-Nickel Substrate by Chemical Vapor Deposition: Wrinkle Formation. *Adv. Mater.* **2009**, *21*, 2328–2333. [CrossRef]
36. Reina, A.; Thiele, S.; Jia, X.; Bhaviripudi, S.; Dresselhaus, M.S.; Schaefer, J.A.; Kong, J. Growth of large-area single- and Bi-layer graphene by controlled carbon precipitation on polycrystalline Ni surfaces. *Nano Res.* **2009**, *2*, 509–516. [CrossRef]
37. Chang, S.W.; Nair, A.K.; Buehler, M.J. Nanoindentation study of size effects in nickel–graphene nanocomposites. *Philos. Mag. Lett.* **2013**, *93*, 196–203. [CrossRef]
38. Kim, Y.; Lee, J.; Yeom, M.S.; Shin, J.W.; Kim, H.; Cui, Y.; Kysar, J.W.; Hone, J.; Jung, Y.; Jeon, S.; et al. Strengthening effect of single-atomic-layer graphene in metal–graphene nanolayered composites. *Nat. Commun.* **2013**, *4*. [CrossRef]
39. Kuang, D.; Xu, L.; Liu, L.; Hu, W.; Wu, Y. Graphene–nickel composites. *Appl. Surf. Sci.* **2013**, *273*, 484–490. [CrossRef]
40. Kim, D.J.; Truong, Q.T.; Kim, J.I.; Suh, Y.; Moon, J.; Lee, S.E.; Hong, B.H.; Woo, Y.S. Ultrahigh-strength multi-layer graphene-coated Ni film with interface-induced hardening. *Carbon* **2021**, *178*, 497–505. [CrossRef]
41. Safina, L.R.; Baimova, J.A.; Mulyukov, R.R. Nickel nanoparticles inside carbon nanostructures: Atomistic simulation. *Mech. Adv. Mater. Mod. Process.* **2019**, *5*. [CrossRef]
42. Safina, L.R.; Krylova, K.A. Effect of particle size on the formation of the composite structure in Ni-graphene system: Atomistic simulation. *J. Phys. Conf. Ser.* **2020**, *1435*, 012067. [CrossRef]
43. Safina, L.L.; Baimova, J.A. Molecular dynamics simulation of fabrication of Ni-graphene composite: Temperature effect. *Micro Nano Lett.* **2020**, *15*, 176–180. [CrossRef]
44. Safina, L.; Baimova, J.; Krylova, K.; Murzaev, R.; Mulyukov, R. Simulation of metal-graphene composites by molecular dynamics: A review. *Lett. Mater.* **2020**, *10*, 351–360. [CrossRef]
45. Wu, C.D.; Fang, T.H.; Lo, J.Y. Effects of pressure, temperature, and geometric structure of pillared graphene on hydrogen storage capacity. *Int. J. Hydrog. Energy* **2012**, *37*, 14211–14216. [CrossRef]
46. Krylova, K.; Baimova, J.; Mulyukov, R. Effect of deformation on dehydrogenation mechanisms of crumpled graphene: Molecular dynamics simulation. *Lett. Mater.* **2019**, *9*, 81–85. [CrossRef]
47. Krylova, K.A.; Baimova, J.A.; Lobzenko, I.P.; Rudskoy, A.I. Crumpled graphene as a hydrogen storage media: Atomistic simulation. *Phys. B Condens. Matter* **2020**, *583*, 412020. [CrossRef]
48. Dimitrakakis, G.K.; Tylianakis, E.; Froudakis, G.E. Pillared Graphene: A New 3-D Network Nanostructure for Enhanced Hydrogen Storage. *Nano Lett.* **2008**, *8*, 3166–3170. [CrossRef] [PubMed]

49. Ranjbar, A.; Khazaei, M.; Venkataramanan, N.S.; Lee, H.; Kawazoe, Y. Chemical engineering of adamantane by lithium functionalization: A first-principles density functional theory study. *Phys. Rev. B* **2011**, *83*. [CrossRef]
50. Yang, R.T. Hydrogen storage by alkali-doped carbon nanotubes–revisited. *Carbon* **2000**, *38*, 623–626. [CrossRef]
51. Khantha, M.; Cordero, N.A.; Molina, L.M.; Alonso, J.A.; Girifalco, L.A. Interaction of lithium with graphene: Anab initiostudy. *Phys. Rev. B* **2004**, *70*. [CrossRef]
52. Medeiros, P.V.C.; de Brito Mota, F.; Mascarenhas, A.J.S.; de Castilho, C.M.C. Adsorption of monovalent metal atoms on graphene: A theoretical approach. *Nanotechnology* **2010**, *21*, 115701. [CrossRef]
53. Yoshida, A.; Okuyama, T.; Terada, T.; Naito, S. Reversible hydrogen storage/release phenomena on lithium fulleride (LinC60) and their mechanistic investigation by solid-state NMR spectroscopy. *J. Mater. Chem.* **2011**, *21*, 9480. [CrossRef]
54. Waqar, Z. Hydrogen accumulation in graphite and etching of graphite on hydrogen desorption. *J. Mater. Chem.* **2007**, *42*, 1169–1176. [CrossRef]
55. Łoś, S.; Duclaux, L.; Letellier, M.; Azaïs, P. Study of Adsorption Properties on Lithium Doped Activated Carbon Materials. *Acta Phys. Pol. A* **2005**, *108*, 371–377. [CrossRef]
56. Stuart, S.J.; Tutein, A.B.; Harrison, J.A. A reactive potential for hydrocarbons with intermolecular interactions. *J. Chem. Phys.* **2000**, *112*, 6472–6486. [CrossRef]
57. Humphrey, W.; Dalke, A.; Schulten, K. VMD: Visual molecular dynamics. *J. Mol. Gragh.* **1996**, *14*, 33–38 [CrossRef]
58. Girifalco, L.A.; Weizer, V.G. Application of the Morse Potential Function to Cubic Metals. *Phys. Rev.* **1959**, *114*, 687–690. [CrossRef]
59. Katin, K.P.; Prudkovskiy, V.S.; Maslov, M.M. Molecular dynamics simulation of nickel-coated graphene bending. *Micro Nano Lett.* **2018**, *13*, 160–164. [CrossRef]
60. Galashev, A.Y.; Katin, K.P.; Maslov, M.M. Morse parameters for the interaction of metals with graphene and silicene. *Phys. Lett. A* **2019**, *383*, 252–258. [CrossRef]
61. Petucci, J.; LeBlond, C.; Karimi, M.; Vidali, G. Diffusion, adsorption, and desorption of molecular hydrogen on graphene and in graphite. *J. Chem. Phys.* **2013**, *139*, 044706. [CrossRef]
62. Ganz, E.; Ganz, A.B.; Yang, L.M.; Dornfeld, M. The initial stages of melting of graphene between 4000 K and 6000 K. *Phys. Chem. Chem. Phys.* **2017**, *19*, 3756–3762. [CrossRef] [PubMed]
63. Los, J.H.; Zakharchenko, K.V.; Katsnelson, M.I.; Fasolino, A. Melting temperature of graphene. *Phys. Rev. B* **2015**, *91*, 045415. [CrossRef]
64. Ye, Y.; Ahn, C.C.; Witham, C.; Fultz, B.; Liu, J.; Rinzler, A.G.; Colbert, D.; Smith, K.A.; Smalley, R.E. Hydrogen adsorption and cohesive energy of single-walled carbon nanotubes. *Appl. Phys. Lett.* **1999**, *74*, 2307–2309. [CrossRef]
65. Pinkerton, F.E.; Wicke, B.G.; Olk, C.H.; Tibbetts, G.G.; Meisner, G.P.; Meyer, M.S.; Herbst, J.F. Thermogravimetric Measurement of Hydrogen Absorption in Alkali-Modified Carbon Materials. *J. Phys. Chem. B* **2000**, *104*, 9460–9467. [CrossRef]
66. Schimmel, H.G.; Kearley, G.J.; Nijkamp, M.G.; Visser, C.T.; de Jong, K.P.; Mulder, F.M. Hydrogen Adsorption in Carbon Nanostructures: Comparison of Nanotubes, Fibers, and Coals. *Chem. A Eur. J.* **2003**, *9*, 4764–4770. [CrossRef]

MDPI\
St. Alban-Anlage 66\
4052 Basel\
Switzerland\
Tel. +41 61 683 77 34\
Fax +41 61 302 89 18\
www.mdpi.com

Materials Editorial Office\
E-mail: materials@mdpi.com\
www.mdpi.com/journal/materials

www.ingramcontent.com/pod-product-compliance
Lightning Source LLC
LaVergne TN
LVHW070710100526
838202LV00013B/1066